The Way We Build

The Way We Build : A Journey Through the Spaces of Hyundai Card

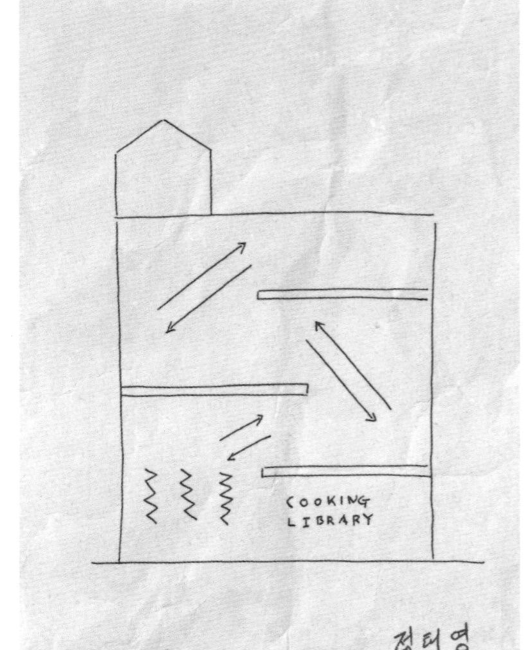

Ted Chung

Pre-face

Hyundai Card's space initiatives serve as

guideposts for Seoul's contemporary culture

Foreword

The world of financial services and credit card transactions is one of the most transient and virtual sectors of today's global economy. Every second, a multitude of transactions is processed in the digital computer networks of the international financial system. It is therefore not immediately evident why a company such as Hyundai Card would champion actual, 'analogue' spaces for cultural activities in Korea's capital city and beyond. As a matter of fact, these 'spaces'—dedicated to themes such as contemporary design, travel, cooking, or vinyl records and music—are contributing to no small degree to Seoul's changing image as one of the most dynamic and innovative metropolises on the planet. Moreover, these spaces have become intrinsically linked to Hyundai Card's public perception, which, in the world of the finance industry, equals that of the company's hometown.

It would be too shortsighted to understand Hyundai Card's space initiative as merely a Public Relations campaign in pursuit of a favorable public perception. The spaces, while catering primarily to the many customers who hold a credit card from the company's wide assortment of products, have become hubs of Seoul's contemporary culture. They are not primarily made for selling products—even though rare records can be found for instance in the particularly popular (and open to the public) VINYL & PLASTIC store—, but they are places of encounter and learning. It is no coincidence then that several of Hyundai Card's spaces are designated as 'libraries'; and as such they are as much well-curated repositories of knowledge on specific aspects of cultural production (design, travel, cooking) as they are a bold statement in support of one of the most archaic vessels carrying information: the book. While gurus of the digital age have repeatedly prophesied the demise of centuries-old book culture, this remarkable initiative testifies to the ongoing relevance of the book, and, in a wider sense, to the culture of knowledge production and cultivation in an exemplary manner. Nothing seems more urgent in our current times, where knowledge is increasingly contested by unfounded opinions and superficial 'likes'.

But the Hyundai Card spaces are much more than sterile containers of books and knowledge. All of them invite visitors to a multi-sensory experience and to a broad range of activities tied to the specific themes of the spaces, whether it is concerts, exhibitions, or cooking classes. The theoretical knowledge available in the books on the shelves is thus directly activated and made accessible. And it is perhaps here where the social significance of the project becomes most manifest. Korea underwent a rapid modernization relatively late, having been one of the poorest countries in Asia after World War II. But the development in the last three or so decades has been impressive and sustained. In a culture that is so rapidly changing, Hyundai Card's space initiative provides essential reassurance. It serves as a guidepost for how to adapt to the new, cosmopolitan lifestyle that the country has so vehemently embraced. In this, historic examples of comparable impetus come to mind. One thinks of the Berlin based AEG electricity company and its chief designer Peter Behrens, who at the beginning of the 20th century worked on a strong corporate design that would have a significant and lasting impact on Germany's early modern visual culture. To be such a cultural beacon is the aspiration of the 'space' initiative. Its various aspects are assembled and documented here for the first time—in the form of a book.

Dr. Martino Stierli
The Philip Johnson Chief Curator of Architecture and Design
The Museum of Modern Art (MoMA), New York

Writer's Comment

The 'Father Said' advertisement series that pioneered a new future of advertising and achieved popularity without celebrity appearances, the Super Concert of the unparalleled lineup featuring Paul McCartney and Coldplay, the M and X series that set a new paradigm of credit cards with ground-breaking designs, and the spatial project that built libraries containing the identity of Hyundai Card when everyone else shouted 'digital'—what is truly surprising and unprecedented worldwide is that all these achievements were actualized by one company. Further, the fact that all these projects sprouted from Hyundai Card's corporate culture and branding naturally triggers admiration. The architectural and spatial projects of Hyundai Card lead to greater insights when you understand them comprehensively as to how they have progressed within the corporate strategy, not as individual cases. Hyundai Card did not stop at simply constructing buildings with beautiful looks. Unlike the usual process of designing first then considering the utility later, Hyundai Card started only after fierce discussions, sometimes for years, on why it should create the particular building. That is, 'Why' has been the most vital question for Hyundai Card. <The Way We Build> is a holistic book that narrates the path of Hyundai Card beginning from 'Why' and exploring the answers relentlessly. In all architectural and spatial projects of Hyundai Card, it is difficult to identify a prominent style or physical unity. Rather, the spirit that Hyundai Card pursues is clearly revealed in the concept of 'house within a house', transparent low-iron glass, utilization of pure materials like iron, and tenacious content planning capability. On the 8th floor of Building 2 in the Yeouido Complex, the most recent work of Hyundai Card's architectural project, you can read the confidence surpassing even the obsession in expressing unity and identity. This is because the building reflects the recent transition in Hyundai Card's corporate culture, focusing more on the comfort of employees and the living space beyond the spatial completeness. Before Hyundai Card, there has never been a case where a company's architectural and spatial projects have impacted the community beyond the office building and branding space. While creating this book, we sought various advice from experts overseas. Yet, the common answer was that such as case is extremely difficult to find. At the core thereof, of course, is Vice-Chairman Ted Chung. The three of us—as the authors and production managers of this book—met and interviewed Vice-Chairman Ted Chung dozens of times over the past two years. In doing so, we had the opportunity to hear the beginning of the architectural and spatial projects of Hyundai Card in detail, as well as ongoing projects. Vice-Chairman Ted Chung was the de facto creative director of this book. He is a rare executive who personally proved that the architecture and design of a corporation are also the business area of the CEO. Based on his innate sense added with efforts, his acumen is always one step ahead, on top of his superb intuition. Now is an era where architecture and space itself are a medium and means to express brand philosophy and spirit. In this era, Vice-Chairman Ted Chung played a key role in raising the level of Korean architecture by making decisions ahead of anyone else. No one can deny this achievement. As Vice-Chairman Ted Chung candidly puts it, the architectural and spatial projects of Hyundai Card continue to evolve while reflecting on the results. Hyundai Card, which once chased logical design, declares that it will seek new aesthetics to pursue after pronouncing the end of minimalism. The declaration further raises expectations for its next endeavors.

Most of the books on architecture have followed the architect's perspective or the critic's interpretation. For <The Way We Build>, there were virtually no books to use as reference or comparison. A client ordering an architecture project rarely develops a story. That is because no company has ever devoted its zeal to create a space in every direction for over 10 years. Without a clear goal concerning a space interlocked with the corporate strategy, this story could not have begun. Included here are UX/UI designs from signage to map-changing planning like regional branding or urban regeneration. Because the goal is sharp, the efforts to realize it becomes meticulous, and the imagination is maximized. The bold, empty space of the Music Library, the opening walls of the Castle of Skywalkers, and attempt to bring vitality with a spiral staircase cutting through the floor of a rented building—these are not mere sagas. This is because Hyundai Card believes that a good space is essential to positively change the lives of those who use it. The diverse voices and materials contained in this book, including those of architects, industry officials, and users, will elucidate why financial companies are intensely obsessed with architectural perfection.

Each of the authors has been working as an editor-in-chief of a lifestyle magazine, editor-in-chief of a design magazine, and architecture columnist, respectively. In doing so, our relationship with Hyundai Card has been developed over a long period, covering various architectural, design, and cultural projects. Despite our confidence that we knew more about Hyundai Card than others, our path to complete this book was by no means easy. More time and effort were required to research, explore, and archive all materials of all architectural and spatial projects of Hyundai Card, interview persons involved, and interpret and turn the results into writing. However, the job was meaningful. It was more profound because an opportunity to have such a closer look into the important company that has impacted the entire Korean society is rare. In a world mediated by smartphone screens, space is compressed into a single picture, and the speed of it being shared is faster than ever. Superficially decorated spaces, spaces that feel familiar in the eyes are increasing exponentially. However, meeting a space perfected with a single principle down to its details is becoming harder. Here, you will find great insights simply by reading the principles of spaces Hyundai Card created and the process of realization thereof. <The Way We Build> is a book that encapsulates the spaces Hyundai Card has built and is the beginning of the architectural and spatial projects of Hyundai Card that will continue to evolve and progress.

Jiho Park
CEO of Inspiration 102
Eunkyung Jeon
Director of the Monthly DESIGN magazine
Yoonkyung Bae
Adjunct Professor, Journalist

Interview

Ted Chung, Vice-Chairman of Hyundai Card · Hyundai Capital · Hyundai Commercial

The architectural and spatial projects Hyundai Card has carried out encompass the Yeouido Complex for its executives and employees, Hyundai Capital offices in various countries, the Library series involving membership features, brand spaces such as Studio Black, Vinyl & Plastic, and Castle of Skywalkers, as well as regional projects like Gapado. This globally unprecedented magnitude is hard to be believed as the work of a financial company. At the heart of Hyundai Card's architectural and spatial projects was Vice-Chairman Ted Chung, who is deeply interested in architecture and design and makes leading decisions with unparalleled perspectives. It is widely known that he has inspired many as a business executive. In the early 2010s, when everyone was racing towards digital, Vice-Chairman Ted Chung intuitively recognized the value of space—the most powerful language of brand expression—and has been driving relevant projects. And here, he shared the key moments in the architectural projects of Hyundai Card.

"Architecture for the exterior, architecture to show taste and sophistication is not the space Hyundai Card pursues. Instead, we concentrate on the content. That is why it generally takes several years to design and prepare the content alone."

It is an era in which space itself has become a medium and one of the key methods of brand expression. So far, Hyundai Card and Hyundai Capital have exemplified this well ahead of others via various architectural and spatial projects. You have often emphasized that the philosophy of the company and the brand must be upright before building a space.

In the digital age, in addition to TV commercials, all-around exposure is essential. So, you can say it is an era of selling minds, not products. Paradoxically, the more the digital is strengthened, the greater the importance of a space connoting the corporate spirit. That is why the space of Hyundai Card is inevitably the hero and root in the digital age.

The Library series began opening its doors in the order of travel, music, and cooking since the design in 2013, and it has become a key asset of Korean modern architecture beyond a company-operated space. How did the Library series come to its existence?

The Library series has been the subject of my serious interest from the beginning. (laughs) It is no exaggeration to say that it is quintessential of the brand spirit of Hyundai Card. The Library was conceived to reveal what Hyundai Card likes, its essence, and the value it pursues. Marketing was not the starting point. If spatial marketing was the clear goal from the beginning, I might have compromised appropriately without investing such level of effort into it.

A total of four libraries have been completed. Was building only four libraries the original plan?

At first, I thought of only a design library. Then, the plan gradually expanded. Think of it like this. Hyundai Card has written four books so far, of which Design Library can be considered as the first book written on the favorite topic. Doesn't everyone invest the most effort into writing the first book? That is why the first one is more meaningful than anything else. The second is the Travel Library. On this one, I confess that I wrote a sellable book from a marketing standpoint on a topic that anyone would like. And the third book, the Music Library, was written because I had another topic I wanted to write about, while the last book, Cooking, was written because it was a book everyone would enjoy. (laughs) The location where the book would be written was also important. The Design Library was a book that I wanted to write in Bukchon, and for the Music, it was Itaewon. In particular, for the Music Library, the staff were unsure whether to mark it as Hannam-dong or Itaewon. So, I said, "Jazz or rock is Itaewon. Have you ever found music in Hannam-dong?" For the Travel and Cooking as well, I thought that they should be written in Gangnam, where sales would be active.

I would like to hear your thoughts as a client of an architectural project. Many might wonder, what do you focus on when selecting architects, space designers, and artists to work on projects?

Whatever others say, it will depend on my taste in the end. That is why I chose the architects to work with in most of the time. When a person has 100 yet reveals only 20 above the water while hiding 80, I tend to be in awe of the revealed blade of 20. When a person with 50 shows all 50, I consider that as an exaggeration, not a hidden blade. That is, instead of people who show all their cards, I like talented people who have more than what I can see, more under the water, showing only a part of what they have. If I must talk about the direction, I want a refined architect like Kazuyo Sejima, who designed the classical concert hall exclusively for Hyundai Card. After working with many architects and designers, I learned that half of the design is completed on the drawing and the other half is finished on site. The response by good architects and designers comes from deep within. Whether we agree or disagree, it would be correct to say that this is not a realm of having or not having skills, but a difference in the fit: 'is the person a good fit for us?'

There is a saying that a good client creates a good project. You have been a client in numerous architectural projects. In such projects, there are moments when you must make big and small decisions on the interior, landscaping, art, etc. How do you coordinate and make important decisions in such instances?

Above all, it is important to organize what you want by yourself. I never say, "just make it good." This is because if the client is ambiguous and swayed, the architect also loses his/her way. The direction must be clear. The determination must be swift. I rarely use conceptual words. When the relationship among architecture, interior, art, and landscaping should be organized, for instance, I draw the 'lines' between key matters. That is why project management and the client of the project are crucial. The person who makes requests knows best what he/she wants to do. There is no right answer; however, I think that the architect may suggest the direction of the interior, while the interior designer can finish it. Taking the Cooking Library as an example, the architecture was done by 101 Architects, and Blacksheep, which did the interior design for Jamie Oliver Kitchen, was responsible for the interior. In between, I structured the relationship. Also, for the Busan Office Building, 101 Architects handled even the interior. The important decision there was to let the architect's spatial senses complete the interior entirely. To what extent a service provider should do—such a decision is on the client. No matter who you work with, the client needs to be alert and play a key role. Thus, in many projects, I am also part of the team as the client and share the credits of the project. One should not avoid responsibility and there is no need to be humble.

You have worked with various architects for over 10 years. You have often worked with Gensler, Spackman Associates, and 101 Architects in the projects.

I have one weakness: I can have smooth conversations only when the other person knows me well. It seems that I have a higher entry barrier for working with or meeting others. Instead, once I build a relationship with someone, I rarely let go of it. (laughs) Instead of meeting new people, introducing oneself, and explaining our style every time, the synergy is greater to work with people who understand the values Hyundai Card pursues. Gensler, Spackman Associates, and 101 Architects are good examples of such partners. In the case of Spackman Associates, they have a higher understanding of Hyundai Card's corporate culture as we have been working together for many years since the early 2000s, starting with the interior design of the Yeouido Complex.

With Gensler, you have worked on a variety of projects, such as the Yeouido Complex and office buildings overseas, as well as the interior of the Music Library.

It is less about Gensler but a lot with Philippe Paré, who is now in charge of Gensler Paris. When I contacted Gensler for the first time to design the Yeouido Complex, the person who came to see me was Philippe. He possesses a high level of architectural knowledge and cultural literacy. He has pride but no self-conceit. After his first work, we continued to collaborate in various ways, heightening the mutual understanding. That is how we became life partners who help each other grow.

In addition to Gensler, 101 Architects, which worked together in the Design and Cooking Libraries, the Yeongdeungpo office building, the Busan office building, the Yeouido Complex, and the Gapado Project, has also collaborated significantly with Hyundai Card. According to Wook Choi, you have a high level of intuitive understanding of architecture and are fluent at creating new paradigms based thereon. Further, you actively give opinions, providing feedback very quickly.

Wook Choi is a person whose stubborn world can be felt by others. And he never meddles with a world he does not have. He also interprets and expresses my intentions superbly well. For example, when we started the Cooking Library, I drew a simple sketch of the building. Then, he completed all the structures. He even included my suggestions such as 'I don't want the fuss of please-look-at-our-building in Cheongdam-dong, but I want the building to offer various experiences once you enter, even if the look may be ordinary', 'The finest value in the future is a meal in the greenhouse of a farmhouse', and 'I hope the smell of bread flows bottom-up.' These could be achieved thanks to our longtime collaboration.

What kind of client are you to architects?

I am unsure as I have never asked the architects. But I can assure you one thing: a client who understands the value of architecture and design well. I think it is because, although I listen to architects' opinions, I make clear judgments and decisions.

You emphasize the relationship between corporate culture and the workspace. That seems to be the reason for constantly trying new changes in the interior space of the office buildings.

Some spaces trigger intended behaviors well, some do not. An outsider expecting a grandeur marble or artworks visits the company and even says, "It's not as great as I expected." (laughs) Our pride stems from Hyundai Card being the only company that invests great effort to match corporate culture and identity. If you are not very interested in corporate culture, it will be difficult to find anything special in our office buildings. Many have better facilities and interior in their office buildings. However, it has never come to my mind to make the Yeouido Complex more stylish. I only focus on what would make the employees more comfortable, what parts would change the ambiance, and which details would trigger new ideas. That is why I don't even dream of erecting a new office building. We will change the space as we continue to evolve. Of course, the principle of facilitating free and informal communication with an open-plan office will always remain intact.

When I visited the Frankfurt office building, the most impressive was the spiral staircase that pierced through three floors. The third and fourth floors of the Yeouido Complex are also connected by stairs. I can see the will of communication via the architectural connection. Is there any reason behind the focus on the stairs?

I researched stairs to a great extent. As ordinary stairs take up a lot of space, a circular staircase was devised, ascending in a spiral. As such, even in the same headquarters, more than one floor can be used, and they can freely communicate with other headquarters. The elevator is only for commuting—it is quite cumbersome to use the elevator during usual work hours. Especially, waiting for the elevator to go up one floor is the most irritating. Above all, it represents the will to eliminate the isolation of floors. There are stairs in the Music Library and the Understage. The Cooking Library also has weaved floors. The desire for exchange and communication—Hyundai Card's DNA—is expressed in the design.

Is there another principle that you ensured to be reflected in the architecture as preventing the separation of floors for communication?

Not a principle, but an obsessive idea existed that lines must be removed from every creation of office space. For Hyundai Capital America and the Frankfurt office, we worked very hard to make the lines a single line by matching the lines of the desks, lighting, and windows. I was quite obsessed with reducing lines because if even one desk was misaligned, another line would be created. Then, after building the Beijing office, I realized that conversations among employees far less frequent as the lines were aligned Since the office is larger than Hyundai Capital America and Frankfurt office and has curves, aligning lines made the space look empty. Afterward, I brushed off my obsession with lines. With Phillippe of Gensler, I reflected, thinking 'I was so obsessed with lines that I made it difficult for the employees.' (laughs) For the London office created afterward, I worked with more freedom. We are also changing the Yeouido Complex in this context. I sometimes think: Is perfection looking 100 years ahead possible? We continue to make regrets but we also make progress. Another answer would be being flexible and changing things little by little as necessary.

The corporate office building itself reveals the types of thinking the organization does. Thus, managing the identity of architecture and space across all businesses of Hyundai Card is also the CEO's realm.

Many companies want good architecture. However, few invest serious considerations like Hyundai Card does. I hope our cases serve as an example to show that architecture and design play an important role by interlocking with the corporation's internal strategy instead of simply being good architectural works. If you go to certain corporations' office buildings, there are spaces built to show-off, only caring about visitors' perspectives, those emphasize only the company's perspectives, deprioritizing the daily life of employees, or those that lack connection with the essence of the business. I hope Hyundai Card can be a reference point for such places.

Hyundai Card is well known for its pursuit of logical design. However, it was profound that it stated the end of minimalism through the 'Da Vinci Motel' project in 2019. It seemed like hinting at the aesthetics to pursue in the future.

Minimalism, which originated from Apple, has somehow deteriorated into a haven for the untalented. We liked the fresh aesthetic of minimalism in the first place. We did not follow minimalism itself. We have concluded that minimalism no longer offers a unique aesthetic. However, there is still no opposite term of minimalism. That is why I think this realm is a boundless ocean. (laughs) You can interpret the 'Da Vinci Motel' project, which built tone and manners by combining the kitsch word 'Motel' with 'Leonardo da Vinci', as the first flag. In a way, creating a more popular space, the Vinyl & Plastic, and trying to collaborate with various brands and creators can be considered as part of such efforts. These days, I do a great deal of research on the new aesthetics we are pursuing.

In the city of Seoul, the buildings created by Hyundai Card occupy a unique status. What are your thoughts on the positive impact a company's headquarters and buildings have on the city?

Every time I hear a story like this, I am happy and grateful. Many visitors from foreign companies or countries still ask us whether they can visit the Yeouido Complex or the libraries. And I keep getting feedback that many are surprised that architectures of such a level stand in Seoul. I am delighted to have set a precedent that a corporate building, not a public building, can positively influence the society. Meanwhile, architecture for the exterior, architecture to show taste and sophistication is not the space Hyundai Card pursues. Instead, we concentrate on the content. That is why it generally takes several years to design and prepare the content alone. Whenever we built a library, the process of looking around and discussing all relevant spaces worldwide was quite extensive. Architecture and construction are thereafter. It differs from the usual way of discussing the shape of a building on Day 1. And the focus was on the staff in the case of the office building, the visitors in the case of the library, and the people who have been living there and the history in the case of the traditional market and Gapado. That is, we did not try to replace what was there. I wanted to show that the architecture with such efforts exists in Seoul as well.

Hyundai Card Headquarters

A space of innovation where renewal and

experiment are constantly taking place

The Headquarters Complex of Hyundai Card in Yeouido ("Yeouido Complex") is an innovative space of unending renewal and experiments. Began with Building 1, the Yeouido Complex has expanded to include Building 2, directly facing Building 1 with the street in between, and nearby Building 3. Since 2002, the renewal of the space has continued, from the lobby to all floors. Stereotype-shattering meeting rooms including the auditorium and lecture rooms are located throughout the Yeouido Complex. The Design Lab, the key department

responsible for numerous confidential projects of Hyundai Card, is in the lobby of Building 1. Also included in the Yeouido Complex is the Card Factory, which exhibits the entire card production process. In addition, convenience facilities continue to evolve, including the Box, The Kids, Open Studio, Open Radio, and Cafe & Pub. Hyundai Card is also experimenting with new office building designs tailored to its innovative operations, such as Workspace.

Convention Hall, Auditorium & Lecture Room

p. 30

The nexus of the Convention Hall is swiftly movable tables and chairs. The Auditorium was built on the theme of "timeless". The Lecture Room was inspired by and designed after a theatrical auditorium of a partner company, Santander.

1.1

Design Lab & UX Lab

p. 44

An encounter with Jean Nouvel at the Milan Design Week hatched the Design Lab. Inside of the Design Lab has exposed the finishing materials of the floor and walls, unveiling the atmosphere of a warehouse. The UX Lab was built separately on the second floor as an extension of the Design Lab.

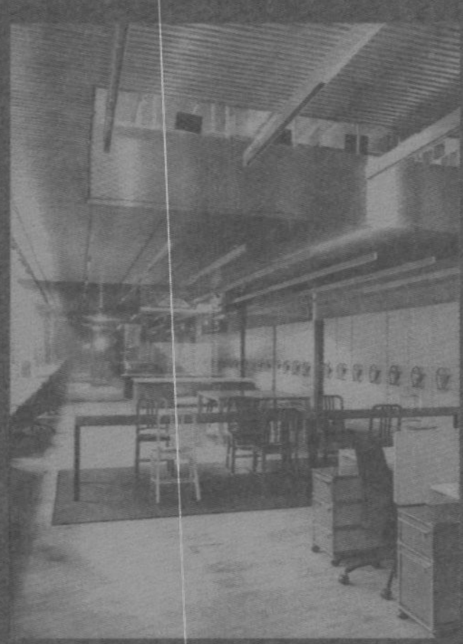

1.2

Card Factory & Traffic Monitoring Center

p. 50

The Card Factory is designed to convey the unique sensibilities of industrial capital and the joy of nostalgia to visitors. At the Traffic Monitoring Center, you can grasp all transactions of Hyundai Card at a glance.

1.3

the Box, Café & Pub

p. 58

While renewing Building 2 of the Yeouido Complex, the vast lobby was secured, then the Box was created, stretching out toward Building 1. The Café & Pub is the only place where the public can access the Yeouido Complex.

1.4

Workspace

p. 66

Following the Digital Hyundai Card transition, diverse variations have been made in the general office space to transform Hyundai Card into an agile organization. One example is connecting the third and fourth floors with stairs piercing through the floors. The interior has no partition and is designed as an open structure.

1.5

1.1 Convention Hall, Auditorium & Lecture Room

Convention Hall
Location: 10F HQ1 3, Uisadang-daero,
Yeongdeungpo-gu, Seoul
Area: 580m²
Design: 2016.03 ~ 2016.06
Construction: 2016.06 ~ 2016.09
Interior: Gensler
Contractor: Dawon ID&C

Auditorium
Location: 1F HQ2 3, Uisadang-daero,
Yeongdeungpo-gu, Seoul
Area: 260m²
Design: 2009.01 ~ 2009.09
Construction: 2010.02 ~ 2010.08
Interior: Spackman Associates
Contractor: Dawon ID&C

Lecture Room
Location: 1F HQ1 3, Uisadang-daero,
Yeongdeungpo-gu, Seoul
Area: 166m²
Design: 2013.08 ~ 2013.11
Construction: 2014.01 ~ 2014.03
Interior: Spackman Associates
Contractor: Dawon ID&C

The front of the Convention Hall embodies the 'boundary without boundaries' inspired by the work of James Turrell.

Located on the 10th floor of Building 1, the Convention Hall is a place that represents the endlessly evolving Yeouido Complex in keeping with the growth of Hyundai Card. The Auditorium and Lecture Room may be small but have given birth to key ideas with profound impact on the Company. The historical significance of these spaces for Hyundai Card is considerable.

1.1.1
The Way We Build

The Convention Hall on the 10th floor is where continual transformation occurs, soaked in the history of Hyundai Card. Originally, it was a conventional place with many windows opening outwards along the right-side wall. The architectural firm Gensler, however, renewed the space in a major way in 2018, turning it into its current form. The key requirement was that the platform must be eliminated. A high platform typically found in corporate auditoriums creates awkward moments. The platform burdens the speaker trying to climb it up, simultaneously triggering the audience to be rigid. Vice-Chairman Ted Chung wanted to eliminate this awkwardness. Thus, climbing up was replaced by walking out to. What Gensler asked Hyundai Card was whether all windows could be removed. Hyundai Card gladly agreed as the windows had no utility.

The most challenging aspect of the design was the front part for presentations. Adding a frame could create a daily distraction while leaving the wall empty would make the space dull. The project leader Phillippe Paré suggested the idea of 'boundary without boundaries' inspired by the work of James Turrell that offers the experience of watching the sky with a frame. Since the front part of the space is a boundary unlike a boundary, the ceiling was intended to be an edge, unlike an edge.

The standing bench that people can lean on was the idea of Vice-Chairman Ted Chung. It was an attempt to break the dichotomy of sitting or standing. By escaping the situation prone to be rigid or uncomfortable in the Auditorium, the intention was

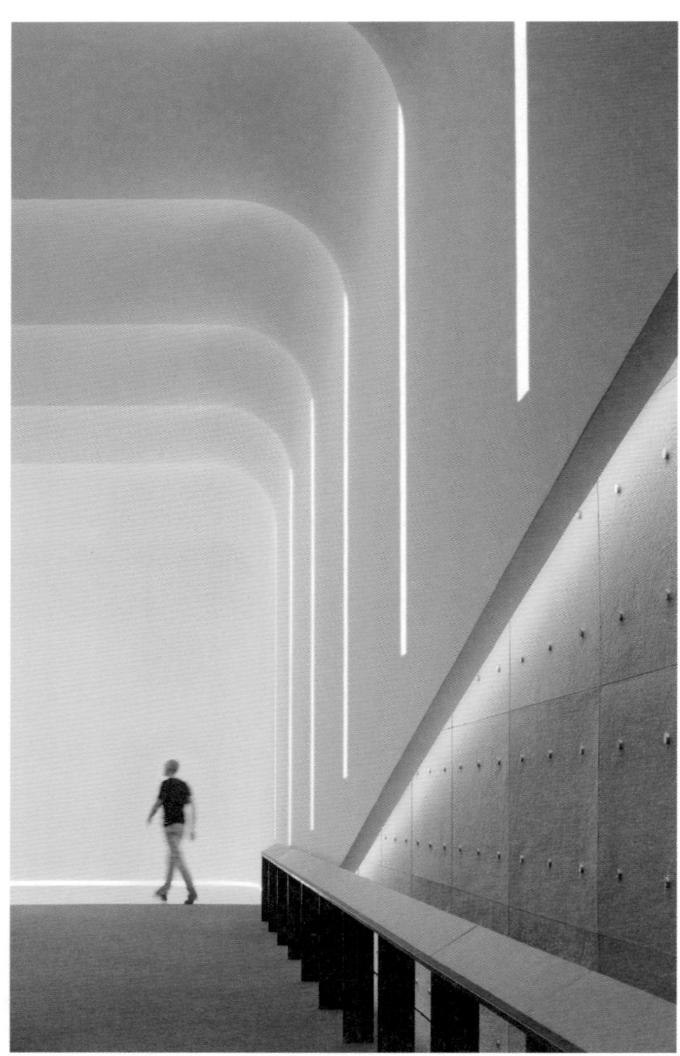

The Convention Hall lacks high platforms usually found in auditoriums of other companies. 'Climbing up' was replaced by the concept of 'walking out to.'

Since the Auditorium has glass on three sides, it could make the sound be reverberated or muted. Thus, to resolve this issue, sound-absorbing material imported from Denmark was used. The Convention Hall won the 2019 AIA award in the interior architecture category. The state-of-the-art system was introduced to enable the seats to be hidden completely into the wall and automatically taken out as necessary.

"Vice-Chairman Ted Chung asked us to imagine a completely new concept of a multipurpose auditorium. Hyundai Card gave a lot of freedom to come up with a perfect solution. This is exactly what the perfect client would be like. Hyundai Card presents us with a clear vision for the project with careful balancing so that we can do what we do best as designers. The Convention Hall is an ideal project that exemplifies the ideal relationship between a client and an architect when it comes to designing and actualizing a new space."

Philippe Paré, Director of Gensler Paris – Convention Hall Space Design

1.1.2 Point of Space

Finished in a modern and practical form, the Convention Hall won the 2019 American Institute of Architecture (AIA) award for interior architecture, which is given annually to innovative architectural interiors. It is a place where cutting-edge technology, the finest sound system, and innovative lighting system harmonize to exhibit the extreme design sophistication worthy of Hyundai Card. The space is not that large compared to other usual convention halls. However, the continuous parentheses of the white walls with rounded corners harmonize with the reflective flooring materials and lighting, creating a sense of boundless space. Various activities such as training, presentations, and conferences can be held here. Particularly, the space is for casual communication in usual times with the state-of-the-art system that enables the seats to be hidden completely into the wall and automatically taken out when needed. To prevent the authoritative form of chairs arranged at the same height, time was invested to complete the theatrical composition. Thus, the front ten rows have the same height and rows thereafter are sloped naturally.

When Hyundai Card purchased Building 2 to expand its office complex, the inside was tight-spaced and the lobby on the first floor had a low ceiling. As such, in planning the Auditorium and The Box, more than half of the second floor was removed to turn it into a double-height ceilinged floor, securing a suitable view. Spackman Associates designed the key spaces of Building 2 including the Auditorium and The Box as a glass box. It was intended to symbolize the transparency of the financial company and exude vitality by naturally exposing the flow of employees. Because the Auditorium has glass on three sides, there is a possibility the sound may reverberate or be muted. Thus, sound-absorbing material made of fabric combined with a metal frame called 'soft cells' from Denmark was installed on the ceiling and the back wall for the first time in Korea. This material is easily mountable with magnets or hooks and boasts ease of maintenance and repair. For the seating, the Series 7 Chair by Arne Jacobsen was installed. The chair is unique in that it is fixed to the floor

"Like the Box, the three sides of the Auditorium are also made of transparent glass. To this end, finding a sound-absorbing material was the key. After several months of extensive research, we finally found a product in Denmark that was made of metal yet had the perfect sound absorption ability. For the lecture room, furniture selection was a crucial project. We found an optimal product that rotates perfectly 360 degrees and smoothly returns to place with chairs attached to the table. However, the problem was that the 50 seats had to be evenly spaced. As the product was a set of two, the gap with other sets had to be widened. In order to match the detailed needs of Hyundai Card, we coordinated with the manufacturer for several months. As a result, we were finally able to match the spacing among the seats perfectly. In the process of applying Hyundai Card's meticulous standards, we could also grow significantly."

Mary Spackman, Principal Designer of Spackman Associates – Auditorium and Lecture Room Space Design

with a single pole. The fabric on the bottom and the back of the chair enables comfortable seating. The backrest fabric has a unique wave pattern, which is intended to always create movement in the space even when no one is there. The Lecture Room was designed for the dual purpose of one-way lectures and two-way communications. Spackman Associates searched all over the world to find a table combined with chairs that rotate 360 degrees so that people can have conversations with a person behind them. The 50 moving chairs automatically return to their original positions when the sitter stands up. Tables were arranged in an arc rather than a straight line for smooth conversation among people around the tables, and the distance between them was also studied to find the most convenient distance for conversation. The final decision was made by accurately calculating even the height of the stairs, making mockups of heights in 40 cm, 50 cm, and 60 cm with cardboard to measure perfect angles. The front frame for presentations was finished with a white wall without a special device to prevent visual obstruction.

The 50 moving chairs installed in the lecture room rotate 360 degrees, enabling unrestricted communication with anyone nearby.

1.2 Design Lab & UX Lab

Design Lab
Location: 1~2F HQ1 3, Uisadang-daero, Yeongdeungpo-gu, Seoul
Area: 580m²
Design: 2013.11 ~ 2014.02
Construction: 2014.02 ~ 2014.06
Interior: Ateliers Jean Nouvel
Contractor: Dawon ID&C

UX Lab
Location: 2F HQ1 3, Uisadang-daero, Yeongdeungpo-gu, Seoul
Area: 780m²
Design: 2015.03 ~ 2015.06
Construction: 2015.06 ~ 2015.09
Interior: Gensler
Contractor: Kesson

The encounter between Jean Nouvel and Vice-Chairman Ted Chung at the special exhibition hall of the 'Office for Living' during the Milan Design Week shaped the Design Lab in its current form.

The Design Lab is one of the symbolic spaces of Hyundai Card's Yeouido Complex. This key department in charge of numerous confidential projects of the Company is symbolically located at the entrance of Building 1. The UX Lab, the key driver of the Digital Hyundai Card project since 2015, was formed on the second floor separately.

1.2.1
The Way We Build

The Design Lab is one of the core departments and is responsible for numerous confidential projects of Hyundai Card. The space also symbolizes the status of Hyundai Card represented by 'design'. When the original plan envisioned the Design Lab on the first floor of Building 1, some opposed, arguing it must be deep inside of the building where outsiders cannot access easily. Vice-Chairman Ted Chung, however, disagreed. He thought that the first floor with high story height, where visitors are first welcomed, should be left for departments embodying Hyundai Card, not for general offices. Around that time, an encounter with Jean Nouvel at the Special Pavilion of 'Office for Living' during the Milan Design Week inspired the birth of the Design Lab. Vice-Chairman Ted Chung, who was considering how to develop the concept of 'warehouse' worthy of Hyundai Card while researching places like Brooklyn, recalls the encounter as follows: "At first, I thought our encounter would end superficially after about 10 minutes of exchanging greetings. Usually, it is not easy for a French designer and a Korean businessman to communicate in English. But as we start talking about 'office design', our conversation began to heat up to an hour or two. At first, I was worried that it would be too much for a world-famous designer like Jean Nouvel to take on small office space, but inwardly, I started thinking that I would like to 'show-off' properly." As they shared the concept and vision at the first meeting, the process thereafter was smooth. A month later, he visited Jean Nouvel's office in Paris and received the first design results. Little has changed since this design during its implementation. An ideal structure that is open yet with perfect confidentiality has been realized where the middle lattice space is visible from the entrance without exposing the office space.

To the design, Hyundai Card simply added VITRA furniture. Jean Nouvel prefers furniture of a more dignified style that uses more wood, but the furniture naturally fitted with the concept of 'warehouse-type office', which included a design of portable filing cabinets. Hyundai Card did not request even minor detail corrections. This was directly applying the principle unique to Hyundai Card—i.e., arbitrary and partial changes ignoring the overall concept can cause serial issues. The UX Lab was the work of Gensler. The outcome was not flawless. There were too many staffs in the department and the necessary elements had to be 'squished' into the space. Nevertheless, the space is evaluated as quite a success in terms of practical use after applying differentiated elements in reference to the HCA method, such as cascading conference rooms and height-adjustable desks.

The interior of the Design Lab, the double-height ceilinged structure, has exposed finishing materials of the floor and the walls. Additionally, a large rectangular plastic case is placed on the shelf, emphasizing its warehouse-like atmosphere.

"I have always thought that modern offices are too monotonous and dull. Functionality was emphasized for office furniture and furnishings; however, those were repetitive and lacked individuality. For modern dwellers spending most of their time in the office, it is crucial to have an office with a unique personality that enables unrestricted attitudes and exchanges. When I met Vice-Chairman Ted Chung at the Milan Design Week, I determined that he was the CEO who perfectly understood my desired concept. A design lab is a space where a wide variety of works and activities take place. Sometimes you must hide your work or must reveal them. Flexibility must exist to combine or divide the space. Ultimately, we have created an ideal place where people are free to choose their location at any time."
Jean Nouvel, CEO of Atelier Jean Nouvel – Design Lab Space Design

1.2.2 Point of Space

In front of the entrance to the Design Lab, a glass wall is present with embedded transparent LED elements with a height and width of over 2 m. From the beginning, Jean Nouvel wanted to create a space that is see-through from the outside. The technical barriers for this design were solved by embedding LED elements on a transparent and thin acrylic bar one by one. Passing through the entrance, one sees a transparent glass wall resembling a grand wave dividing the space. It is a space used as a meeting or reception room, and a whole glass from Lasvit, a Czech interior design brand, was utilized. Each glass plate weighs 700 kg with a rugged surface.

Inside the double-height ceilinged Design Lab are exposed finishes of the floor and the wall, conveying the feel of a warehouse. On the first floor, archiving shelf mobile racks moved by turning handles cross the office space and hallway. As a large rectangle plastic case is placed on the shelf, the ambiance

Once one enters the entrance, a transparent glass wall resembling a huge wave compartmentalizes the space at large. The whole glass from Lasvit, Czech Republic was utilized.

of a warehouse becomes even denser. All walls, including the second floor, facing the outside or lobby are made of clear glass. The glass used for the windows is a double-layered special glass invisible from the outside. And the shades are tightly installed. One side of the shades by the French brand Langlaff is a matte white material, while the other is a mirror. When the shades are opened on the mirror side in morning times, the sunlight comes in bright enough that there is no need to turn on the lights.

The UX Lab was built separately on the second floor as an extension of the Design Lab. As a forceful driver of the 'Digital Hyundai Card' initiative since 2015, the UX Lab is a stand-alone space. As digital became a social trend, the importance of UX (User Experience) and UI (User Interface) was emphasized in the design realm. Thus, the UX Lab was set up independently to oversee digital design. The UX Lab is without partitions blocking between desks. Since sharing and communicating ideas in real-time is crucial for digital design work, the space was designed to be open.

1.3 Card Factory & TMC

Card Factory
Location: 10F HQ3 18, Uisadang-daero, Yeongdeungpo-gu, Seoul
Area: 1,369m^2
Design: 2012.03 ~ 2013.12
Construction: 2014. 07 ~ 2014.12
Interior: 101 Architects, Wook Choi
Contractor: Dawon ID&C

TMC
Location: 9F HQ3 18, Uisadang-daero, Yeongdeungpo-gu, Seoul
Area: 587m^2
Design: 2012.03 ~ 2013.12
Construction: 2015.01 ~ 2015.03
Interior: 101 Architects, Wook Choi
Contractor: Dawon ID&C

Wook Choi, Head Architect of 101 Architects and the designer of the Card Factory, took the British factories in the 1920s as the motif. The Card Factory has become a must-visit course for industry stakeholders during their visit to the Yeouido Complex.

At the Card Factory, one can see the entire process of card production immediately. TMC functions as the brain of Hyundai Card, enabling visitors to have a bird's eye view of the transactions of Hyundai Card, which averages 3.7 million cases per day.

1.3.1 The Way We Build

The Card Factory stemmed from the idea of mechanizing the credit card production process. Vice-Chairman Ted Chung has been emphasizing the crucial importance of creating uniqueness even for companies without visual characteristics. Expressing the identity of a financial company via abstract methods such as a formula is bound to be limited. That principle of Vice-Chairman Ted Chung expanded to the plan of installing a 'factory' inside the Yeouido Complex where the entire card production process is viewable at a glance. However, roadblocks appeared from the initial research stage. Hyundai Card paid a visit to Tokyo in order to have a meeting with a company which hosted a robot fashion show in collaboration with Chanel. The meeting rather rendered disappointment, only making it clear that pure 'beauty' and 'precise mechanics' could not co-exist. Of course, the case of automobile factories is different. Automobile factories run as a single machine with every element connected. The connection and movement, and the clear reason why the machine should be placed in that position make it more likely to advance the aesthetics of the machine. However, the machines producing cards are segmented. Their reason for existence is unclear. In the end, an artificial arrangement is inevitable by hiding and decorating. After the Card Factory was completed, it received raving feedbacks from most of the visitors. Yet, this is the reason why Vice-Chairman Ted Chung still thinks more could have been done on the project. Nevertheless, the decision to relocate the Factory from Sangam-dong to the Yeouido Complex was valid. The relocation served as an opportunity for better efficiency in utilizing the machines of the Sangam Factory, which was merely spacious and inefficient at the time. Also, it significantly reduced the likelihood of counterfeits of leaked blank cards. Building 3, where IT-related departments are gathered, is a space where you can personally experience how Hyundai Card newly transforms after it declared its transition to a digital company. The Traffic Monitoring Center (TMC), equivalent to the brain of Hyundai Card, is on the 8th floor, where an average of 3.7 million transactions per day can be grasped at a glance. To visit the Card Factory, you must pass through the TMC surrounded by circular glass with a diameter of 14.6 m. This composition, in which the Card Factory as the collective of state-of-art digital and analog is connected, embodies

Installed on the 8th floor of Building 3, where IT-related departments are gathered, is the Traffic Monitoring Center (TMC) where an average of 3.7 million transactions of Hyundai Card per day can be identified at a glance..

M3
M2
M
VISA

PACKING CARD

1.3.2 Point of Space

Hyundai Card reinvented financial capital, which is discussed only in numbers, by amalgamating the perspective of industrial capital—factories. Inspired by a space that symbolizes the 19th century's industrial revolution as well as the original form of today's factories, the Card Factory was designed to offer visitors the unique sentiment of industrial capital and the joy of nostalgia. The ceiling is installed with large-sized lightings that reinterpreted factory chimneys, while elevators, stairs, and furniture made of metal also create the unique texture of a factory. When a customer applies for a credit card, its production is automatically completed in one day and a message is sent to the customer requesting a receipt. Before arriving and collecting the card here, customers can visit the lounge along the iron railings as those installed at automobile factories, enjoy coffee, and have a panoramic view of the Card Factory. When mechanical arms pull out the card and place it on the conveyor belt, customers can watch the entire process of the Card Factory moving automatically, from how small robots collect the cards to how the robots stack them in a unit called 'magazine'. In the TMC, 65-inch curved monitors are hung densely along with the circular space. Insurmountable data including card approval status, call center connection status, traffic, DB access, network, and server usage status are constantly displayed on the monitor. The service operations are indicated by the colors of a traffic light: red (danger), yellow (warning), and blue (normal). The media table installed with an 84-inch large touch screen located in the center of the TMC is the core business control area. There, all screens are shared in real-time, and countermeasures are planned.

In the Card Factory, you can observe the entire process of robotic arms pulling out cards, placing them on conveyor belts, then stacking them in units called 'magazine' after collection.

1.4 the Box, Café & Pub

the Box
Location: 1F HQ2 3, Uisadang-daero,
Yeongdeungpo-gu, Seoul
Area: 283m²
Design: 2009.01 ~ 2009.09
Construction: 2010.02 ~ 2010.08
Interior: Spackman Associates
Contractor: Janghak E&C

Café & Pub
Location: 1F HQ1 3, Uisadang-daero,
Yeongdeungpo-gu, Seoul
Area: 148m²
Design: 2017.05 ~ 2017.10
Construction: 2017.11 ~ 2018.01
Interior: Spackman Associates
Contractor: Dawon ID&C

The motif of the overall design of Café & Pub stemmed from the fashion show image of Thom Browne, which Vice-Chairman Ted Chung had in mind for 10 years.

the Box, an employee-only dining and café space, and Café & Pub, which is open to the public, are located on the first floor of Building 2 and Building 1, respectively. The spaces allow you to get a glimpse of Hyundai Card's efforts to invest in concept and detail even when creating a restaurant or café.

1.4.1 The Way We Build

Before mentioning the Box, the correlation between Buildings 1 and 2 of the Yeouido Complex must be discussed. After Hyundai Card moved into Building 1, which was owned by Kia Motors, the business experienced rapid expansion. As such, Hyundai Card purchased Building 2 across the street as well. Buildings 1 and 2 are like twins. With their calm finish without flashiness, the buildings never become old-fashioned even after many years. Unlike buildings in developing countries where elements are added continuously to stand out, the design is in line with the buildings of developed countries where the emphasis is placed on management, allowing them to retain their dignity even after decades. As such, the tone and manner of Hyundai Card could be transplanted into the buildings without difficulty. Vice-Chairman Ted Chung affirms: "Usually when purchasing office buildings, many focus only on the price. But, from the perspective of operating a company, the most important part is whether synergies can be created upon expanding the office building as the company grows. Disconnected and fragmented companies are likely to face headwinds eventually due to impediments in communication. If we hadn't bought Building 2 because of the price, the loss would have been hundreds and billions of dollars or more. As an added bonus, we also had the effect of having the public road between the two office buildings as a space for Hyundai Card." As Hyundai Card moved into buildings constructed a long time ago, there were limitations. A ceiling height of 3 m is often ideal. However, the ceilings of the Yeouido Complex were about 2.6 m high. The goal was to raise the height by more than 20 cm overall. Considering the total area of the ceiling, the height could be raised by 10 cm only when all pipes and wiring are rearranged. Of course, it was a task that required a significant amount of money, time, and effort. Still, Vice-Chairman Ted Chung had an unwavering will. He argued that the openness the height provides has a far greater influence on the emotions and creativity of employees than good furniture and equipment may. The initial interior

While renewing Building 2, the lobby with vast openness was secured. Then, the Box was created, stretching outward to Building 1.

CAUTION

concept of the Box was a sports bar. The large monitor inside is also inspired by a British pub where people can enjoy watching soccer together. Since Hyundai Card hosts numerous events for employees and invitees at its office buildings, naturally, the idea of a space with a stage was generated. The final concept was decided as a dynamic fashion show runway. Although it is now no longer present, the Box originally had a device called 'Wailing Wall'. When Vice-Chairman Ted Chung visited the New York Times Building during the Insight Trip, he was impressed with the real-time articles being displayed on the on-site monitor. By reinterpreting the device to fit Hyundai Card, he came up with the idea of delivering honest voices of customers in real-time. While employees are eating in the Box, 60 small LCD monitors displayed unedited opinions of customers even including swearing. This led to the process of listening to customers' opinions with transparency and finding improvements throughout the Company. Café & Pub is the only space in the Yeouido Complex that is open to the general public. It is rated higher than the Box in terms of practicality. Café & Pub was designed to emphasize the identity of an office despite being a café, according to the popularization of shared offices, rather than highlighting the limited function as a café. The overall design motif stemmed from the digital image of Thom Browne, a fashion brand, which Vice-Chairman Ted Chung had in mind for 10 years. That is, the office image of the 1970s published by Thom Browne in the mid-2000s became materialized into space. The mood and interior concept of Café & Pub were from the images of steel desks, simple hangers, and men sitting one after another in suits and wearing iron glasses.

At the center of Café & Pub is a fire place. The key furniture, such as tables and chairs, was inspired by a digital image published by Thom Browne in the mid-2000s that had the motif of offices in the 1970s.

COLDPLAY

1.4.2
Point of Space

Building 2 was originally used by a fashion company. Its first floor had cramped stores, giving a feeling of stuffiness due to its low ceiling height. Spackman Associates in charge of the renovation secured a high ceiling height by leaving only part of the second floor intact. In addition to the lobby with openness, the Box was created to extend outward toward Building 1. The exterior wall is made of transparent glass to emphasize the transparency of the financial company. The interior was completed in the form inspired by a fashion show runway, with tables and chairs arranged across the center stretching out along the spotlight toward Building 1. Currently, the space has tables and chairs scattered all over. The inside where the counter is located is in one platform higher, which allows small performance to be held like a mini stage. Thus, it was designed to hold more diverse events by utilizing the front yard as well. Café & Pub has traces of the inspiration after Thom Browne's digital images in the desks, chairs, hangers, and other furniture and furnishings. The same reason is behind using mainly metal materials. Spackman Associates invested various efforts to create a vintage feel by finding hooks from the 1970s to complete the coat stand and applying a dark brown color to the panel using a car paint method. To prove that the vintage tone and manner are not rough, high-quality materials were utilized for the wall lamps and fireplace. Added here was the emphasis on the ceiling pipeline inspired by Soho House, New York. Also, the desks and chairs used in the Hyundai Capital London office building (HCUK) were brought in. Finally, the stylish ambiance of the Office Building was completed by installing a fireplace that gave a warm and luxurious atmosphere as the last element.

In Café & Pub, desk and chairs used in Hyundai Capital's London office building were brought in. The furniture is mainly made of metal.

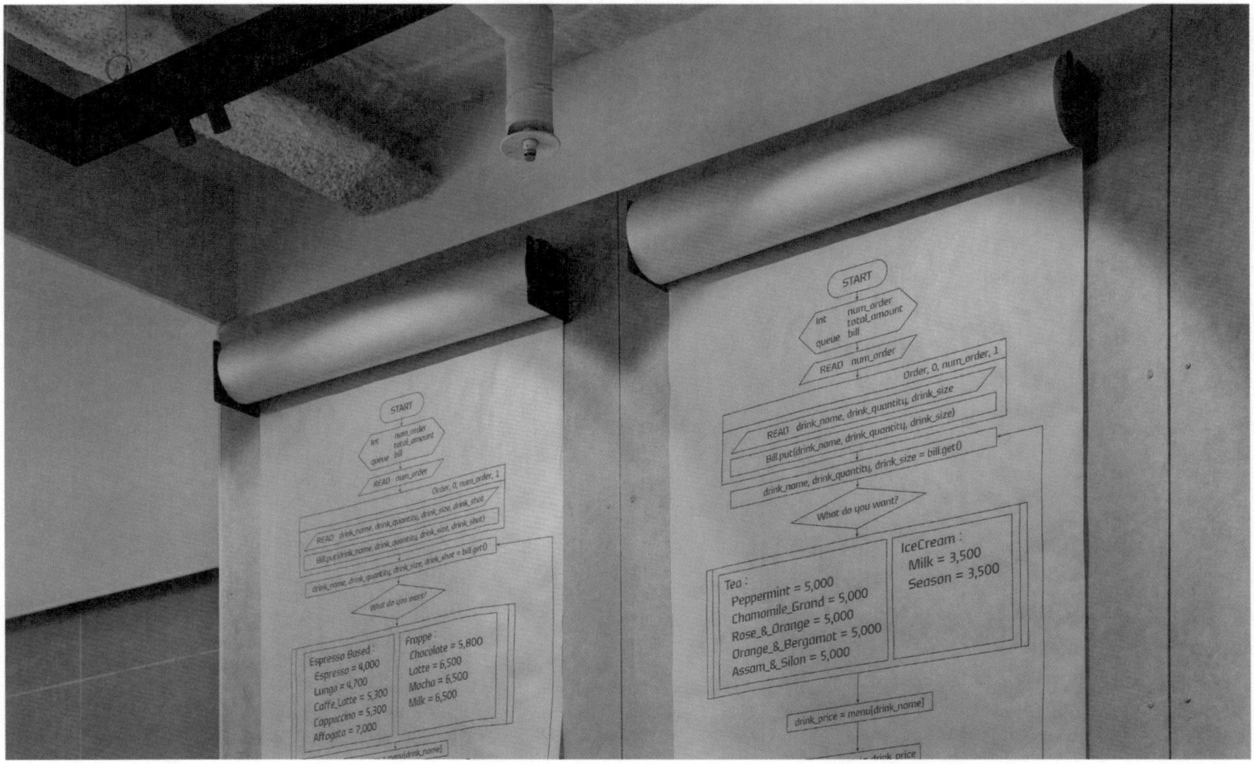

1.5 Workspace

Workspace
Location: 3~4F HQ1 3, Uisadang-daero, Yeongdeungpo-gu, Seoul
Area: 3,000m²
Design: 2019.03 ~ 2019.06
Construction: 2019.06 ~ 2019.09
Interior: Gensler
Contractor: Dawon ID&C

The Yeouido Complex of Hyundai Card pursues a new office building design even for general office spaces as well as special-purpose spaces to transform into an agile organization. By applying slightly different concepts to each floor instead of keeping the consistent design, the plan is to enable employees to experience a variety of environments.

The inconvenience of having to use an elevator to move one floor has been eliminated by making stairs that connect the third and fourth floors. The interior was boldly exposed and finished with metallic materials. The inside is structured as an open space without partitions.

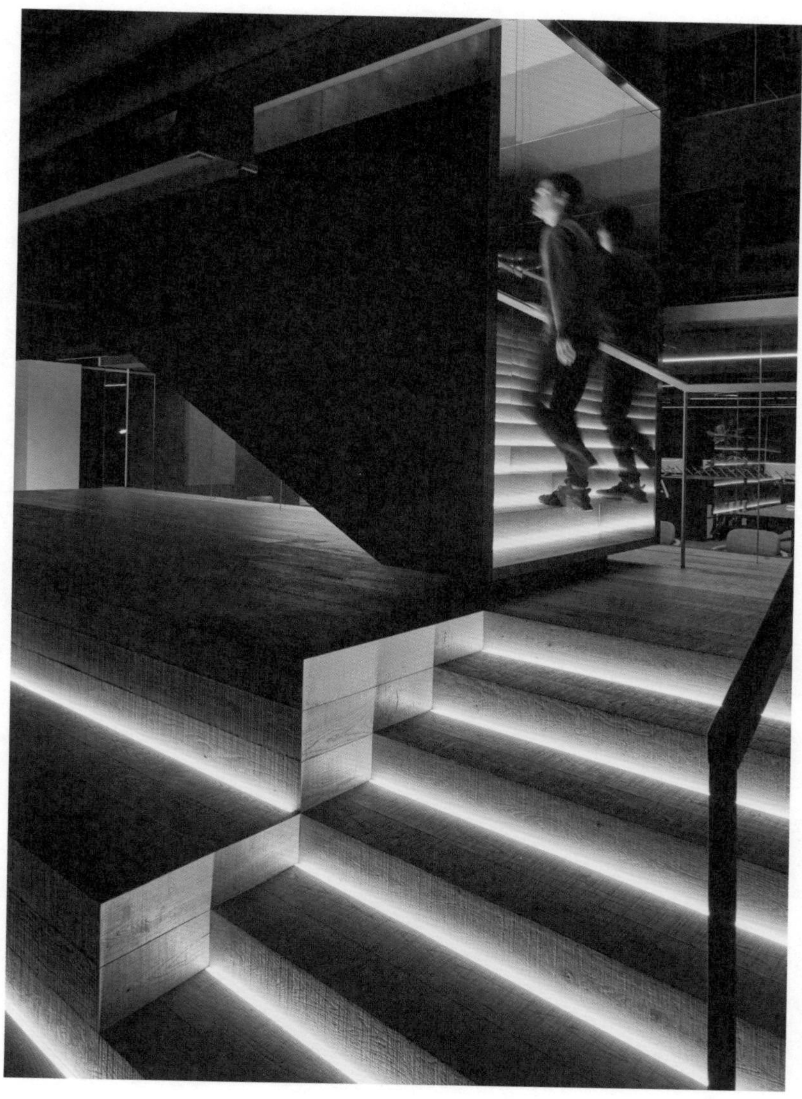

1.5.1
The Way We Build

Since 2002, the Yeouido Complex has undergone a continuous transformation. In the early days, it focused on creating a space to define and express the corporate culture Hyundai Card pursued. After 2015, in particular, as the corporate culture has moved toward a more liberal direction, not only special spaces like the Convention Hall and Café & Pub but also general office spaces experienced diverse trials and changes. That is, it is pursuing a new design for its office building to transform into an agile organization. The old practice of requiring CEO approvals whenever teams were created and disbanded was eliminated. Instead, an agile system that allows teams to be created and disbanded freely under the authority of the head of the department was introduced. As such, the company needed more open-structured offices suitable to the innovative operation of the organization. Accordingly, the 3rd to 4th floors and 8th to 9th floors of Building 1 were renewed first under this policy. Instead of forcing employees to move to a different place inconveniently by constructing or changing the office building in its entirety, Hyundai Card approached it by sequentially changing necessary elements. Another intention was to allow employees to enjoy visiting other types of offices whenever they go to other departments.

The first concept for the new office building was set by Gensler. Dedicated desks were abolished, giving employees their own lockers and portable desks. As the individual executive offices were eliminated, the meeting rooms and common spaces could be expanded. Also, the office layouts were able to be changed frequently. This meant that it became possible to flexibly operate and arrange the organization from time to time according to the situation. The second concept was handled by Spackman Associates. While keeping the concept of the new, agile, and swift organizational culture reflected in the renewal of Building 1, A warmer, more comfortable, practical, and efficient space was created on the 8th floor of Building 2. Without being restricted by a specific style, decorations were removed, and the space consisted solely of basic shapes and finishes. Vice-Chairman Ted Chung suggested using chalks on black chalkboards instead of markers on whiteboards. The suggestion was to breathe the warmth of nostalgia into the office space. The experiment to induce a more comfortable and open vibe by Hyundai Card is still ongoing.

Wall Talkers, on which employees can easily write and erase, were set up in various spaces to enable immediate meeting or sharing whenever ideas come to mind. On the 8th floor of Building 2, which was most recently renovated, nostalgic black chalkboards and calks are installed on each wall.

LTV 70%
Risk : 100% ↓

1.5.2 Point of Space

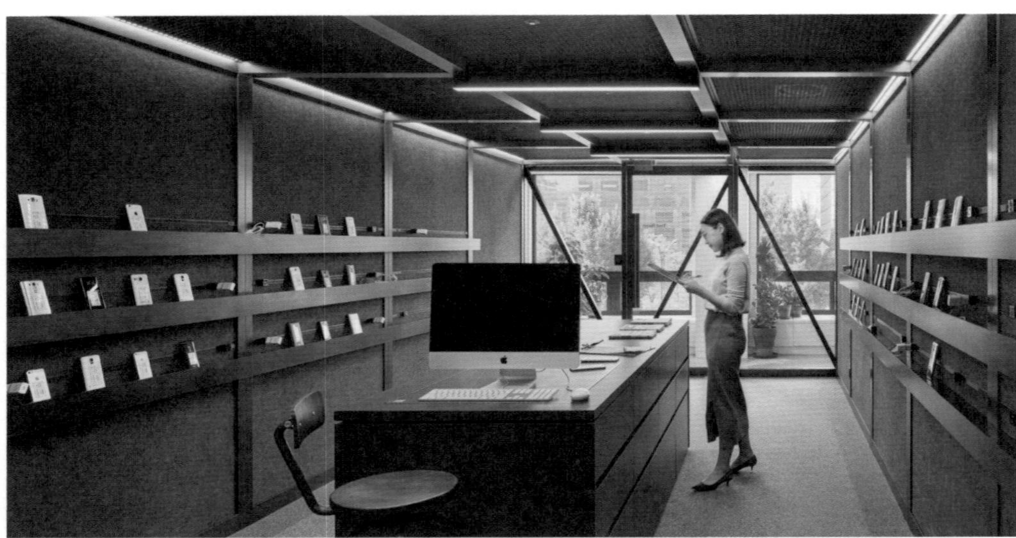

Hyundai Card has recently transformed itself into an agile organization where teams can be freely created and disbanded under the authority of the department head. The Yeouido Complex is undergoing continuous renewal accordingly.

On the third floor of Building 1, which was the first to be renewed, the ceiling seems to be higher compared to other floors thanks to its openness. Whereas the existing ceiling was finished with gray tones, it is now finished with metallic materials while boldly revealing the interior. The interior is an open structure without partitions. In the middle, various seats are arranged for resting or recharging as wanted. A Wall Talker, on which you can easily write and erase, is set up on the wall to enable immediate meeting or sharing whenever ideas come to mind. All meeting rooms have TV monitors installed to quickly connect laptops to share details and derive ideas. In line with the eco-friendly trend, a large sink was installed in an easily visible space so that employees can conveniently wash their tumblers. Plant elements were placed in various spaces to give employees emotional stability, and yellow plastic curtains were installed in each section to create spaces connected as well as separated. The most unique aspect of the third and fourth floors is the staircase piercing through the center. This was particularly due to the inconvenience of taking elevators to move one floor. It is reminiscent of the Frankfurt office building (HCBE), where three floors of a high-rise office building are connected by a spiral staircase. However, unlike the Frankfurt office building that used cold metal, Building 1 used wood materials and appropriate lighting, giving warmth and eye-catching effect. The 8th floor of Building 2, which was most recently renewed, strived to create a warmer and more comfortable ambiance while reflecting the agile system. Top plates of desks are often made of wood or laminate. However, linoleum was selected as an indoor flooring material. This is because it is an eco-friendly material that provides a comfortable and warm sentiment. Although office floors are usually carpeted to reduce noise, the same reason is behind using corks in hallways and lounges of Building 1. Analog black chalkboards and chalk installed on each wall also contribute to creating warmth. In the case of the meeting room installed on the third and fourth floors of Building 1, it was criticized for being inconvenient to have sensitive conversations because the room was placed in between desks of employees, despite being good for impromptu meetings. Thus, on the 8th floor of Building 2, the meeting room was placed on the window side with a great view so that employees could have meetings comfortably. The lounge, where employees can work and have meetings, was evaluated as having a rather low usability as it was a large open space. Therefore, on the 8th floor of Building 2, the lounge was removed to secure more space for desks. Also, separately, a small space with a built-in L-shaped sofa was prepared for concentrating on work alone or for two people to have a quiet conversation.

"Vice-Chairman Ted Chung is a client who has an in-depth understanding of the power of space that physically expresses a company or brand. For the Workspace, we wanted it to have a humble attitude of a startup, returning to the beginner's mindset of the company. Employees can perform their best in a space that best suits their psychological state. We strived to create a quiet and cozy space for concentration, an open space for frequent exchanges, and a flexible space for agile reactions to the ever-changing work environment."

Philippe Paré, Director of Gensler Paris – Workspace Space Design

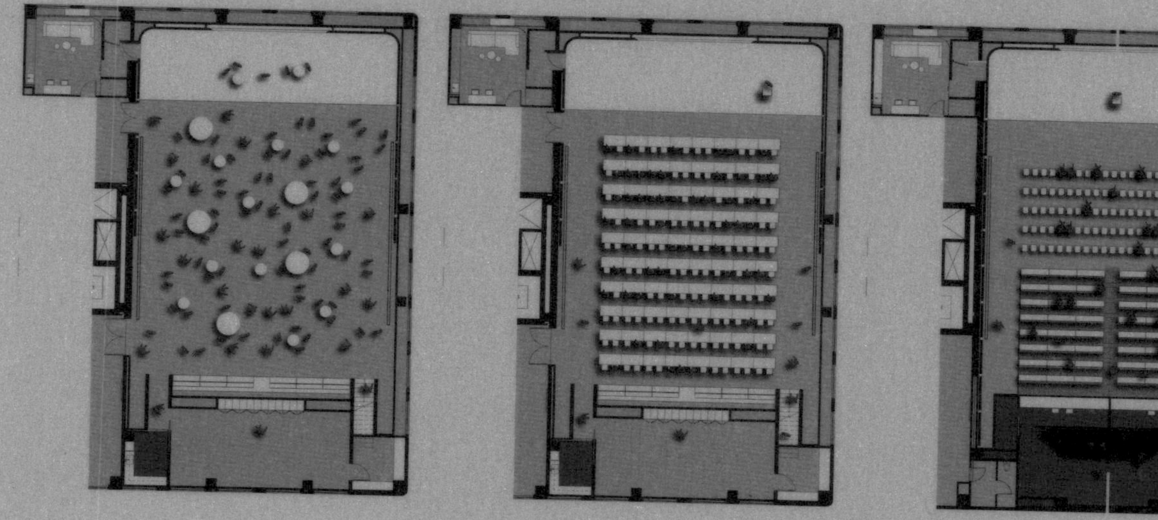

Level 1 - Event Layout

Level 1 - Training Session Layout

Level 2 - Lecture Layout

Section A

Section B

Convention Hall Layout
Convention Hall Section

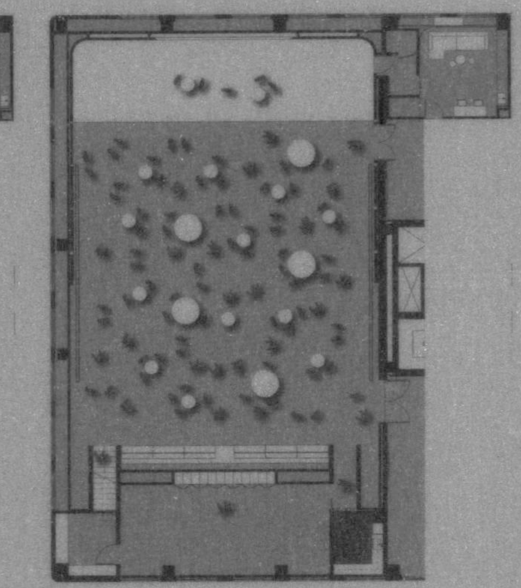
Level 1 - Event Layout

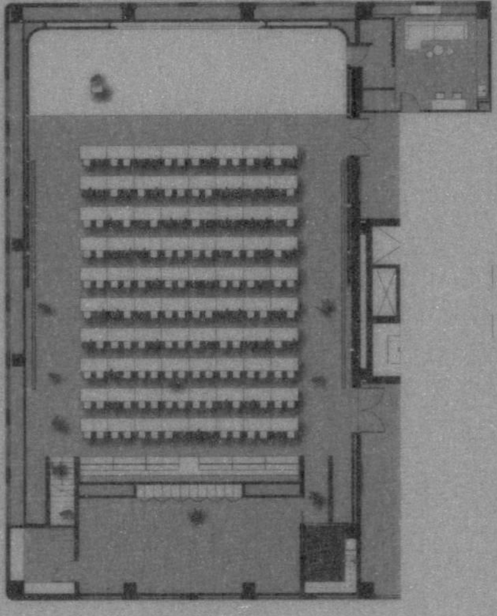
Level 1 - Training Session Layout

Level 2 - Lecture Layout

Section A

Section B

Convention Hall Layout
Convention Hall Section

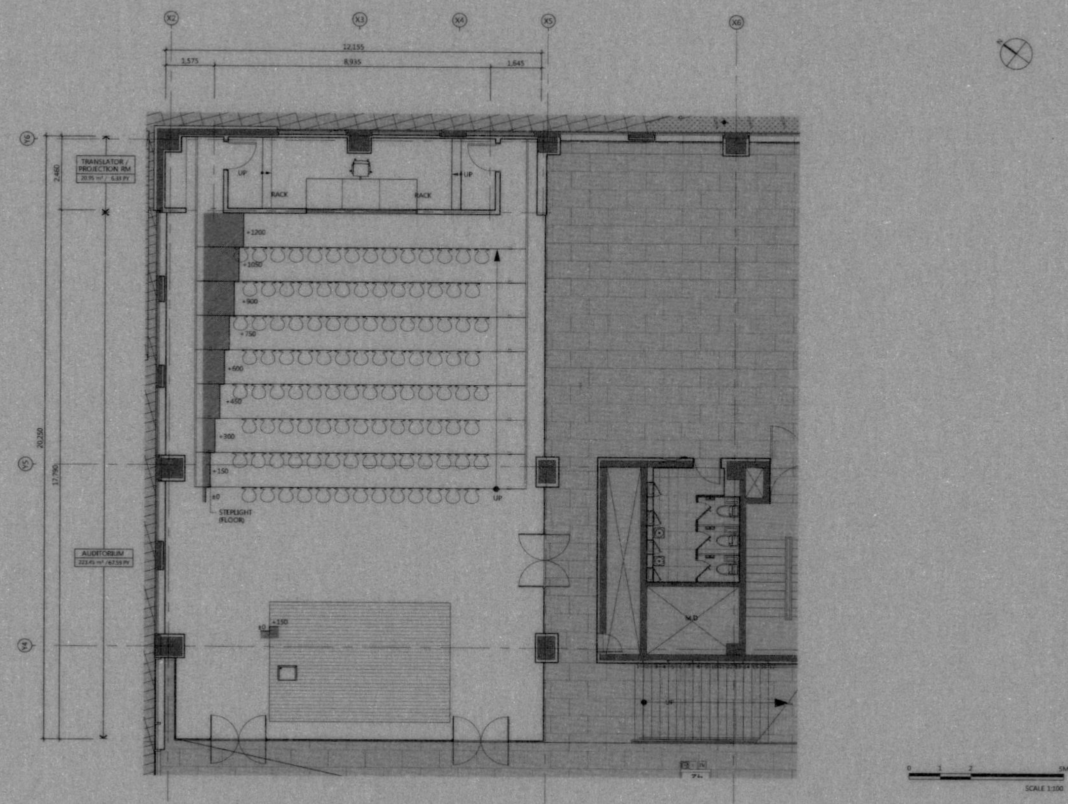

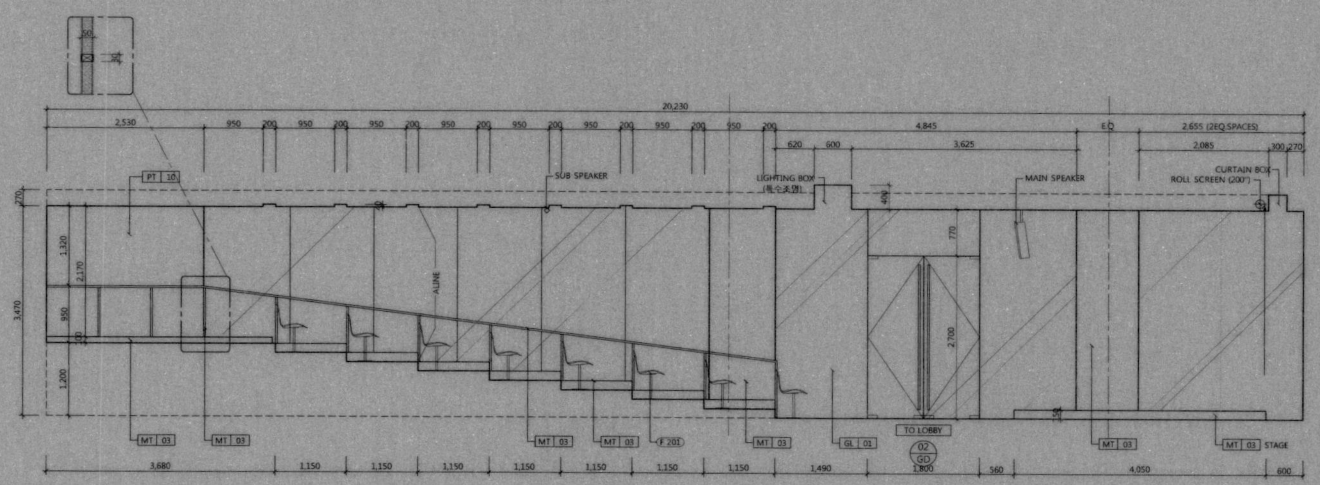

Auditorium Floor Plan
Auditorium Section

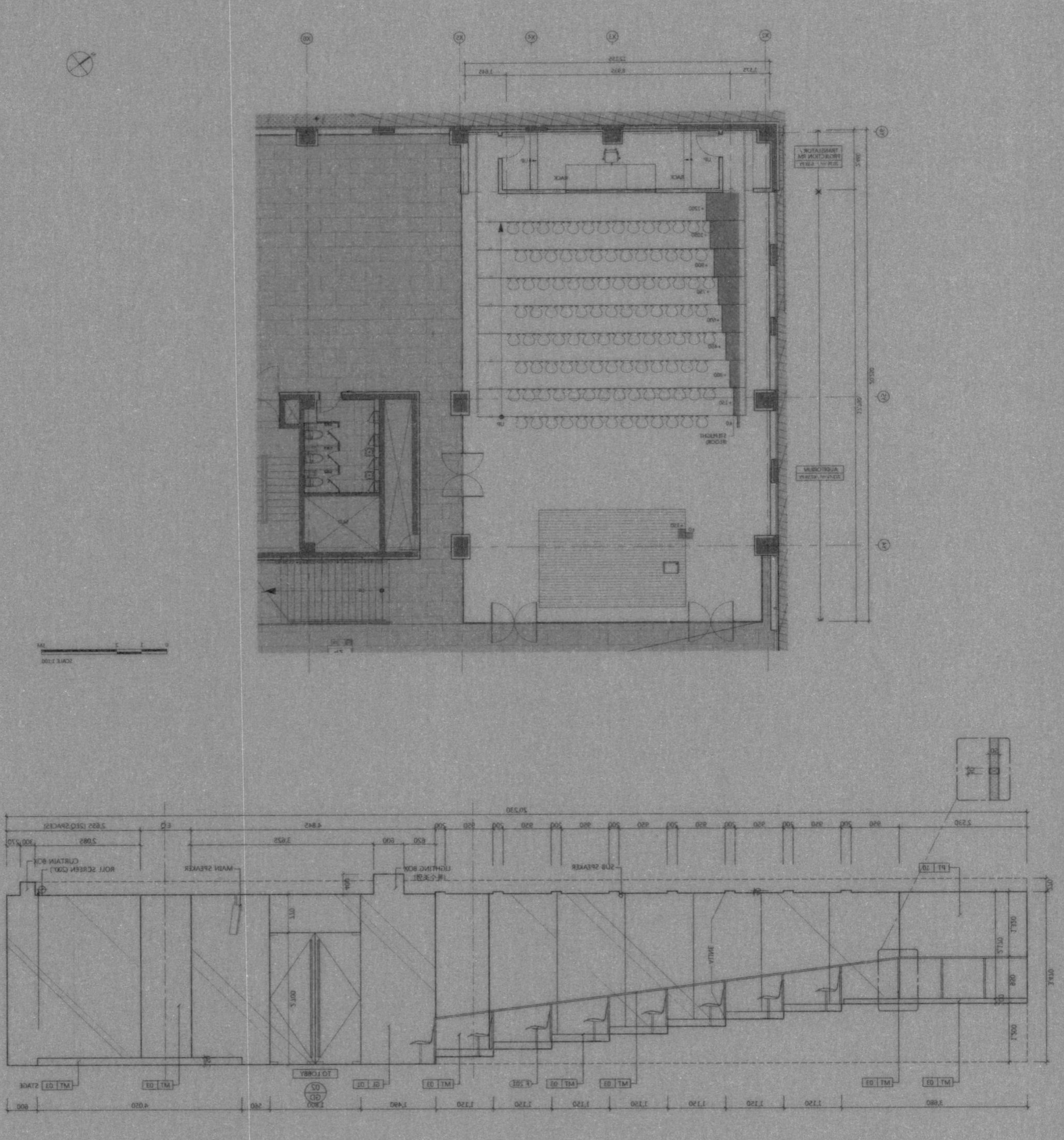

Auditorium Floor Plan
Auditorium Section

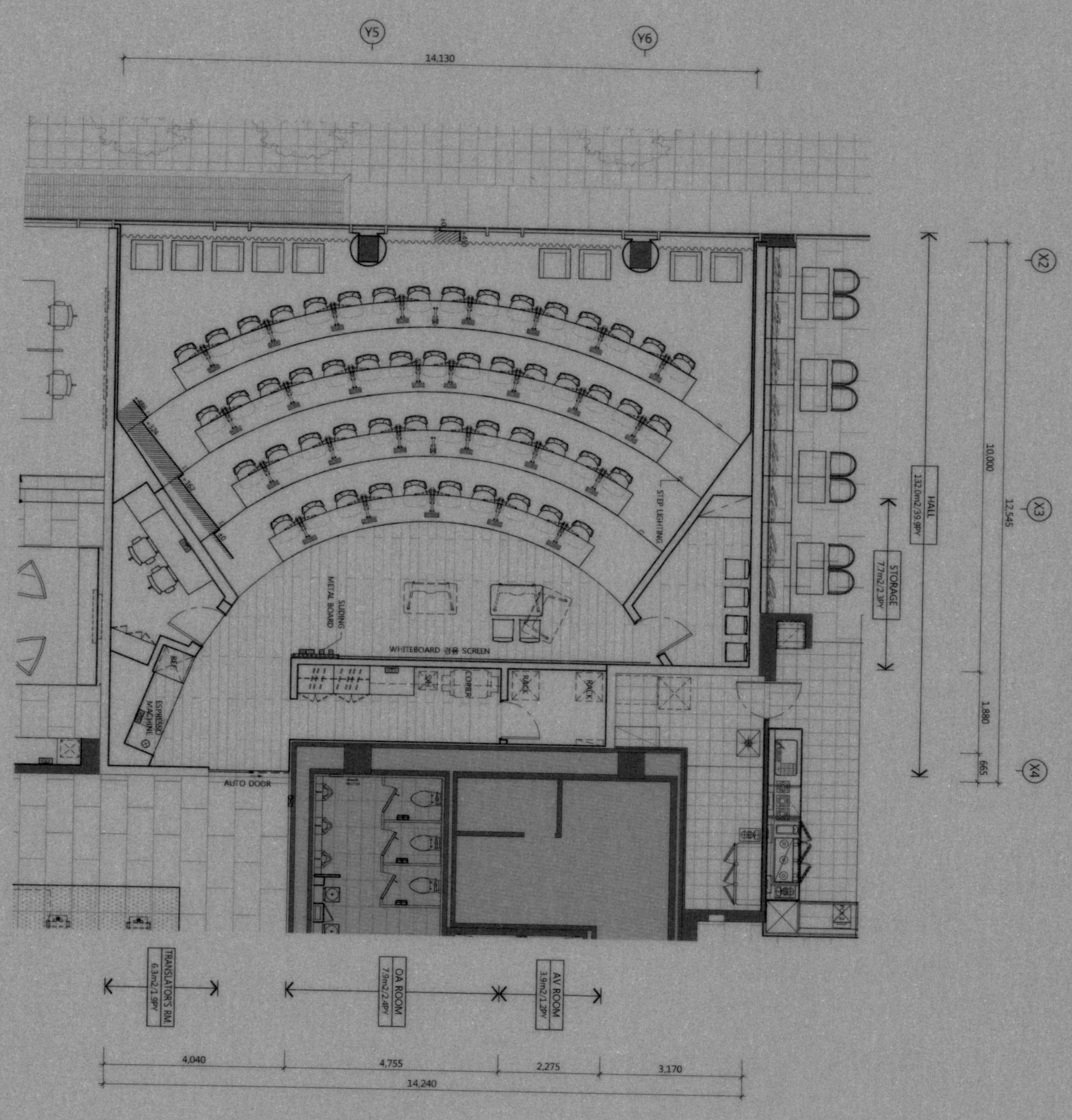

Lecture Room Floor Plan

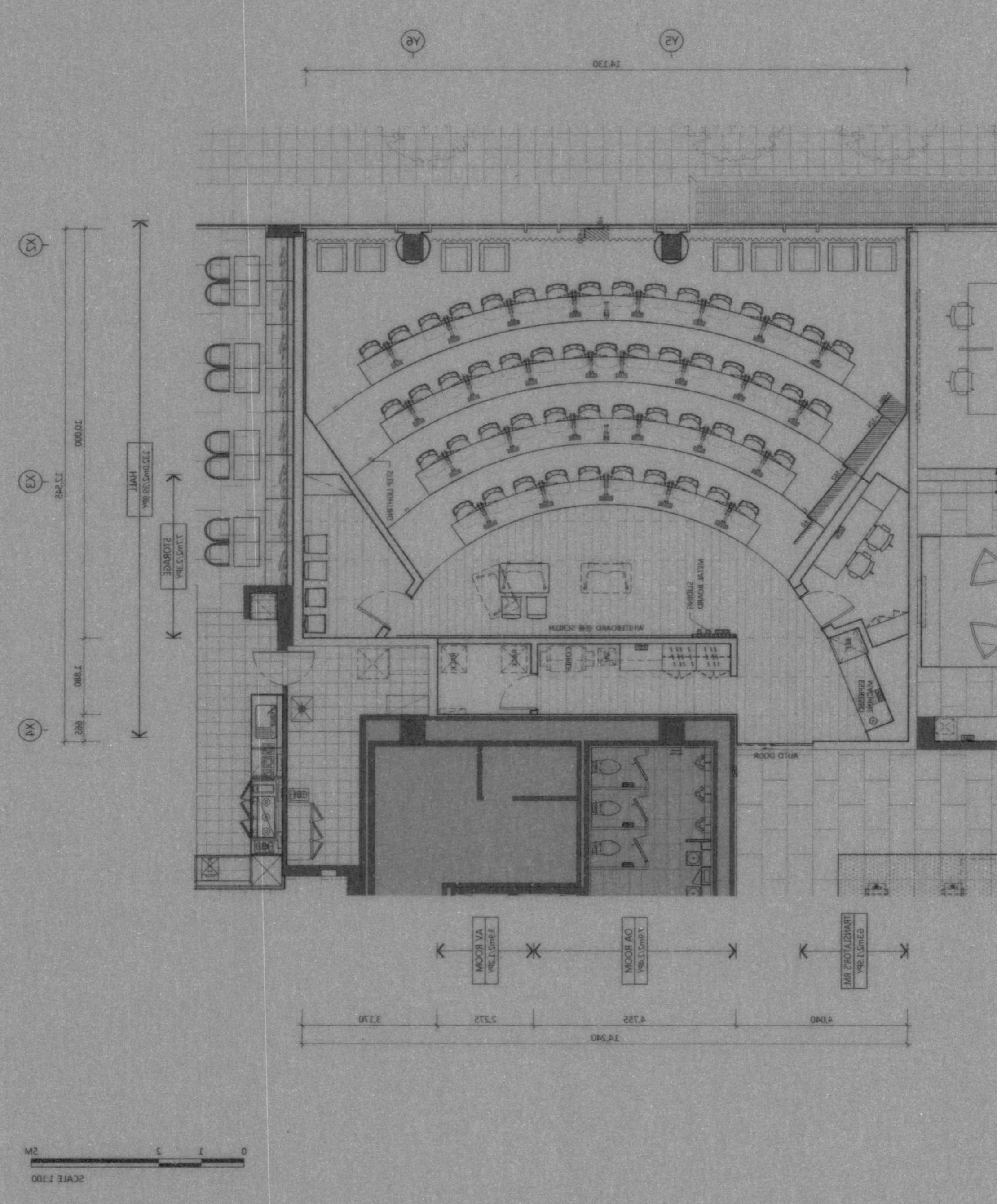

Lecture Room Floor Plan

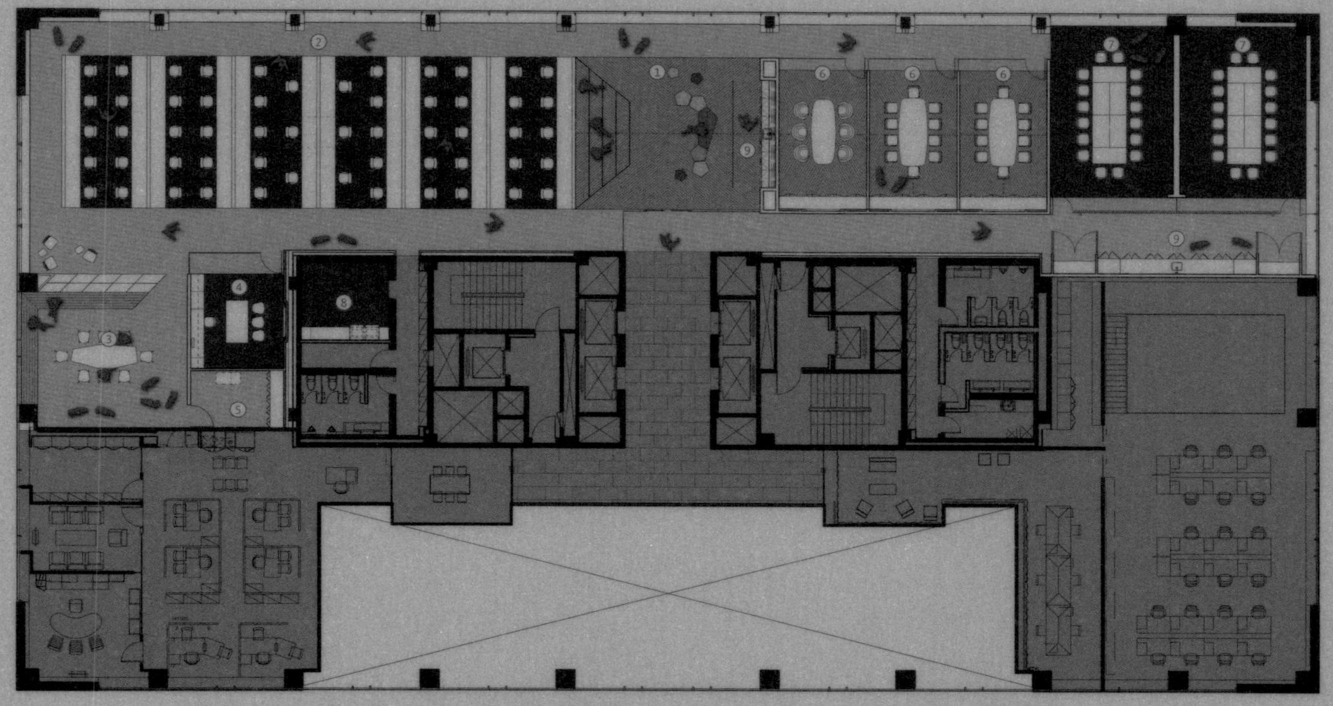

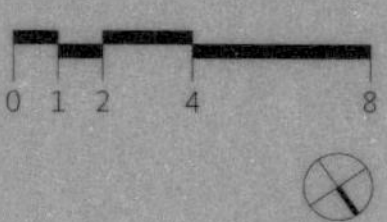

UX Lab Floor Plan

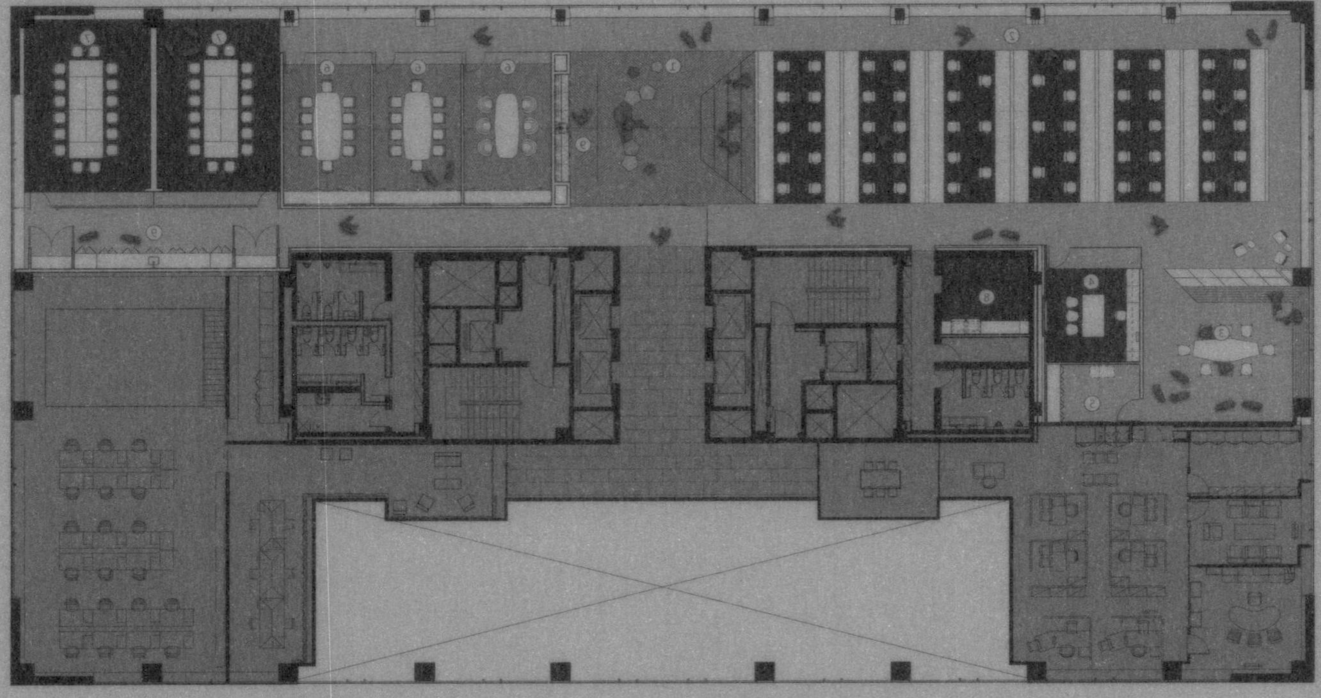

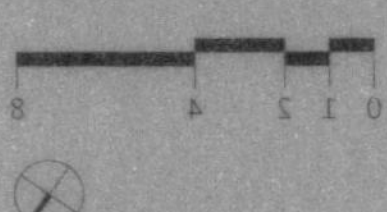

UX Lab Floor Plan

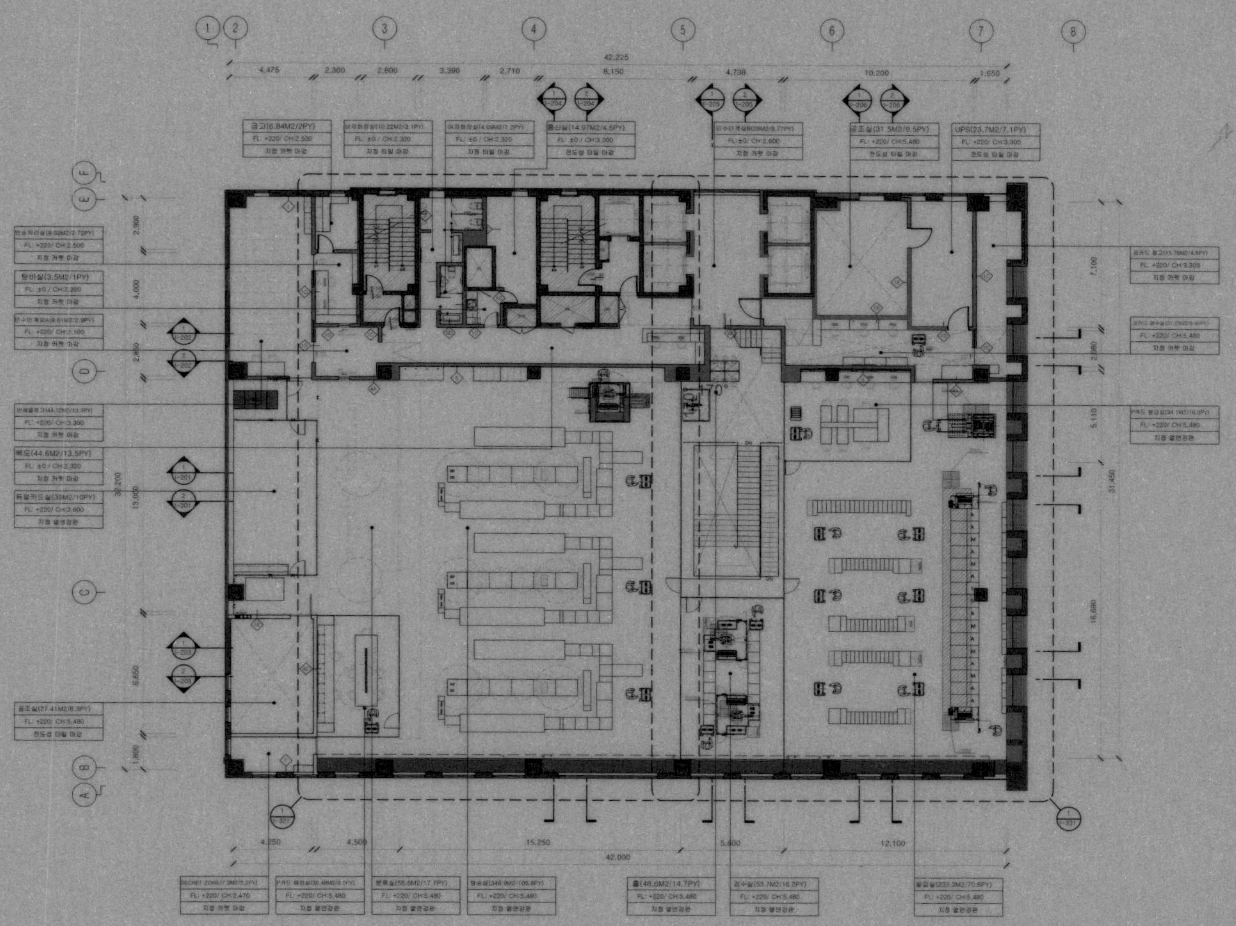

Card Factory Floor Plan

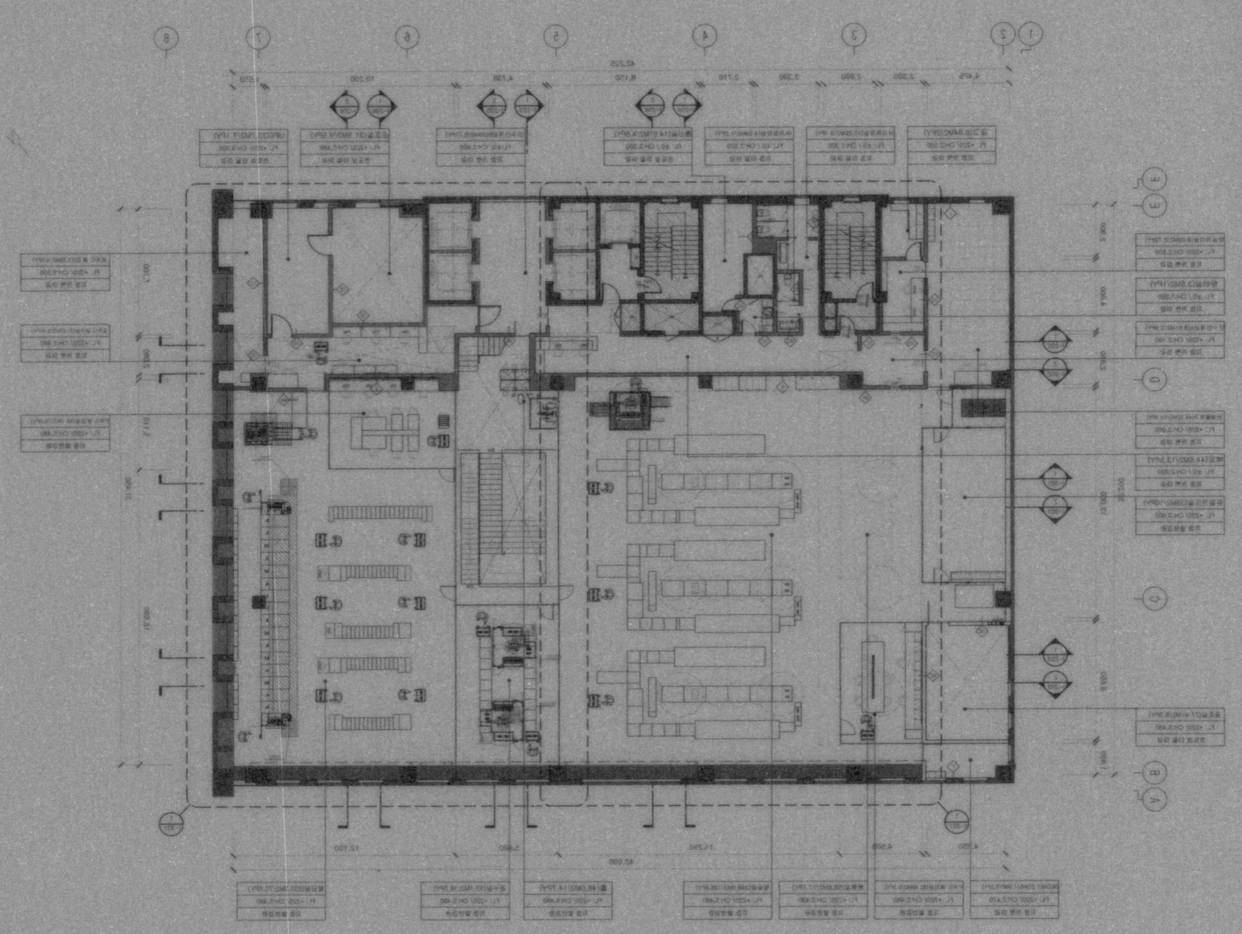

Card Factory Floor Plan

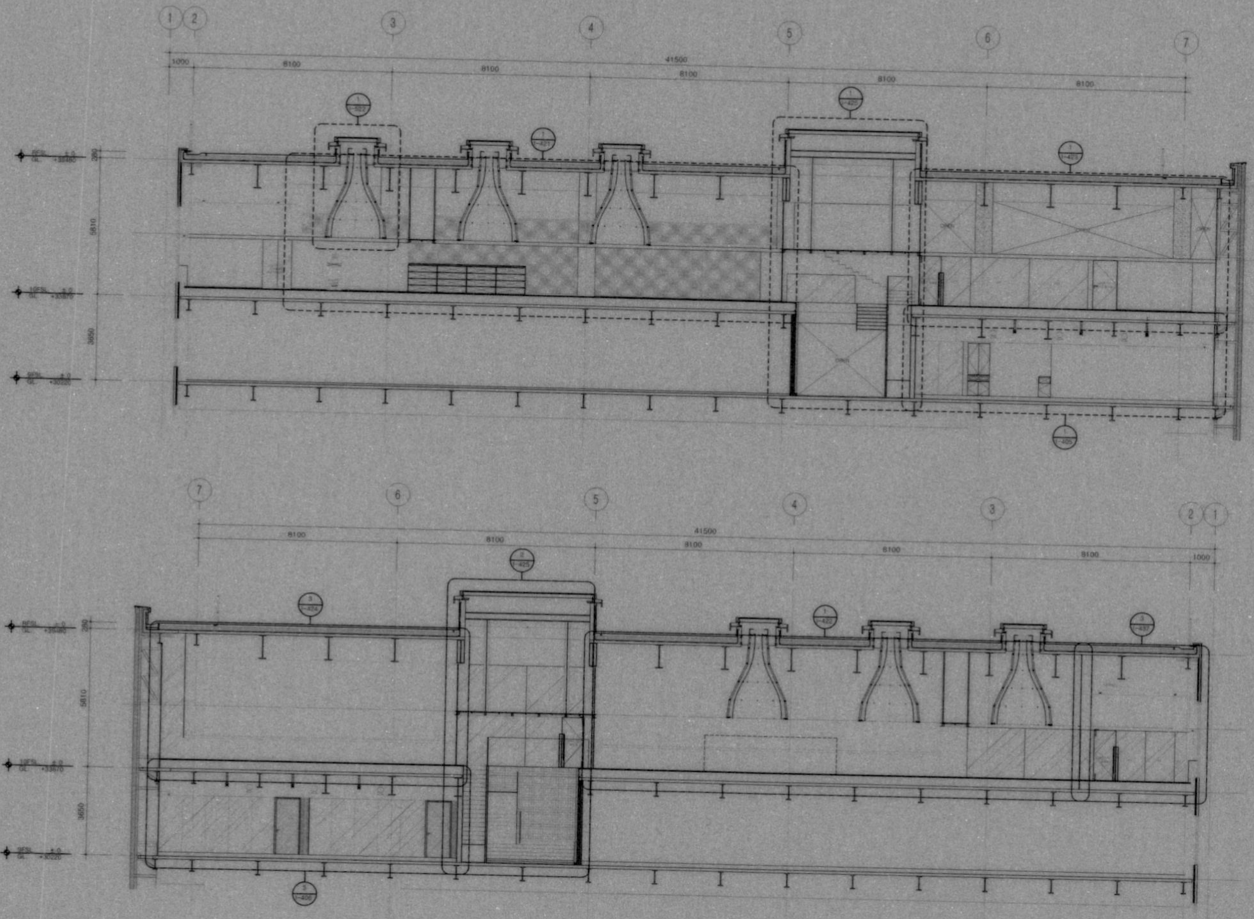

Card Factory Section

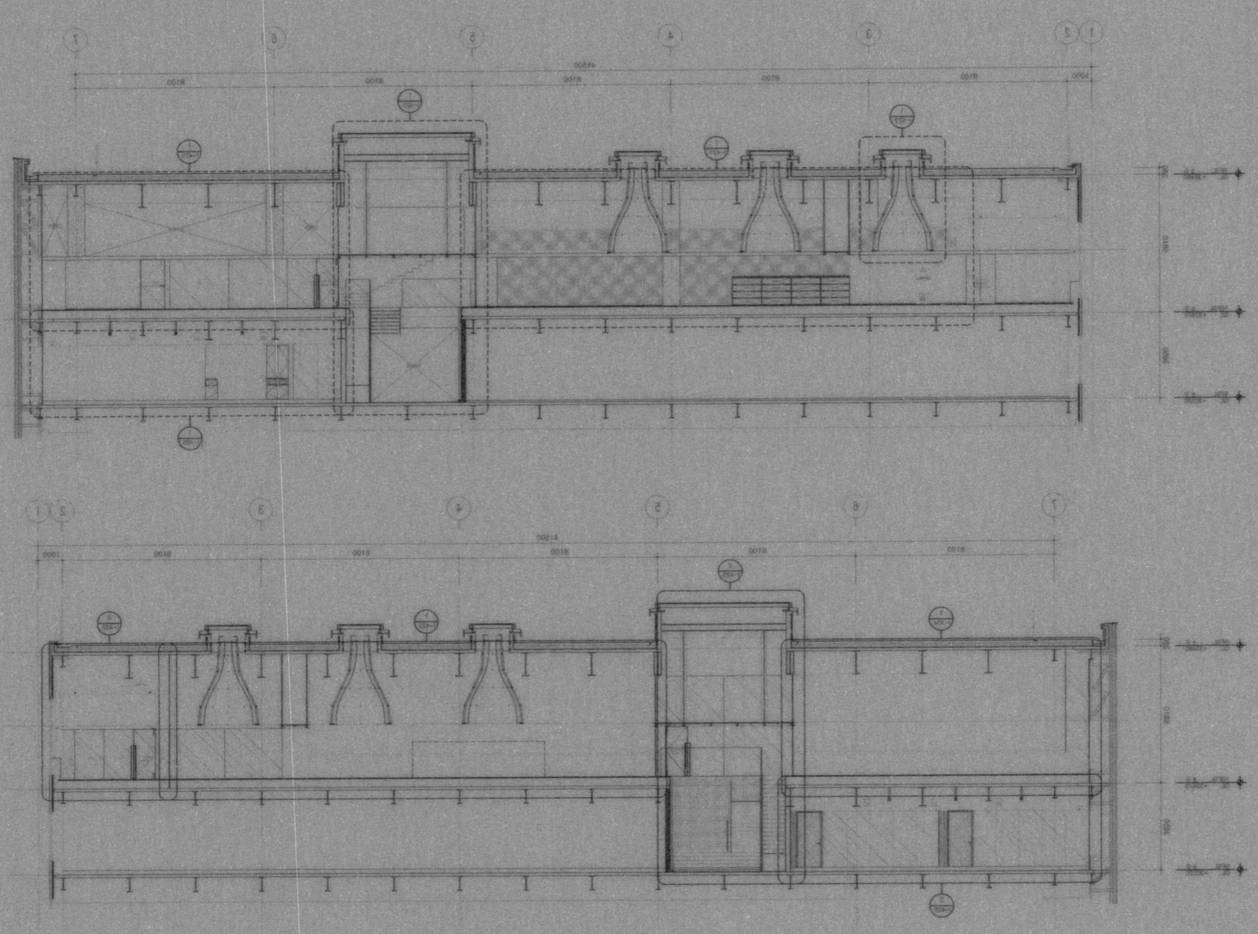

Card Factory Section

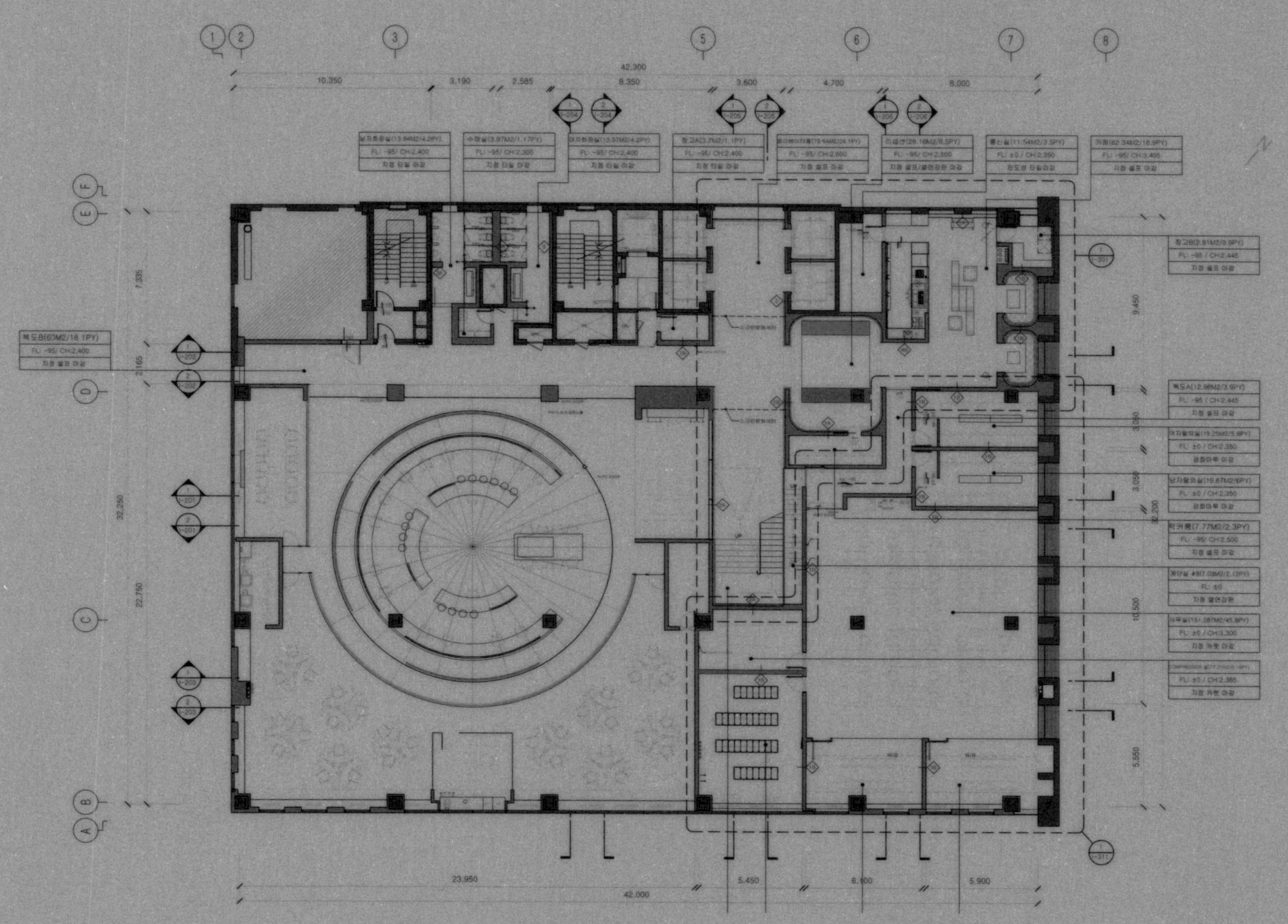

TMC Floor Plan

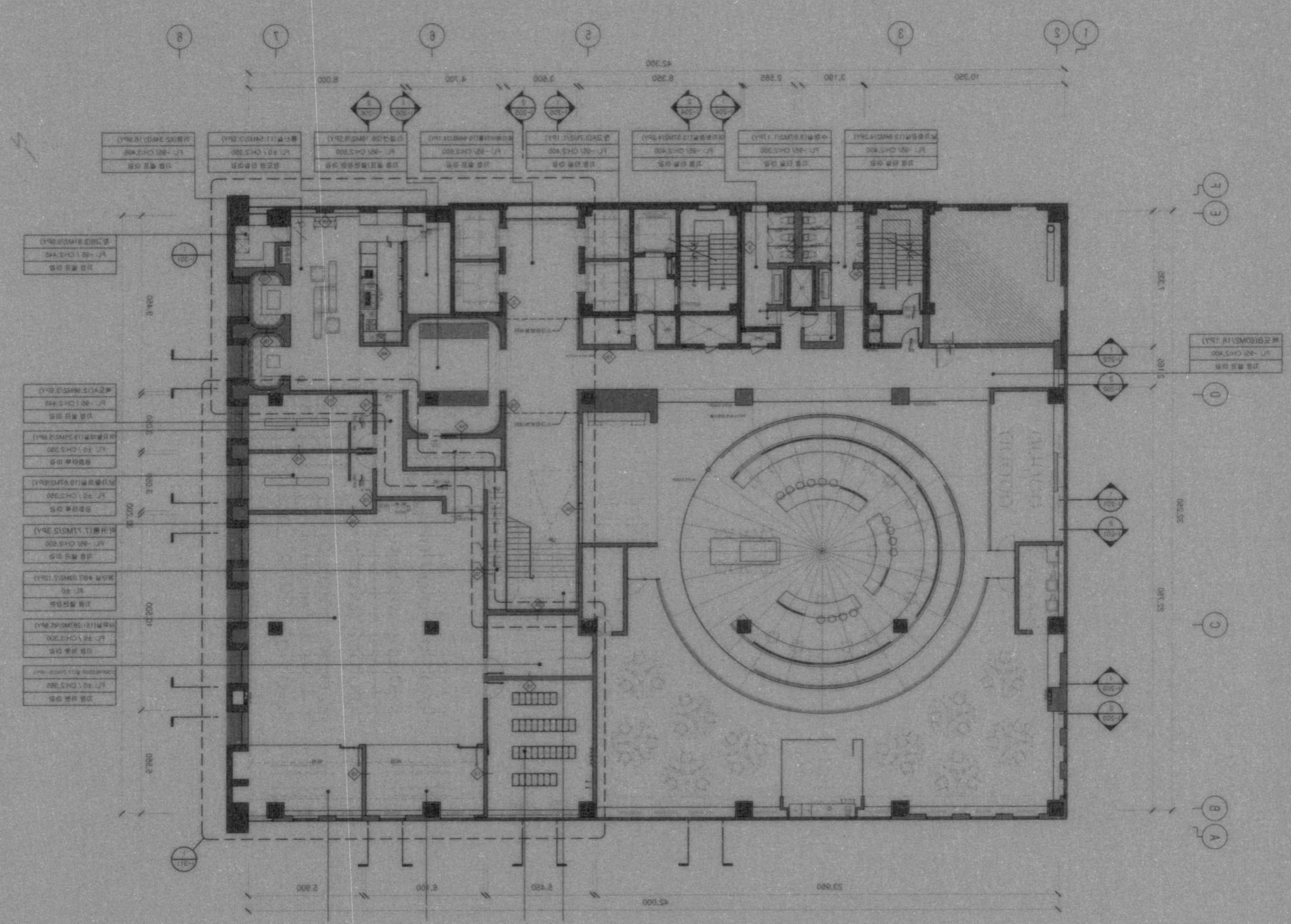

TMC Floor Plan

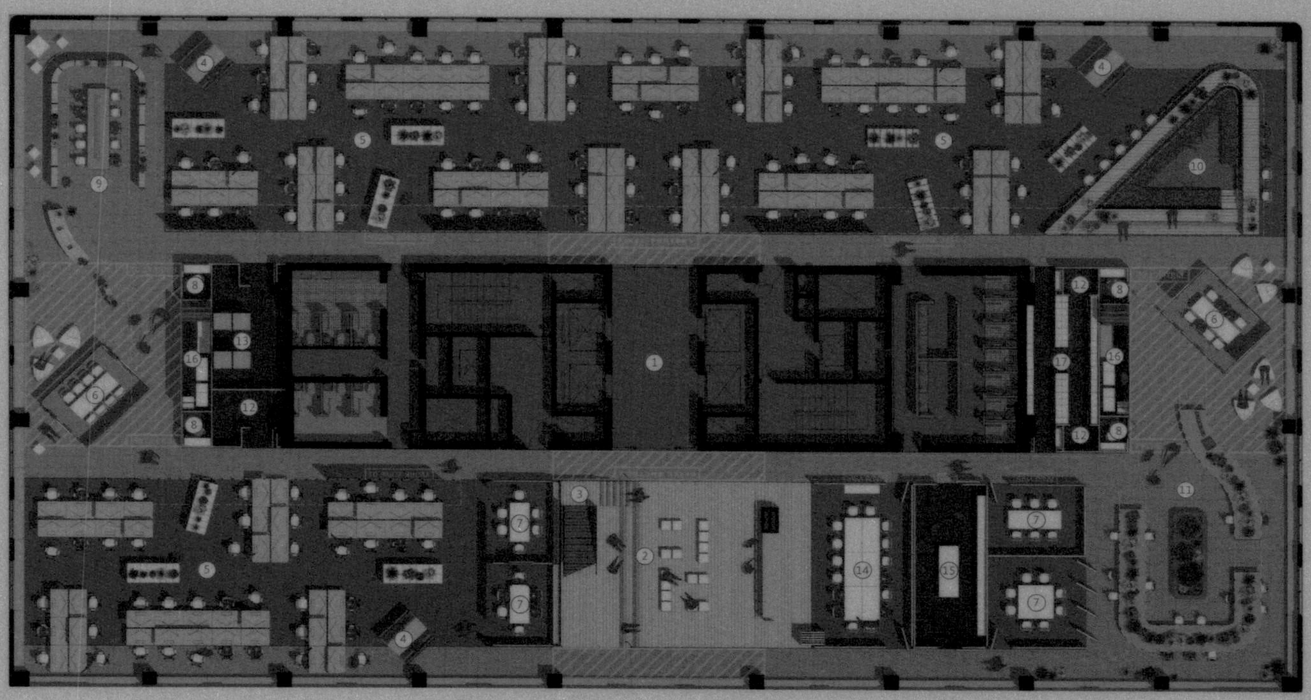

Program

① Lift Lobby
② Lounge
③ Connecting Stair
④ Mobile Meeting Pods - 4 Person
⑤ Flexible Work Zone
⑥ Project Pods / Zone - 8 Person
⑦ Meeting Rooms 6 / 8 Person
⑧ Phone Rooms
⑨ Library / Quiet Work
⑩ Relax Lounge
⑪ Garden
⑫ Storage
⑬ Comms Room
⑭ Conference Room
⑮ Test Room
⑯ Copy and Print
⑰ Lockers

Head Count

Workstations	140	Conference Room
Lockers	140	Phone Rooms
Meeting Rooms	4	Lounge
Project Pods	2	Tea Point
Mobile Meeting Pods	3	Copy and Print
Test Room	1	Alternative Work

Key

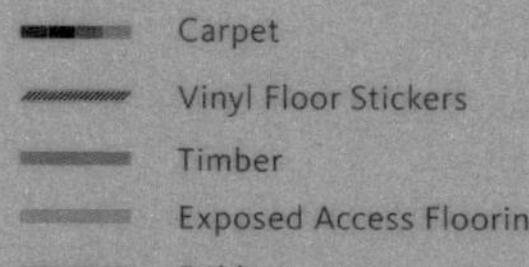

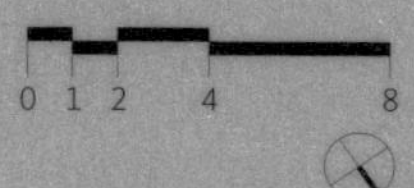

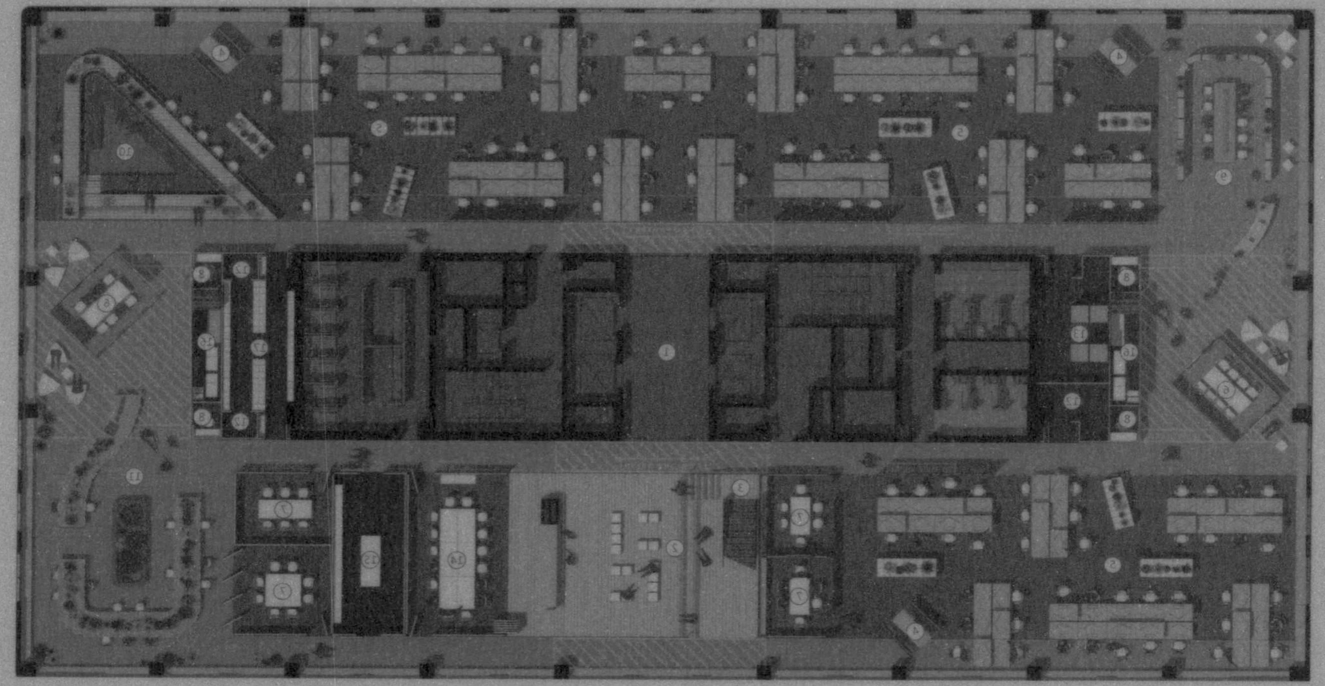

Program

1. Lift Lobby
2. Lounge
3. Connecting Stair
4. Mobile Meeting Pods - 4 Person
5. Flexible Work Zone
6. Project Pods / Zone - 8 Person
7. Meeting Rooms 6 / 8 Person
8. Phone Rooms
9. Library / Quiet Work
10. Relax Lounge
11. Garden
12. Storage
13. Comms Room
14. Conference Room
15. Test Room
16. Copy and Print
17. Lockers

Head Count

Workstations	140	Conference Room
Lockers	140	Phone Rooms
Meeting Rooms	4	Lounge
Project Pods	2	Tea Point
Mobile Meeting Pods	3	Copy and Print
Test Room	1	Alternative Work

Key

- Carpet
- Vinyl Floor Stickers
- Timber
- Exposed Access Flooring
- Rubber

0 1 2 4 8

3F Workspace Floor Plan

Offices Embodying Company Disciplines

Harmonious working spaces rapidly expanding

as local
and global
financial
centers

After entering Yeouido in the early 1990s, Hyundai Card and Hyundai Capital increased their headquarters buildings to three. Thereafter, the Yeongdeungpo office building was built in 2013, and the Busan office building in 2017, gradually expanding the reach of Hyundai Card and Hyundai Capital within Korea. The Yeongdeungpo office building, the first new construction project, was the testbed for follow-up projects. The Yeongdeungpo office was intended to embody the everlasting identity of Hyundai Capital consistent across all locations. The network expansion was also vigorous overseas. Centering on its headquarters in Yeouido, Hyundai Capital has established corporate branches in the U.S., Canada, Germany, United Kingdom, China, Brazil, India, and Australia since the 1980s. The bases built aiming at the world market exhibit various manifestations according to their locations and members.

Hyundai Capital Yeong-deungpo Office

p. 96

Since the Hyundai Card and Hyundai Capital office buildings in Yeouido were purchased, not built from the scratch, the Yeongdeungpo office building was the first office building to be built brand-new. The materials and construction methods of these buildings vary as they were constructed at different times. However, little has changed about the courteous impression that the office buildings intend to exhibit in the city.

2.1

Hyundai Capital Busan Office

p. 102

The Yeongdeungpo office building acts as a mockup for the Busan office building and other office buildings to be built. The Busan office established its core in the fast-changing Seomyeon Rotary. Yet, it kept a distance from boasting itself to the community. Rather, internal stability was ensured, emphasizing the work environment for employees.

2.2

Hyundai Capital America

p. 110

In addition to the two headquarters in Orange County, Hyundai Capital has branches in Dallas and Atlanta. The continuous collaboration with Gensler in charge of design has led to the precise and sophisticated workspace pursued by Hyundai Capital.

2.3

Hyundai Capital Bank of Europe

p. 118

For the Hyundai Capital office building, the base floors were constructed one after another. This design is superior in efficiency and functionality. However, it was limited in facilitating inter-floor communications among employees. Thus the spiral staircase connecting the three floors was installed to trigger encounters, broaden relationships, and strengthen a sense of belonging.

2.4

Beijing Hyundai Auto Financing Co., Ltd

p. 124

Established in 2012, the Beijing office (BHAF) is housed in a state-of-art building designed by world-renowned design firm SOM. The double-layered exterior protects this eco-friendly building from the harsh environment of Beijing.

2.5

Hyundai Capital UK

p. 130

A space with strict order without any error was desired to express the intellectual and sophisticated image of the financial industry. However, after a series of reflections, the company turned its gaze from how the others see it to the humane life of employees inside. This dramatic transition was reflected in the intentional bewilderment.

2.6

2.1 Hyundai Capital Yeongdeungpo Office

Yeongdeungpo Office
Location: 188, Yeongdeungpo-ro, Yeongdeungpo-gu, Seoul
Site Area: 824m^2
Building Area: 493m^2
Gross Floor Area: 4,085m^2
Coverage Ratio: 59.93%
Gross Floor Ratio: 353.40%
Building Scale: B2~7F
Design: 2010.10 ~ 2012.02
Construction: 2012.02 ~ 2013.05
Architecture: 101 Architects, Wook Choi
Landscape: Design Allee
Contractor: Hyundai E&C

Completed in 2013, the Yeongdeungpo office is the first office building that Hyundai Capital built from scratch. The building, which broke away from general types of office buildings, was born after a series of considerations, trials and errors. Diverse materials and finishes, including low-iron glass, sanding glass, and hot-rolled steel plates, were selected for a comfortable working environment.

In the lobby, the liveliness of low-iron glass is juxtaposed with the weight of the hot-rolled steel plate.

2.1.1 The Way We Build

The Yeongdeungpo office building is located near the industrial cluster that has grown based on the Guro International Complex. Since the site has tightly split plots with mostly low-rise buildings, there was a risk that a new building might obtrude the ambiance despite being only seven-story tall. At the first meeting, Vice-Chairman Ted Chung requested: first, the Yeongdeungpo office must be a testbed for future office buildings of Hyundai Card; second, the pillars must be removed from the office space, which must have a ceiling height of 3 m or higher; and lastly, the exterior shall be sophisticated and timeless. The materials and morphological techniques popular at the time were rejected for the exterior of the building. Calm and restrained, the impressions must have neither excess nor deficiency, while a closer look would present a solid and precisely constructed façade. Inside, the goal was to create a space with a variety of meanings for the employees. This did not signify aesthetic achievement; this meant an employee-friendly working environment. A space with an appropriate amount of light, temperature, and sound would improve efficiency while acting as welfare for employees. Of course, removing the pillars and increasing the ceiling height required technical support. The design was the work of 101 Architects. The upper and lower volumes form a single building. The base exudes the solidity of cement. The upper floors are light and modern. The design expresses the foundation of finance and creativity thereon. The materials for the first floor, the entrance of the building, slightly differ from others. For the lobby where the traffic is busy, transparent low-iron glass was utilized for openness and natural communication with the outside. The lobby, which is about 4 m high, faces the main road directly and welcomes visitors with coziness. Four years after the Yeongdeungpo office was completed, the Busan office was built. The Busan office building, designed by 101 Architects as well, has retained the concepts of the Yeongdeungpo office, only expanding it in size. That is, the Busan office utilized the Yeongdeungpo office as a mock-up.

Parts with delicate details connect the floor, ceiling, and glass. The rough cement bricks amalgamate the walls of the neighboring buildings as if they are part of the whole landscape.

2.1.2 Point of Space

When the parts joining materials are highly exposed, the materials may overpower the space. Thus, the design minimized the attaching lines so that the materials combined would look like a single face. The super-floor method, in which concrete is poured at once then the surface is ground, was used for the lobby floor to hide the inter-material boundaries. Thanks to the vast glass curtain wall, light fills the interior. Further, the interior materials are finished to accept the light yet reflect softly, diffusing light tenderly. A distinctive feature of the architectural plan was that the building's core is not at the center, but at the north and south ends. There are elevators and stairs to the north, toilets to the south. When the core is in the center, it splits the workspace. Thus, this shortcoming was overcome by dividing the core. A special structure was adopted to eliminate columns in the center of the office. Thus, the adopted technique supports the slabs with a minimum vertical structure, in place of the general method of using columns and beams. As such, a large space of 16 m x 20 m could be created where the cores on both sides support the weight of the building without columns blocking the view inside. Upon completion, the spaciousness and beauty of the ample space were superb. This was a moving result for Hyundai Card, which endured the hard work of raising the old and low ceiling in the early 1990s after the headquarters moved to Yeouido. The façade consists of six elements. Two types of transoms and mullions were crossed in turn against the concrete wall of the building. Afterward, the sanding glass was fixed. Sanding glass was made by spraying sand at high speed then grinding the surface. From a distance, the building subtly illuminates, giving a soft ambiance. The densely overlapping glass awnings provide a visual effect of turning the building's outline into seemingly multiple lines. The vertical louvers, which ascertain the building's impression, also relate to the working environment. Offices generally use a glass curtain wall. This method, however, decreases the energy efficiency because more of the office space is exposed to the sun, easily raising the indoor temperature. The vertical awnings installed on the east and west facades are not mere decorations. They block direct sunlight in the morning and late afternoon while bringing in the reflected light. The inside of the office is always kept pleasant thanks to the soft and constant indirect light. Within budget constraints, the design had to produce an efficient result by comprehensively considering various functions including lighting and ventilation. Numerous experiments were, therefore, necessary using spatial models. In the end, an optimized module, which did not hinder the building's façade and could block direct sunlight, could be employed.

Mullions and transoms hold the sanding glass firmly, forming a vertical louver. The vertical louvers provide a pleasant working environment by blocking direct sunlight. They also energize the city with rhythmic design.

"When designing the office building, it was not about the region but the essence sought by Hyundai Capital in its office buildings. The concept stemmed from the sky and the sea. The lower floors with people are the sea with numerous ships. The upper floor resembles the sky."

Wook Choi, Head Architect of 101 Architects – Yeongdeungpo Office Building Architecture and Space Design

2.2 Hyundai Capital Busan Office

Busan Office
Location: 8, Seojeon-ro, Busanjin-gu, Busan
Site Area: 1,319m^2
Building Area: 704m^2
Gross Floor Area: 19,362m^2
Coverage Ratio: 53.41%
Gross Floor Ratio: 959.48%
Building Scale: B8~20F
Design: 2012.04 ~ 2014.09
Construction: 2014.10 ~ 2017.05
Architecture: 101 Architects, Wook Choi
Contractor: Hyundai E&C

The Busan office also strived to draw in the diverse landscapes and urban energy around Seomyeon Rotary. Such characteristics can be witnessed in the lobby on the first floor, which is designed to be especially bright and open based on its three-floor height without pillars. The tiered seats provide a space for various activities just like European city plazas.

Four years after the opening of the Yeongdeungpo office, the Busan office was completed in January 2017. The Busan office building has significance as Hyundai Card and Hyundai Capital were integrated into one building to strengthen their local business presence. Also, the building was intended to become an unshakable center with a solid and clean exterior amid the rapidly changing Busan Seomyeon Rotary area.

2.2.1 The Way We Build

This was not the first time that Hyundai Capital worked with 101 Architects to plan the overall concept of a new building. Before, 101 Architects handled the design of the Yeongdeungpo office. The lower floors of the Busan office building had a solid ambiance, while the upper floors had a relatively lighter and modern atmosphere. Simultaneously, the calm order and proportions of the Yeouido headquarters building unswayed by trends were reinterpreted for the Busan office. It was much easier to work on the Busan office having the Yeongdeungpo office as a mock-up. Nevertheless, the new building had noticeable differences such as the increase in scale to eight stories below the ground and 20 above, the proportion of the mass, and the layout. These changes were a response to the regional context of Seomyeon jumbled with multiple industries and users. While the Munhyeon Financial District reinforced its character as the central business district and recently opened citizens' parks sparked the development of high-end residential areas, the Busan office building was distinguished from the rest with its refined appearance and intimate ambiance. Meanwhile, the inside had to be an employee-friendly environment. Value of the space from a real estate point of view carries little relevance to the quality of the space experienced by employees. In order to deliver spaciousness desired by employees, Hyundai Capital's demand was concrete and concise: the ceiling height must be at least 3 m Compared to the legal minimum ceiling height (2.4 m), Hyundai Capital's Busan office boasts a 3.3 m-high ceiling. In addition to the vertical openness, the horizontal aspect was not overlooked. This is a question of where to place the core. In the case of the Yeouido headquarters building, the core is in the center while office spaces are placed along the four sides of the building since it was a purchased structure. Conversely, as in the case of the Yeongdeungpo office building and Building 3 of the Yeouido Complex, when the cores are concentrated to one side, a single and open workspace can be secured. The Busan office was also in line with such concerns and experiments.

The Busan office building has evolved further by applying the post-tension method to the working environment. With a ceiling height of 3.9 m and invisible internal posts, the space seems more spacious than ever. This was to deliver a pleasant work environment to the employees.

태양 공인중개사
매매 임대
갈비
HOTEL

"The lower floor is open like a sea while the upper signifies the sky—this is the same concept as the Yeongdeungpo office building. The lower floors are concerned with pedestrians, while the upper floors are digitized with highly sophisticated construction systems. It does not reject the surrounding landscape as if it encompasses air. Rather, it brings the surroundings in and complies with the local identity. Most lower floors of office buildings often conflict with the road outside. However, the Busan office building seems wide open with a height of about two floors. The structure is simple. We avoided using columns as much as possible. The Yeongdeungpo office has no pillars inside as both cores support the building, and the Busan office also has pillars only at its ends. The upper floors to be functional, the structure to be reasonable, and the overall not to disturb the order of the city—these three were the key desires."
Wook Choi, Head Architect of 101 Architects –Busan Office Building Architecture and Space Design

2.2.2 Point of Space

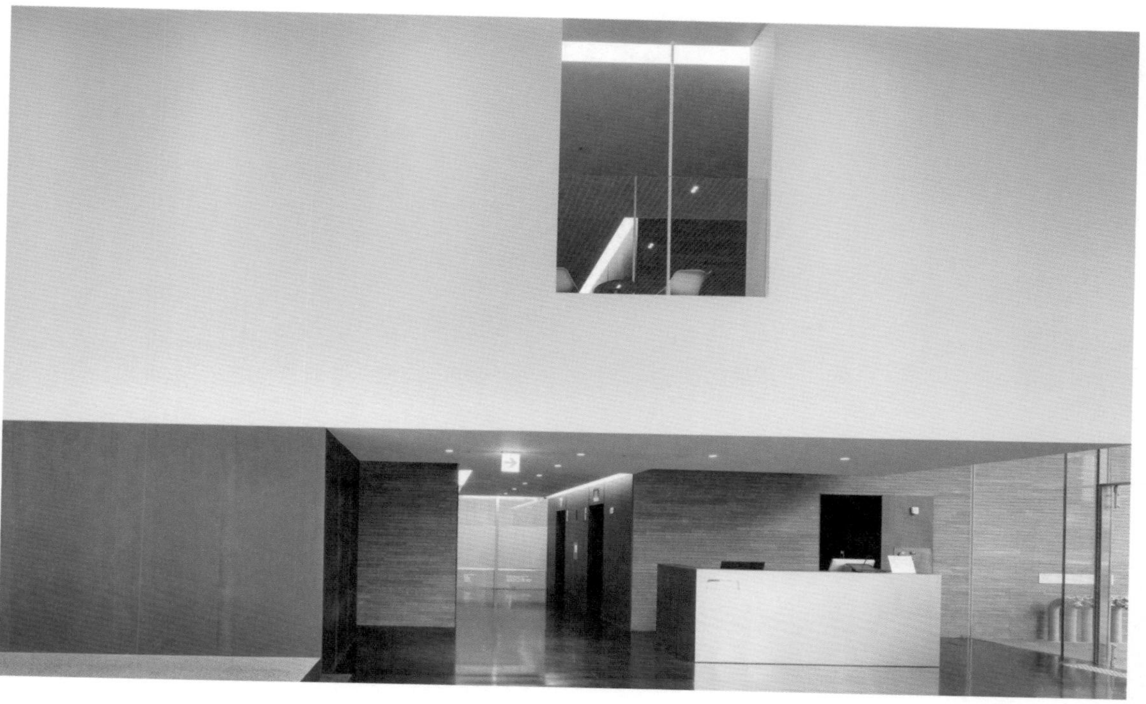

The space is expansive and open as no columns block the perspective in the center. This was achieved by applying the post-tension method. The U-shaped structure of columns and beams repeating every 3.9 m make the space exciting and emphasize the effect of one-point perspective.

There is no endpoint to Hyundai Card's spaces. Large-scale changes are rare, but small-scale, precise attempts and alterations continue to be made, especially in terms of spatial use and functionality. The optimal functioning of the spaces is under constant inspection. That is why attention to specific behaviors or experiences must precede the development of a space. For instance, a rooftop garden with no clear purpose and the awkward and inefficient circulation in the restaurant's kitchen both underwent revision. The same principle was applied to the building's contact with the city. The public space on the first floor was designed to withstand unplanned or unpredictable occurrences, which could be presented by the continuous redevelopment in the surrounding area and to vitalize Seomyeon, an area frequented by younger crowds. The façade is composed of glass curtain walls and vertical louvers, as in the Yeongdeungpo office. The main façade, which catches the eye of outsiders, has increased transparency with low-iron glass. The west side has been finished with opaque glass to trigger the curiosity of passersby. The east side, which receives the most sunlight, has vertical glass louvers hung close together at 975 mm intervals to scatter the light.

The north and south sides, of relatively narrower widths, provide a panoramic view through their horizontal frames. The lines crossing vertically and horizontally at pertinent proportions add an elaborate and well-structured depth to a seeming simplicity. This was to manifest the identity of a financial company by demonstrating the charm of a grid through controlled hierarchy and rhythm. The interior evinces the efforts made to converge the cores on one side. Thanks to the long core located at the west, the workspace could be conceived as a single space of 34.8m x 13.5m. Another way of emphasizing this sense of expansiveness was to remove the pillars. The ceiling beams were secured between the core on the west and the pillars on the east using the post-tension method. Thus, floors over 10m in width could be supported without columns. The U-shaped exposed concrete pattern repeated every 3.9 m is reminiscent of the symbol of the Hannam-dong Understage or the Convention Hall in the Yeouido Complex. The magnificent lobby, 8.4m in height, is graced by glass curtain walls, black rolled iron plates, and a super floor of polished concrete. The synergy in the relationship between Hyundai Capital and 101 Architecture is evident in the use of unadorned materials and materials that reveal their charm over time. The suspended mezzanine and cascading floor convey the impression of a city square and an amphitheater brought inside. Yet, the space still poses questions in terms of its use.

2.3 Hyundai Capital America

Irvine Office
Location: 3161 Michelson Dr #1900, Irvine, CA 92612
Area: 97,519ft^2
Design: 2008.07~2008.12
Construction: 2008.11 ~ 2009.10
Interior: Gensler
Contractor: Clune Construction

Newport Beach Office
Location: 4000 Macarthur Blvd, Newport Beach, CA 92660
Area: 83,866ft^2(1~5 floors only)
Design: 2014.05 ~ 2014.10
Construction: 2014.10 ~ 2015.08
Interior: Gensler
Contractor: CMC Construction

Dallas Office
Location: 6100 W Plano Pkwy, Plano, TX 75093
Area: 66,115ft^2
Design: 2010.01 ~ 2010.06
Construction: 2010.07 ~ 2011.03
Interior: Gensler
Contractor: Gray Construction

Atlanta Office
Location: 4100 Wildwood Pkwy, Atlanta, GA 30339
Area: 100,000ft^2
Design: 2012.03 ~ 2012.08
Construction: 2012.09 ~ 2013.05
Interior: Gensler
Contractor: Gray Construction

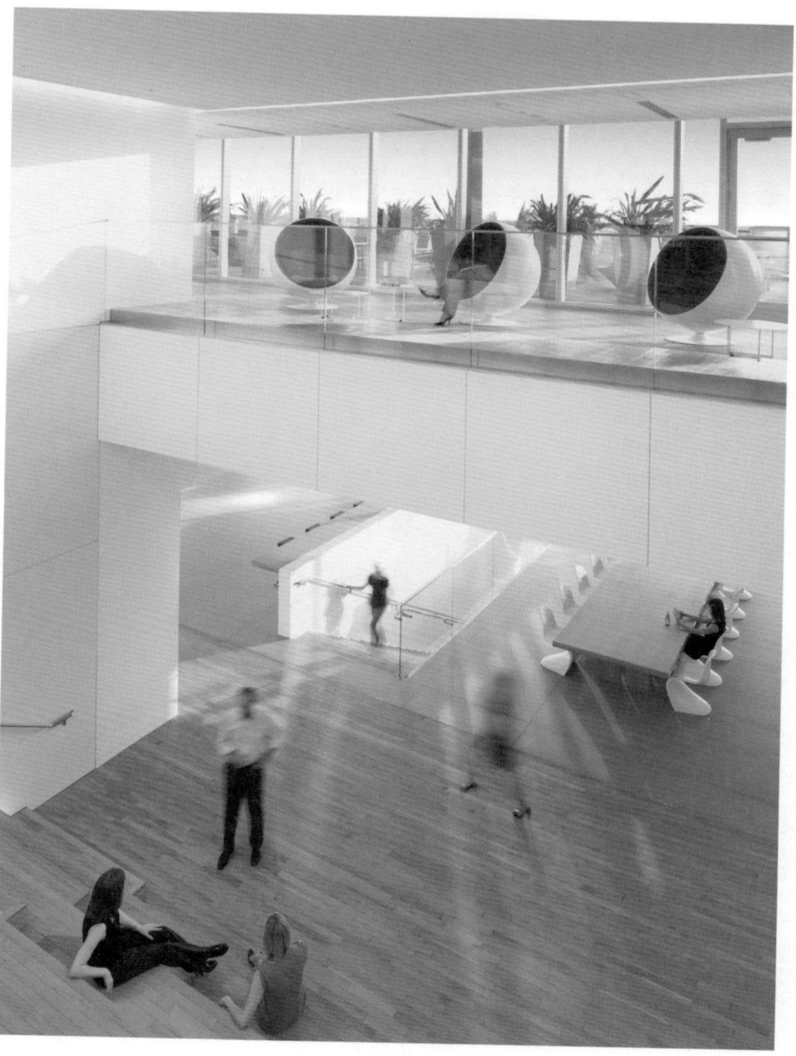

The Irvine office building is characterized by a bright ambiance with white walls and low iron glass. In addition, the three floors are connected around the wide staircase similar to a theater-style auditorium. The stairs are finished with wood, adding coziness. Ball chairs on the upper floor allow employees to have personal time while enjoying the Californian scenery.

Hyundai Capital America (HCA) has grown into one of the top 10 auto finance companies in the US by supporting financial services for Hyundai, Kia, and Genesis brands since its establishment in California in September 1989. As of 2019, total assets amounted to over $30 billion. HCA operates customer support centers in Dallas and Atlanta, in addition to its headquarters in Orange County.

2.3.1 The Way We Build

Hyundai Capital America (HCA) is the first overseas case that applied the spatial ideas behind Hyundai Capital headquarters in Korea. Forming an appropriate culture for the corporation is imperative. To establish corporate culture, the workspace must be strategically aligned. The working environment equals a physical declaration of corporate identity. When Hyundai Capital had just entered the U.S. market, it needed an image of being a leader of the financial world. Another task was to instill the ideas of the headquarters. Above all, an open working environment was its priority. The visual directions such as an intelligent and refined atmosphere, a level-headed and accurate image disapproving even an inch of deviation followed thereafter. HCA is the first project of Philippe Paré, who has overseen Hyundai Card and Hyundai Capital projects at Gensler for a long time. Vice-Chairman Ted Chung recalls his first encounter with Philippe as follows: "Philippe's explanation was extraordinary from the beginning. At that time, Philippe was immersed in studying lines and proportions. He said that overlooking thereof will jumble all elements, ultimately costing more to get things back in order." In fact, Philippe set up one module and used it to multiply or divide other elements down to the smallest detail. For instance, he would align the ceiling light with the center of the window frames and desks, assuring that the axis is not shifted to the maximum extent possible. He is said to be obsessed with eliminating all miscellaneous lines. As such, he integrated worlds that move in different units, such as architecture, interiors, and equipment, into one coherent order. The completed interior gave room by raising the ceiling height to 3 m and setting the baseline at 5-feet intervals. The design is a result of composing the space with numbers and proportions, not with symbolic sculptures of instant references or attention-grabbing decorations. The orderly and transparent glass-enclosed meeting room has given customers confidence. There, the orderliness is so concrete that arbitrary placement of the furniture would be immediately noticeable. People label such office spaces as minimalism; however, Hyundai Capital has never intended so. The two simply have a visual similarity. The module-applied design continued. From Dallas to Atlanta, from Frankfurt to Beijing, the basic principles have remained the same, such as hanging art on the wall or covering the wall with living plants. Then came a moment of transition. It was the realization that cold and dry design overlooks the employees working in the space. The ample light of California is brought deep into the building's interior via the stairs piercing through the three floors. The viewing deck in the middle of the stairs, in terms of design, was great. However, it was more appropriate for particular events and scheduled group gatherings, rather than facilitating daily and casual interactions among employees. Today, Hyundai Catpial's position as a leader of the industry is widely recognized. Space alone does not epitomize corporate identity anymore. Now, the vision and potential of the corporation must be unearthed in employees' daily lives. The look is important; however, the aim must be to offer people a better working experience. Based on continuous experiments, the perception of office space has leaped forward. Vice-Chairman Ted Chung puts it as follows: "Now, we do not esteem modulized layout. Henceforward, pursuing intentional disarray would be more appropriate."

Columns, walls, hallways, desks, and ceiling lights do not allow even an inch of deviation, precisely located at their respective center. This design was to architecturally transfer the intellectual and transparent character of the financial industry. Further, the temporary wall in the ceiling deems the lighting, while the sound-absorbing fabric reduces the noise.

2.3.2
Point of Space

Irvine, Newport Beach

The Irvine branch features white walls and furniture, and transparent low-iron glass partitions. The ample light of California makes this space even more glaring. Employees lead a calm daily life in the office. By using wood as a finishing material, the spaces provide additional warmth. The stadium-like wide stairway located on the center of the 19th floor is suitable for all employees to gather to watch videos or to carry on social exchanges. In particular, as the stairway abuts the middle floor of the internal staircase connecting the 18th and 20th floors, it encompasses all lines of circulation as the square of a medieval city. Nonhierarchic relationships based on open office plans are no longer rare in other companies. However, a vertical circulation among floors still carries limitations since it requires rearranging the ground of respective floors. This was the very beginning of Hyundai Card's continuous challenge on the vertical exchange over floors. Half of the 20th floor was planned as a rooftop garden so that the beautiful scenery of Orange County could be appreciated. Also, for those who enjoy light but hate the sun, the Ball Chair by Eero Aarnio was placed. One immerses oneself in the comfort of the chair while enjoying the scenery of the three floors along the stairway. Newport Beach, as its name suggests, is decorated with the theme of sailing and piers so that employees can relax in a familiar environment.

Overseas offices were designed to reflect their unique local identity. In the Irvine office, employees gather on the theater-style stairs to have a party or watch a video. Lined with wooden beams, the Newport Beach office building is reminiscent of a waterfront pier.

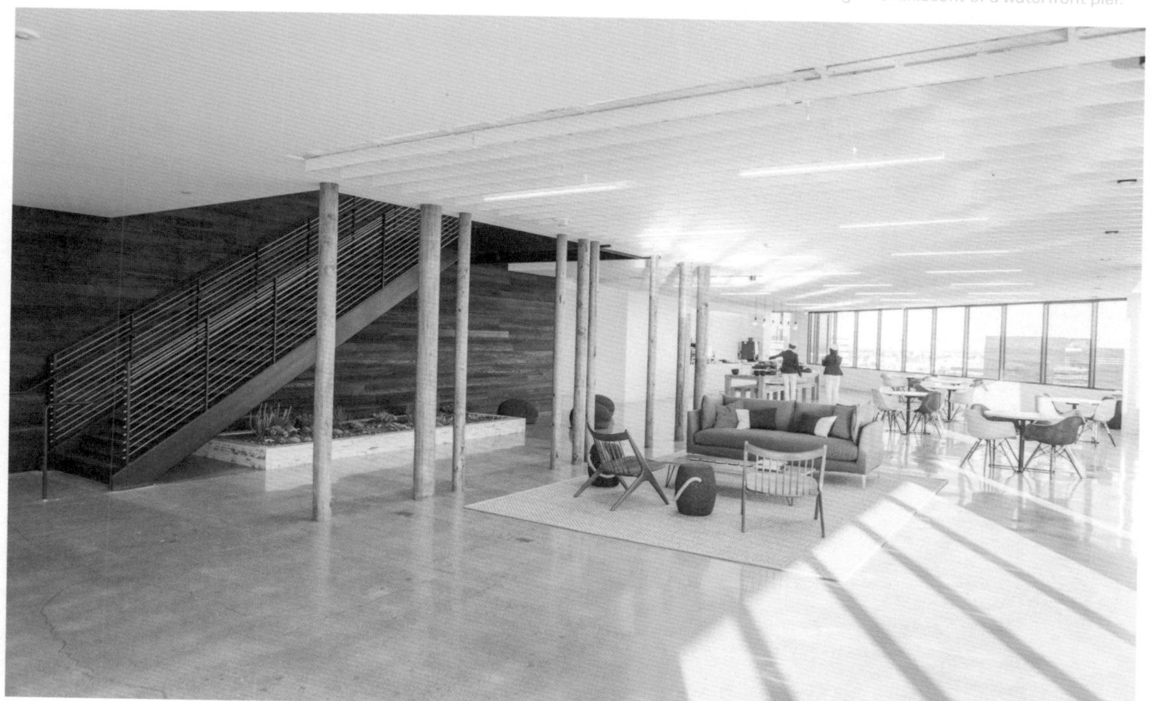

Dallas

The operations centers in Dallas and Atlanta house call centers. The facilities were designed to simply meet the minimum requirements. However, thanks to the company's principle of striving for the best with clear limits, a high-quality space could be born on a low budget. The workstations at the Dallas Operations Center are compact and functional maximizing the efficiency of the space with high employee density. Each desk is aligned along with the center of the windows, and vertical bulkheads hang from the ceiling. This not only is eye-catching but also creates a subtle atmosphere by reflecting the pendant light once again. Also, a functional role of absorbing noise is included. A translucent glass panel installed in the center of the desks protects privacy between employees while leaving room for visual interaction. The view is exceptional from anywhere as there is no building adjacent to the building, allowing the light to permeate throughout the entire office. A space for instant meetings is not far from each department. The meeting spaces have the cheerful name "Think Tank" and have walls painted in bright colors to give them a casual and lively personality. The Number Wall, an artwork hanging in the hallway, and the walls in the break area decorated with various plants also break the silence of bland workspaces and give an impression of an art museum or park. As the large arc-shaped wooden partition envelops the space, the dotted circular skylights illuminate the interior with varying textures of lights.

The Dallas office created a natural and comfortable space with green walls of live plants and skylights accepting natural light. The Atlanta office offers a romantic and warm feeling with the cloud shape on the ceiling. Around 2,000 conical elements composing the complex shape of the cloud have slightly different geometric shapes created with parametric software.

Atlanta

When the Atlanta office building was voted as the Best Interior of 2014 by <The AN (The Architect's Newspaper)>, Kate Orff, the founder of New York-based landscape and urban design studio SCAPE, commented that "The contrast between the abundance of plants and the frugality of the space is eye-catching." As such, the Atlanta office contains the contrast between formal workspaces and causal public spaces. The composition of the workspace is the same as that of the Dallas office but its design is more well-defined with the theme of nature. Three layers of sky, plants, and earth were applied. In doing so, the ceiling of the break area designed after clouds became a distinctive element of the Atlanta office. The 2425 cones made of parachute fabric are similar in size but slightly different due to parametric design. Together, these cones create gentle floats like a flock of clouds. The Green Wall, which is also found in the Dallas office, fills one entire wall of the resting area. The Vegetal Chair by Vitra, featuring a branch-shaped backrest, integrates all these elements. The video conference room across the elevator resembles a giant rock, creating an immersive space that ensures privacy.

2.4 Hyundai Capital Bank of Europe

HCBE
Location: Tower 185, Friedrich-Ebert-Anlage 35-37 60327 Frankfurt am Main
Area: 3,414m²
Design: 2013.10 ~ 2014.04
Construction: 2014.07 ~ 2015.04
Interior: Gensler
Contractor: Omnicon

The Frankfurt office building epitomizes the Hyundai Capital-Gensler collaboration. Amid flowing light, the workspace follows a clean line along with the windows. The spiral staircase that cuts through the concrete floor and runs independently is the kernel that encompasses both design and corporate culture.

Tower 185 has a baroque-style front yard shaped like a horseshoe. Hyundai Capital Frankfurt office, located within the building, also features a spiral staircase reminiscent of baroque passion and bliss.

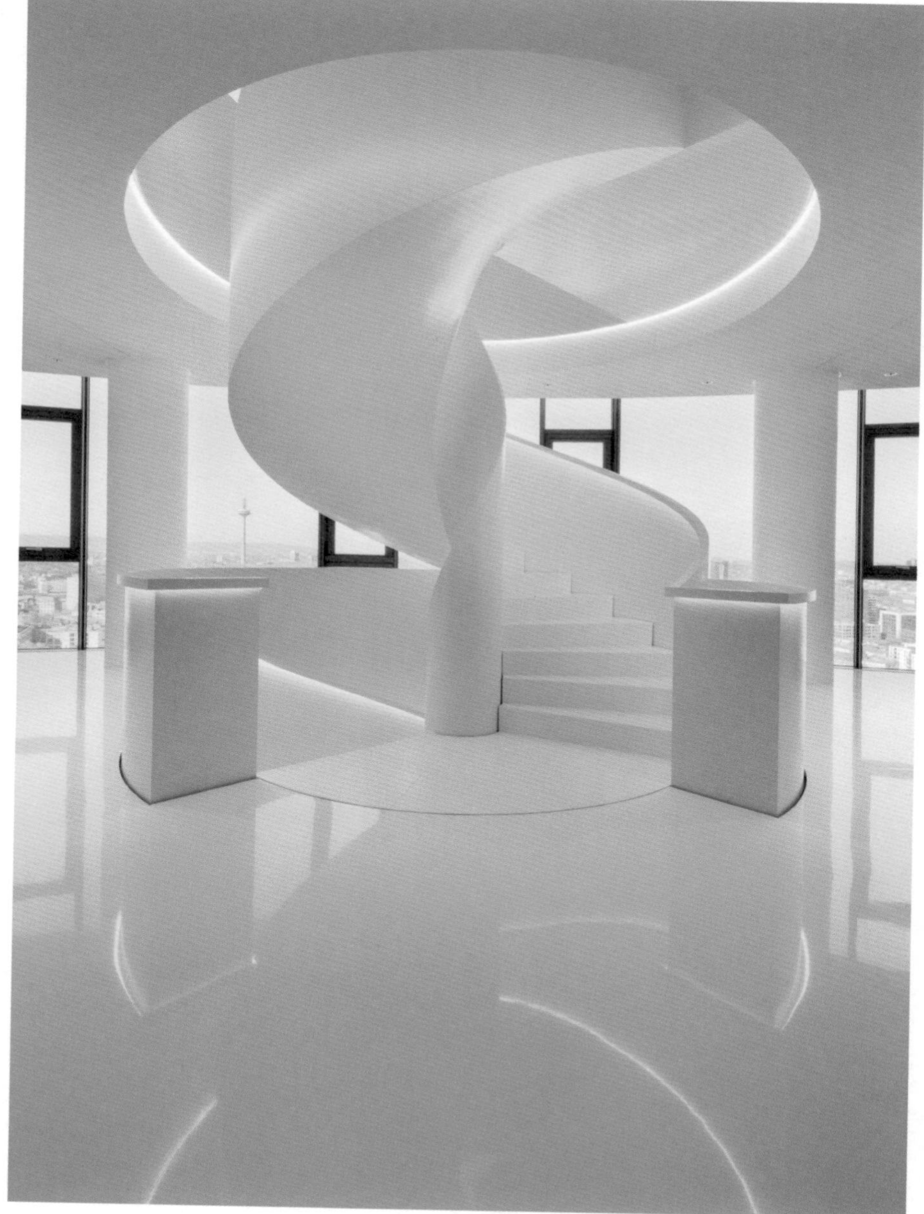

2.4.1
The Way We Build

HCBE (Hyundai Capital Bank of Europe) occupies the 22nd to 24th floors of Tower 185 in Frankfurt, Germany. Completed in 2012, Tower 185 has a baroque front yard in the shape of a horseshoe and consists of a low-rise block and a 50-story tower. This majestic architecture is the fourth-tallest landmark of Germany. The accessibility provided by the location, 5 minutes and 20 minutes away by car from Kia Motors and Hyundai Motors, respectively, was a key reason why Hyundai Capital chose the building. Since Hyundai Capital does not own the entire building, the offices are structured based on the already established architectural order of the building. How desks, partitions, and conference rooms are arranged precisely aligns with the window spacing and the center of the column. The same goes for the floors and ceilings. The hallway and personal workspace are divided around pillars, and the line provides the yardstick for the circulation system on the ceiling. The simple color finish and minimal arrangement of furniture leave room for employees to decorate their workspaces. HCBE clearly divided the sections based on functions: spaces for concentration, spaces for general

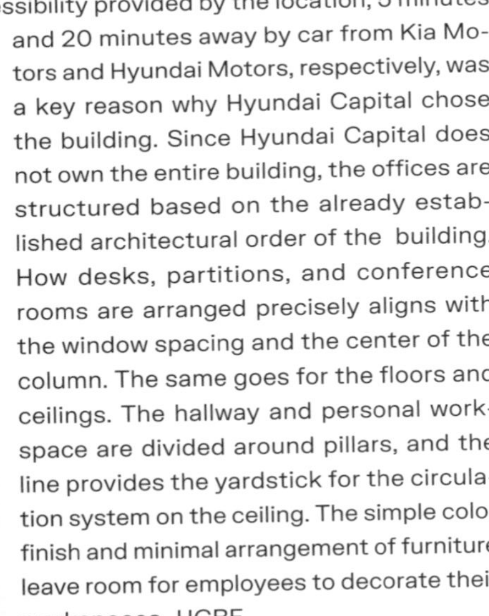

The floors occupied by Hyundai Capital are differentiated from other floors. Beyond the lobby with a dignified atmosphere resembling Studio Black, the employees are welcomed by a bright and cheerful workspace. The Hyundai Capital neon sign on the ceiling is the work of Italian artist Maurizio Nannucci.

working, and spaces for sharing bonds. The division allows each employee to individually set the pace of his or her work. Meanwhile, to solve the vertical separation problem, a spiral staircase that penetrates three floors has been installed. To use the elevator in the core, one must wait for a long time to move one floor. Also, the direct stairs used as a fire escape are dreary. The spiral staircase is thus the solution. The vertical line of human circulation attracts employees from different departments on different floors. Placing amenities around the stairs, for instance, increased casual interactions among employees. On the 23rd floor, there is a lounge-like living room. The lounge is then connected to the study on the 22nd floor, and the game room and the dining room on the 24th. Each represents collaboration, learning, and exchange among employees. In particular, the game room and the study are spaces unusual in other European offices.

2.4.2
Point of Space

Unlike the entrance on other floors, the main entrance on the 23rd floor has been completely renovated. The overarching theme was 'sophistication with profoundness.' Different shades of black—from granite tiles on the floor, glass on the walls, to the MDF on the ceiling—were employed to deliver the theme. Italian artist Maurizio Nannucci created the neon sign on the ceiling using 'Youandi' font. The heart of HCBE is the spiral staircase. Two things were asked of Gensler for the staircase: it must be as thin as possible and must stand alone without structural support. These requests are quite contradictory. Arnold Metals, who collaborates with world-renowned artist Jeff Koons, has answered this thorny request. The smooth and dynamic shape of the steel staircase could be achieved by combining two 5 mm steel plates. Countless holes were made in the outer steel plate so the bonded part would be smooth. To diagnose the structural stability, Hyundai Card made a mock-up staircase with 25% of the actual size. Despite the preliminary work, the on-site construction was challenging. The staircase was split into 16 pieces to be loaded on the elevator then assembled on site. The uppermost pieces were hung on the steel beams installed on the ceiling of the 25th floor. Thereafter, the stairs were assembled one by one downward. After a round-shaped light was installed on the ceiling of the 25th floor, and the staircase was named 'The Path Leading to the Light'. The executive room has small meeting tables in the center, which can rotate to face the wall for people who want to work individually. The transparent partitions allow employees to be oblivious of others when concentrating. Yet, it does not interfere with light entering the room. The hierarchy of superiors-subordinates has been eliminated from the executive room, which can also be used as a meeting space when executives leave. Also highly useful are the auditorium with theater-style seating and the conference room with partitions foldable and hideable up to the ceiling. The Cove Lighting, an indirect lighting system negative molding at the juncture of the wall and ceiling, allowed the light to flow down the walls of hallways. Most of the furniture in the workspace is from Vitra. The meeting point, centered on the spiral staircase, showcases furniture of established designers and artworks. The lobby and reception on the 23rd floor feature the Panton Chair by Verner Panton, Loop Sofa by Arper, and Pink Pumpkin Lounge Chair by Ligne Roset that matches the white finish of the interior. The Cuckoo Clock by German artist Stefan Strumbel hanging on one wall always points to 11:55, triggering two opposing interpretations of 'still five minutes left' and 'only five minutes left' until lunchtime. On the 22nd floor filled with books on design and travel, a herringbone-patterned wooden floor was installed, in addition to the sofas by Chesterfields of England and an antique table by Brazilian architect Jorge Zalszupin. The refined atmosphere continues with the Ball Chair by Eero Aarnio and the Chair by Jean Prouve. As Mutina's ceramic tiles add lightness from the floor to the game room on the 24th floor, the Rocket Stool by Eero Aarnio and the Yello Sofa by Bernhardt Design add emphasis to the space.

The spiral staircase is beautiful and extremely functional in that it elicits diverse communication. The spiral staircase decreased the inconvenience of using an elevator to go up just one floor. All interior elements from desks, meeting rooms, hallways, ceiling lights, to fixtures were positioned based on architectural grids.

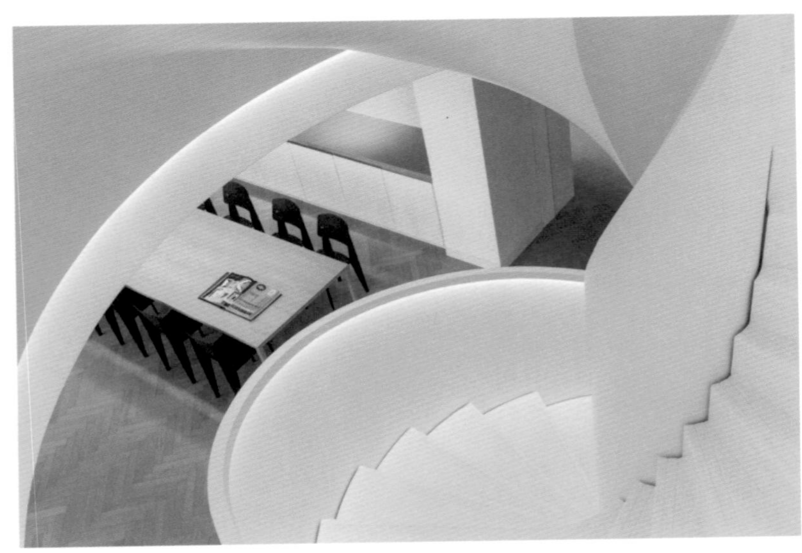

"What a space does is more important than how it looks. We believe that the office space we designed can support accelerating innovation. Designing spaces that promote interaction can result in more communication among employees, more interactions, or a variety of work environments that each employee would love to work in. In such cases, employees can display their utmost ability. For instance, the space to enjoy games was considered unproductive 10 years ago. However, now, it is extremely productive. This is because games strengthen the bond among team members, enhancing mutual trust."

Philippe Paré, Director of Gensler Paris - HCBE Space Design

2.5 Beijing Hyundai Auto Financing Co., Ltd

BHAF
Location: Poly International Plaza Zone 7, Wangjing East Park, Chaoyang District, Beijing
Area: 11,857m^2
Design: 2015.01 ~ 2016.08
Construction: 2016.09 ~ 2016.12
Interior: Gensler
Contractor: Hyundai Engineering

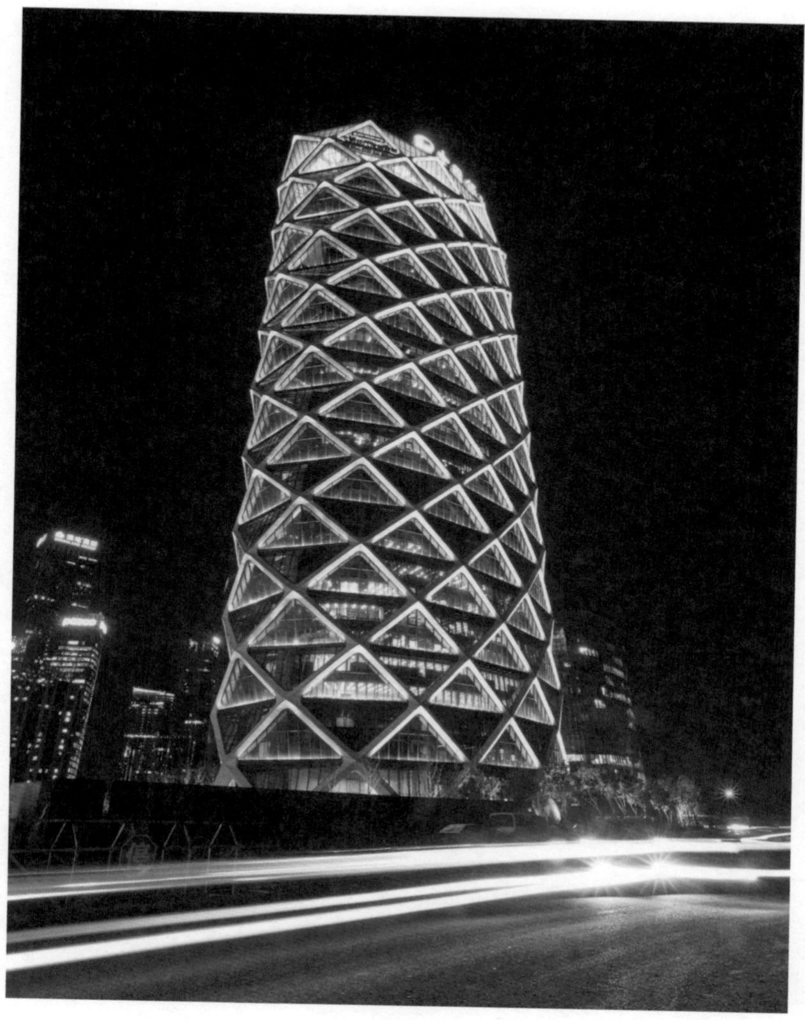

SOM, an American architectural firm, has designed three buildings with Chinese real estate developer, Poly Group. The Beijing office building (BHAF) has occupied Tower 1, the tallest one with 31 floors among the three. The structure encloses an oval flat building with a rigid, rhombus exoskeleton. The design has eliminated internal columns, thereby facilitating a pleasant and flexible design of the interior space.

The Beijing office building allowed Hyundai Capital to encircle its space to the new direction of 'intentional disorder' from perfection and preciseness. That is, Hyundai Capital turned its focus to the day-to-day life of its employees instead of pursuing architectural perfection. This led to the honest self-evaluation: "We continue to regret and also progress."

2.5.1
The Way We Build

Founded in 2012, BHAF (Beijing Hyundai Auto Financing Co., Ltd) opened a new office building in 2017 at the Poly International Plaza between the Forbidden City and Beijing International Airport. The location is an emerging business district in Wangjing, the home of the offices of Mercedes-Benz, Microsoft, Alibaba, Siemens, Uber, and POSCO. It is also a transportation hub, 30 minutes from Beijing Hyundai Motors and Beijing Kia Motors offices and 20 minutes from Beijing International Airport.

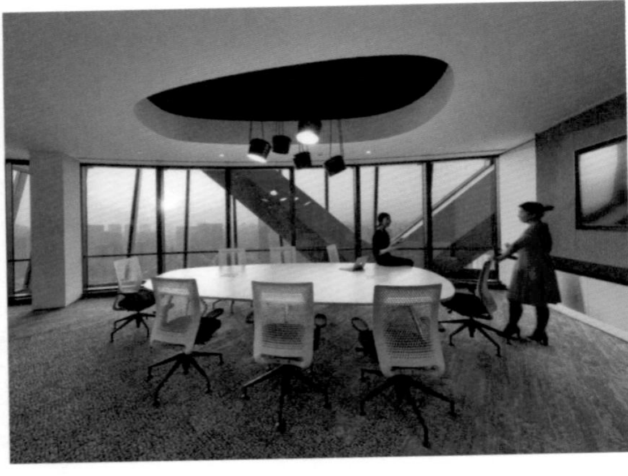

The Beijing office, like the Frankfurt office, has workspaces completed with strictly aligned lines. Since the building is oval-shaped, the inside rotates outward in a large arc. And such a spatial formation creates emptiness.

BHAF is a commanding corporation in charge of a market as large as that of HCA. BHAF's spatial construction follows the design applied to HCA. The expansive floor-to-ceiling glass curtain wall fully accepts natural light. To let the light flow into the interior, no partition has been erected. Furthermore, special equipment to facilitate KGM Architectural Light was applied to make the interior lighting resemble natural light as closely as possible. The vast workspace is visually expansive and exhibits every advantage of large office space. It also articulates the transparency of a financial company. Public spaces have been installed at the center of each floor in the Beijing office building, such as a library, general lounges, and a tea lounge, to instill a sense of belonging and pride in 500 employees. BHAF required no separate

staircase because the building had a magnificent vertical staircase. This revolving staircase was selectively built on certain floors at the building's upper part with a grand atrium. This staircase was an important factor in the selection of this building and the floors therein. The building designed by the world-renowned architectural firm SOM boasts a range of technological advancements in office architecture. Coupled with the interior design of Gensler, which presents strictly aligned lines and orderly space, the office spaces received positive feedback from employees. However, Vice-Chairman Ted Chung disagreed: 'The spaces based on modules and aligned lines were perfect. The metrological or mathematical approach was absolutely beautiful, but I realized that the design was so perfect that it obscures everything else. As the Beijing office building is oval, the space rotates with the radius value, but the curve makes the space appear limitless and bleak. The aligned lines diminish a sense of vibrancy or possibility. The minimalist styling and blankness of composition made the space seem larger. That is when I first realized: our means of expression is icy and dry'. Also suspended was the idea that a clear corporate DNA can integrate all spatial elements despite the company's continued expansion. Unifying space had been difficult not only in its global corporate branches but also in the Yeouido headquarters. A unified identity tends to become obsolete after only two years. The company decided to forego virtual architectural lines and focus instead on the working atmosphere, selecting materials that radiate warmth. As such, it was no longer necessary to insist on furniture by Vitra. Vice-Chairman Ted Chung redefined this as follows: 'This obsession with luxuriousness had to go. I suggested that we should stop expressing luxury via hardware as it emerges from a materialistic corporate culture. Instead, what kind of furniture should we use in our spaces? I thought we could try out Chinese products, so now all are made in China. The key is not a perfect foresight projecting forward 100 years. We continue to regret and to progress'. And the individuality of each branch overseas is respected. Even the Yeouido Complex employs different concepts on each floor.

2.5.2 Point of Space

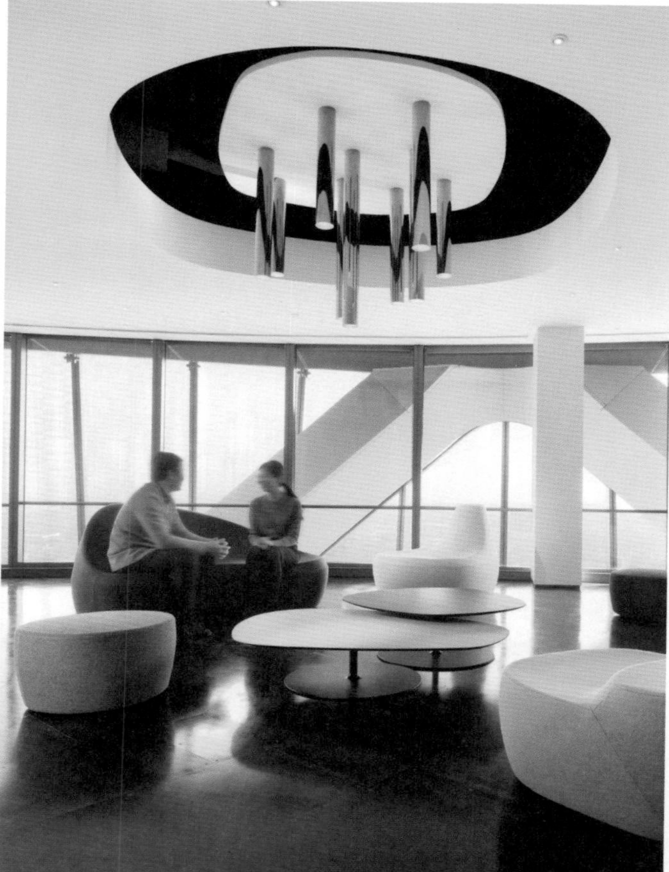

The break area inspired by a small garden and traditional Chinese spaces improves the desolate atmosphere and promotes communication among employees. There is a gap for air circulation between the lattice frame supporting the building and the indoor space. This double-layer structure is a state-of-art construction method that saves energy and facilitates a pleasant environment.

The Poly International Plaza, designed by the U.S. architectural firm SOM, consists of two small towers and a 31-story tower. designed it. BHAF occupies six floors (20th to 25th floors) with the best view in the 31-story tower. The continuous grid pattern of the tower is inspired by the folding structure of the Chinese traditional red lanterns. The exoskeleton structure and office space are each finished with individual glass layers. The outer glass curtain wall is equipped with windows controlled by the building system per section. On summer days, the windows are opened to circulate air via the chimney effect by the double layers. In winter, the insulation is strengthened by trapping the heat accumulated from the outside between the double layers. The elongated elliptical shape from east to west draws in an ample amount of light, minimizing the use of artificial lighting. Accordingly, the building's design contributes to resolving the chronic air pollution in Beijing by minimizing the use of electricity. The interior design, meanwhile, is in line with the California office. Generally, white interior elements dominate, while gold, yellow, red, and green are partially utilized. Compared to the highly polluted air outside, the office spaces inside are as thoroughly hygienic as the cleanroom. To express the dignified ambiance of a traditional Chinese space, blackwood is used on the floor and the pavilion. The black material makes dust more visible, forcing regular management. Inspired by traditional Chinese gardens, a variety of plants are placed between the personal workspace and the hallway. The small waterfall garden with plants flowing down from the ceiling also acts as a place to escape from the outside where the breathing of fresh air is virtually impossible.

2.6 Hyundai Capital UK

HCUK
Location: London Court, 39 London Road, Reigate Surrey RH2 9AQ
Area: 868m²
Design: 2016.03 ~ 2016.08
Construction: 2016.09 ~ 2016.12
Interior: Gensler
Contractor: OD Group

The initial plan for Hyundai Capital's UK headquarters, planned by Gensler, was the same as Hyundai Capital overseas offices. Yet, after establishing the Beijing office, Hyundai Capital changed its course. It began to refuse a tight space without even a room to hang artwork in order to better cope with employees' daily lives. Such a completely changed view on its office building has brought a historical transition to Philippe Paré of Gensler, a long-time collaborator.

The architecture and furniture of the UK office building were steered toward flexible adjustment. The furniture made of OSB is height-adjustable and can turn into a desk or a sofa. The cork stool and the power cable that can be pulled up anywhere contribute to the comfortable and free atmosphere.

2.6.1
The Way We Build

Along with the European financial crisis, the need for a captive finance company was increasing in the automobile sales market to offer appropriate financial products and services reflecting diverse preferences. Accordingly, Hyundai Capital, as the captive finance of Hyundai Motor Group, launched Hyundai Capital UK (HCUK) in July 2012 after establishing a joint venture with Santander, Spain's largest bank, in March 2012. Thereafter, the office was moved to Reigate, Surrey, located a 1-hour and 15-minute drive from central London. In the reception on the third floor of the London Court, plants welcome the visitors instead of solemn reception desks or soft leather sofas. The visitor lounge exposes the offices inside, abolishing boundaries or tension between the office and the outside. Unlike previous office buildings overseas, the UK office has the familiarity of being at home. The warm natural materials and plant decorations are key contributors. However, the main reason is a major change in the furniture arrangement. The initial plan of Philippe did not deviate much from his earlier works. He intended to apply the old way of Hyundai Capital—aligning lines and using achromatic colors. However, Vice-Chairman Ted Chung reacted more honestly and critically than ever. "I do not think that we should continue to use the perfection of numbers or quantitative measures as our yardstick. What is the perfection we are pursuing anyway? We should focus more on people. I want to start telling a different story." His thoughts had accumulated over time. In the case of the California office, the issue was invisible due to the charming natural environment and outgoing personalities of the employees. There, the light and air moved the emotions of the employees. And, thanks to the outgoing ambiance, people were not swayed by the architecture. Conversely, the Beijing office differed. The Beijing office, designed in the same way as the California office, dispirited the movement of employees. It was the result of directly applying the rationale of the California office—leaving the area empty through a meticulous organization—onto the office of different architectural and cultural backgrounds. In both offices, employees were not the main protagonists. Rather, the space itself was prioritized, which was also the case

The hallway leading past the reception has a mural by artist Aakash Nihalani. The humorous optical illusion art contrasts well with the repetitive box shapes of the cabinets. A break area was mainly placed on the east side with ample natural light.

in the California branch. More changes were required and thus realized within the UK office building. The dizzyingly free layout of the UK office was later applied to the interior design of the Yeouido headquarters, establishing itself as a new identity. The attention to employees also extended beyond the architecture. Quite many financial companies are in Reigate. Yet, the city is small and quiet compared to central London. Vice-Chairman Ted Chung was concerned that the employees would not be able to share the urban vitality. He wanted to instill in his employees the pride that they are working for a global company. Then, by coincidence, an idea came to his mind while dining with Korean actor Suro Kim. At that time, he owned Chelsea Rovers FC, a team belonging to the 13th division of the English Football League. The office happened to be located only a 40-minute drive from the team's hometown, Cheswick. As such, HCUK became the official sponsor of the team. After the arrangement, the employees and their children are learning soccer from the players, maintaining a sense of belonging outside the company. Above all, the pleasure of being part of the deep-rooted culture of England was terrific. This was a splendid solution that instilled pride in the employees.

2.6.2 Point of Space

The four-story London Court blends well with the traditional brick-built neighborhood. This is so because, despite being supported by steel frames, the exterior wall is attached with brick panels. Moreover, as the building faces south, an enormous amount of natural light flows into the building. Finished with a glass curtain wall, the east side is exceptionally bright, making the area suitable for employees to relax. The offices are aligned along the central hallway with eight meeting rooms to the north, and private workspaces to the south. By placing easily movable furniture and numerous plants as a greenhouse, the space gives a home-like atmosphere with an independent lifestyle. The detailed areas of the offices are largely divided into the living room, kitchen, game room, meeting room, workspace, and space for lockers. The materials composing the space were neither painted nor covered in order to reveal their natural characteristics. Thus, even a non-interior design expert can immediately notice the materials used. There is no chasing after luxurious ambiance or artificial smoothness. Rather, the naturalness—being worn out, having scratches over time—was desired. Artist Aakash Nihalani, who collaborated in the Vinyl & Plastic Project, underlined the casual atmosphere with a witty mural. The furnishings reflect the preferences and habits of each employee and are flexible enough to be configured into a familiar environment at any time. The height of the desk can be adjusted with a lever, and new combinations can be created by bringing in other furniture. Crafted from Oriented Strand Board (OSB) plywood, the desks for employees create a warm and comfortable atmosphere. OSB is a material made by bonding and compressing crushed wood and is usually used as a material for buildings and furniture. It is economical as leftover or unused construction wood can be used. Also, the unique pattern created by the small, crushed fragments assembled is a design in itself. The desk, 'Hack', was the latest item released by Vitra at that time. The piece itself was like a small architecture, having each part precisely intertwined, and was easy to assemble and move. One could witness how practical and complete it is by simply adjusting the height with the lever. The power cables suspended from the ceiling are visible and reachable from anywhere. Employees can have pop-up idea meetings as the walls of the hallway are made of chalkboards. The Cork Stools of Vitra designed by Jasper Morrison can easily be moved to other spots when needed. The open office plan with minimal participants enables a free layout. The interior is open on all sides except for the glass for the meeting room. Materials and colors of the floor, not the walls, divide the space. The large meeting room adjacent to the lounge has consistently erected glass walls, which can be fully folded into one corner to expand the lounge area.

One side of the meeting room was finished with glass to facilitate visual interaction. The physical boundaries can be eliminated as the glass wall can be folded to one side. The space leads to a living room with cozy sofas and a kitchen with bar tables.

"Companies must not stop at providing a great workspace. Their focus should rather be on how the workspace can be utilized to create corporate culture and community for employees. Employees want choices. They want a space resembling the environment they grew up in. They want to seat according to how they feel, wear the clothes they want, and be isolated when they want to hide. The paradigm has shifted from a workspace requiring employees to adapt to a workspace that adapts to employees' needs. You feel at home in the London office because the design encompasses such ideas."
Philippe Paré, Director of Gensler Paris - HCUK Space Design

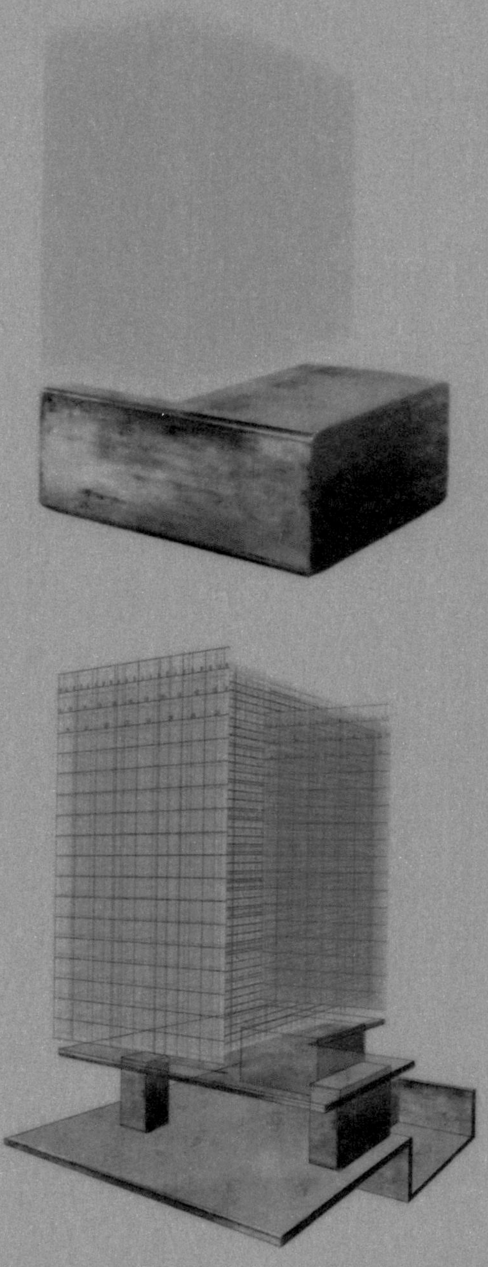

Hyundai Card, Hyundai Capital
Design Concept

Hyundai Card, Hyundai Capital
Design Concept

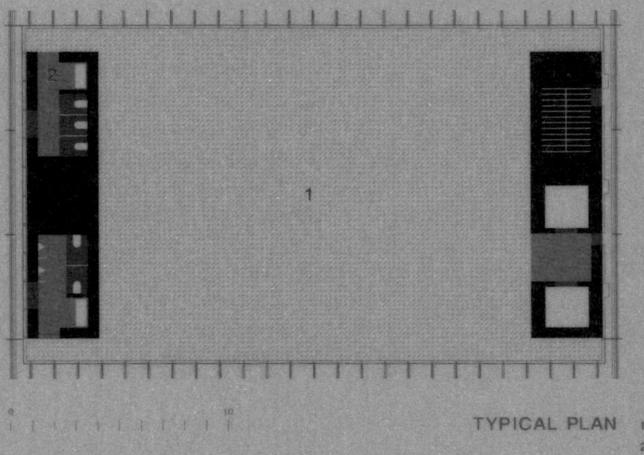

TYPICAL PLAN 1 Office
2 Toilet

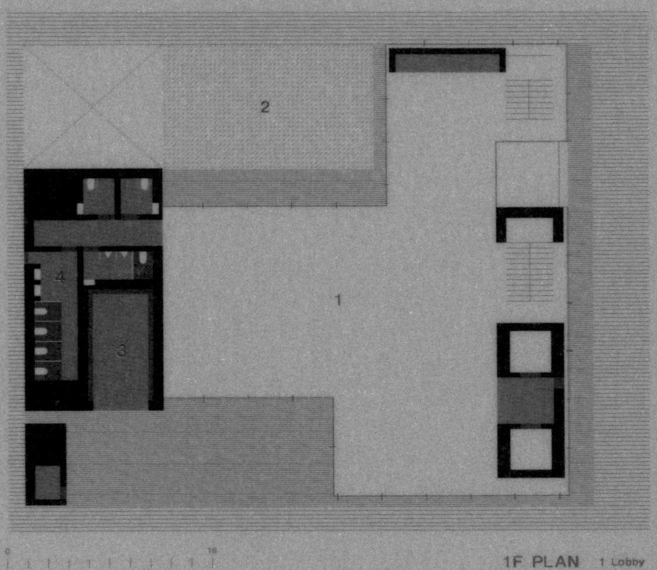

1F PLAN 1 Lobby
2 Garden

Yeongdeungpo Office Floor Plan

TYPICAL PLAN
1 Office
2 Toilet

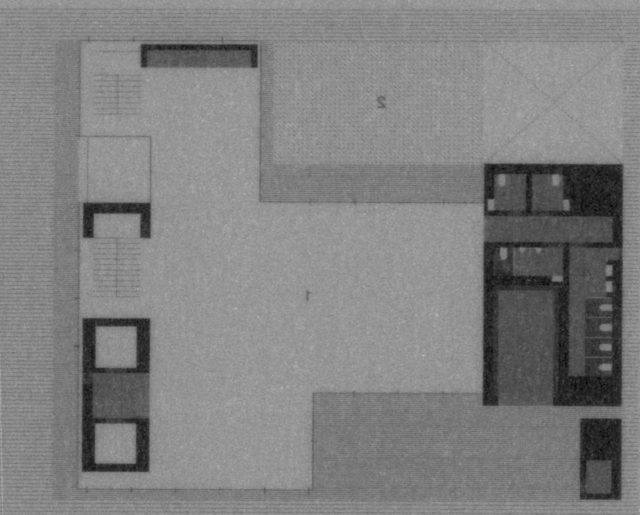

1F PLAN
1 Lobby
2 Garden

Yeongdeungpo Office Floor Plan

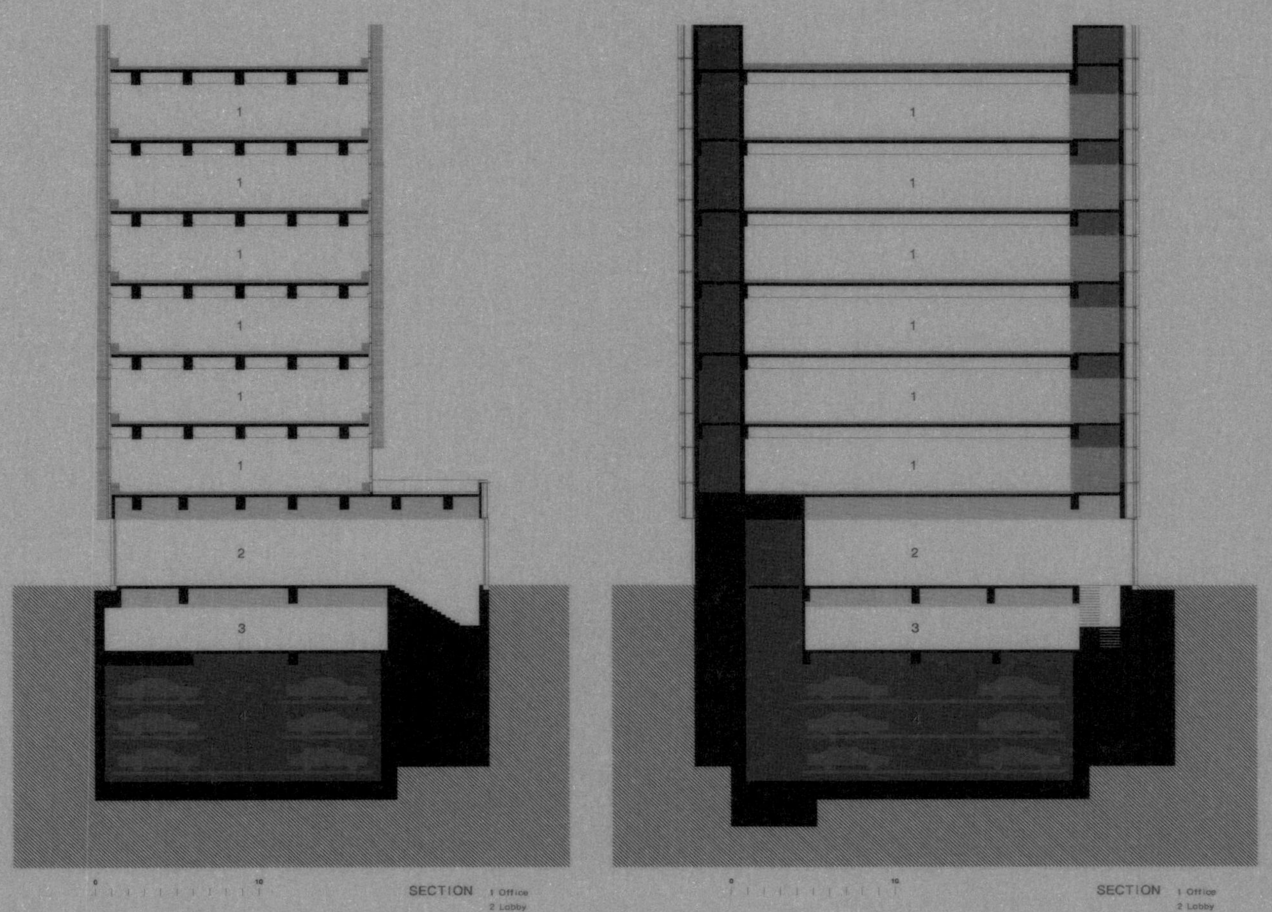

Yeongdeungpo Office Section

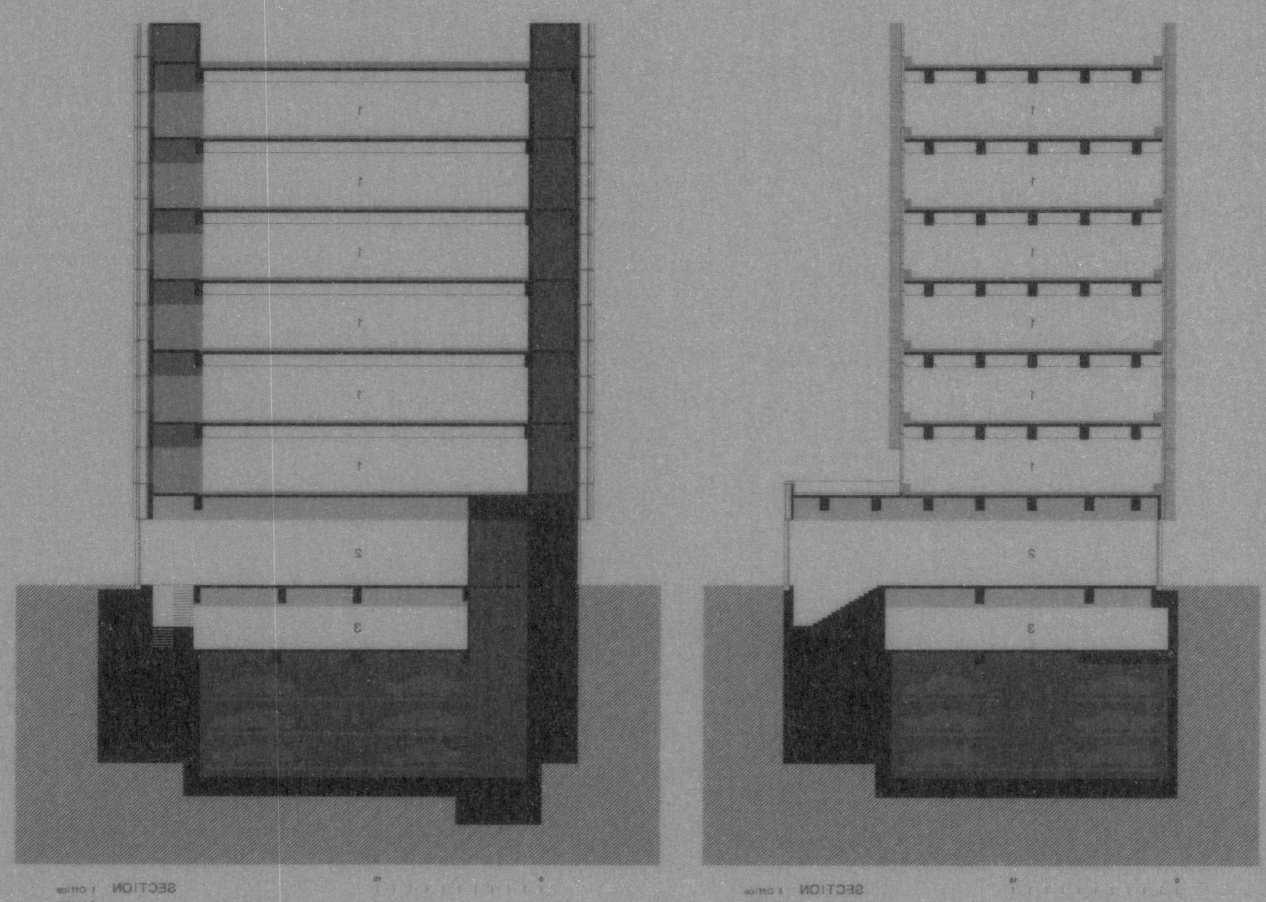

Yeongdeungpo Office Section

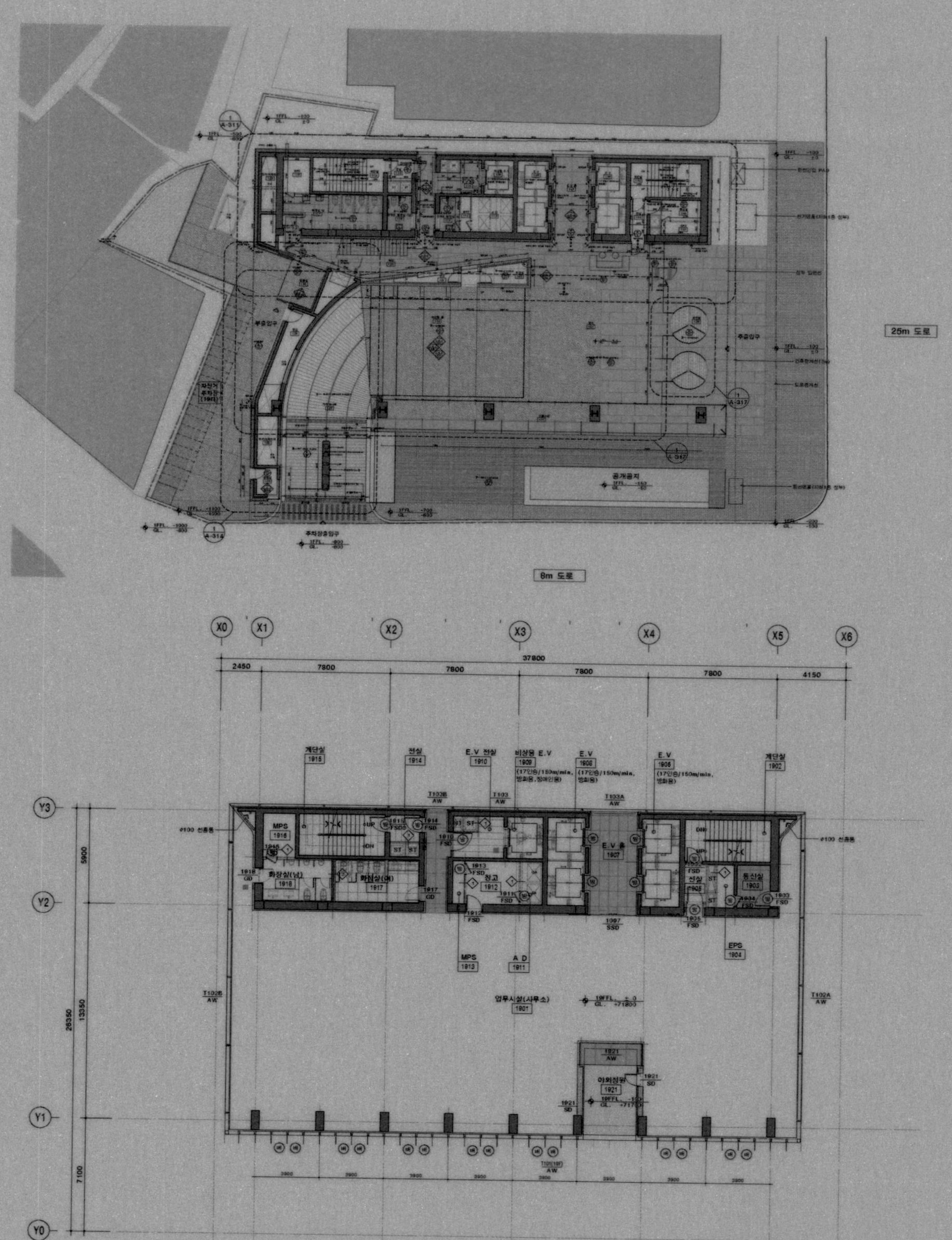

Busan Office 1F, 19F Plan

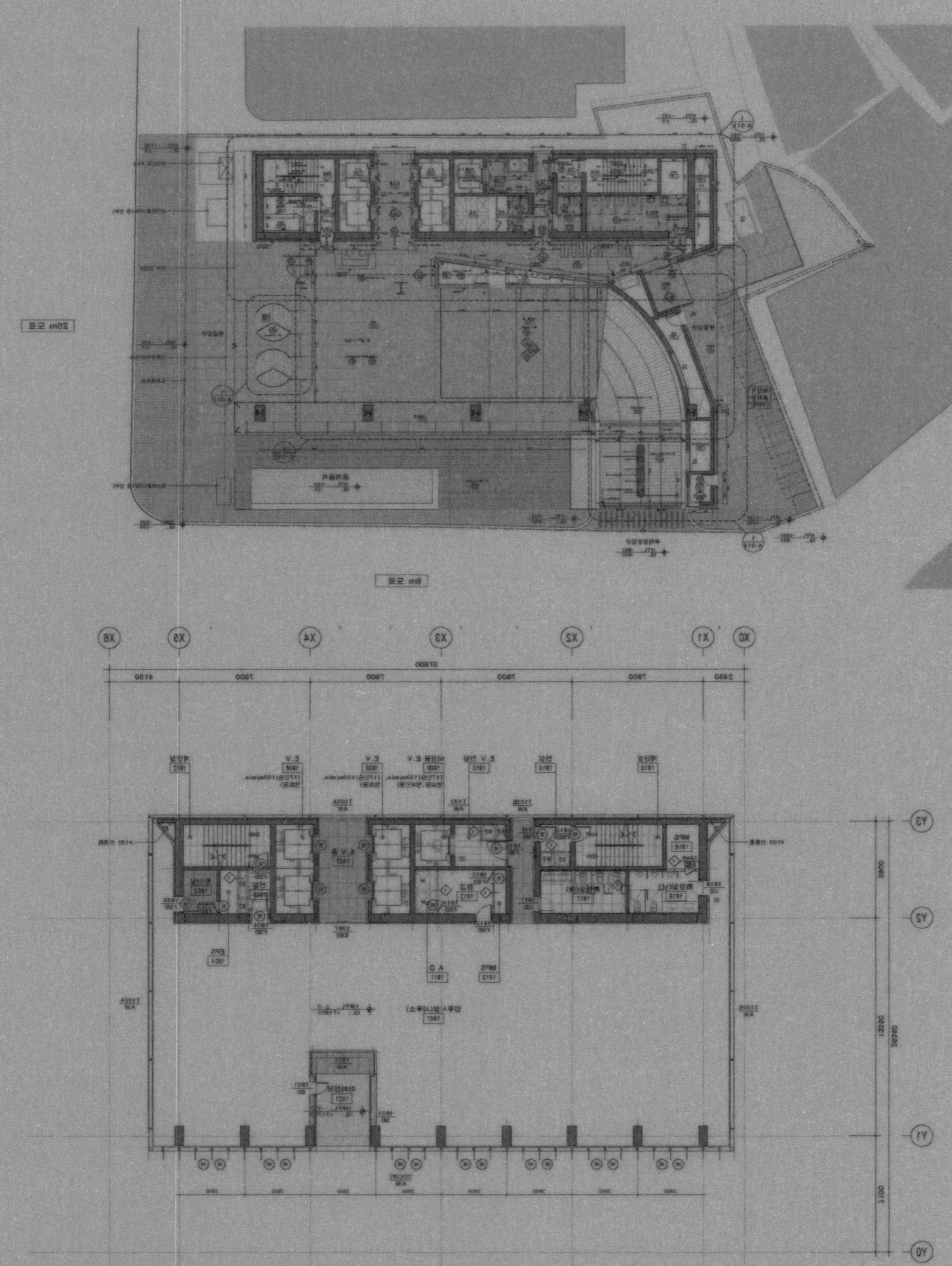

Busan Office 1F, 19F Plan

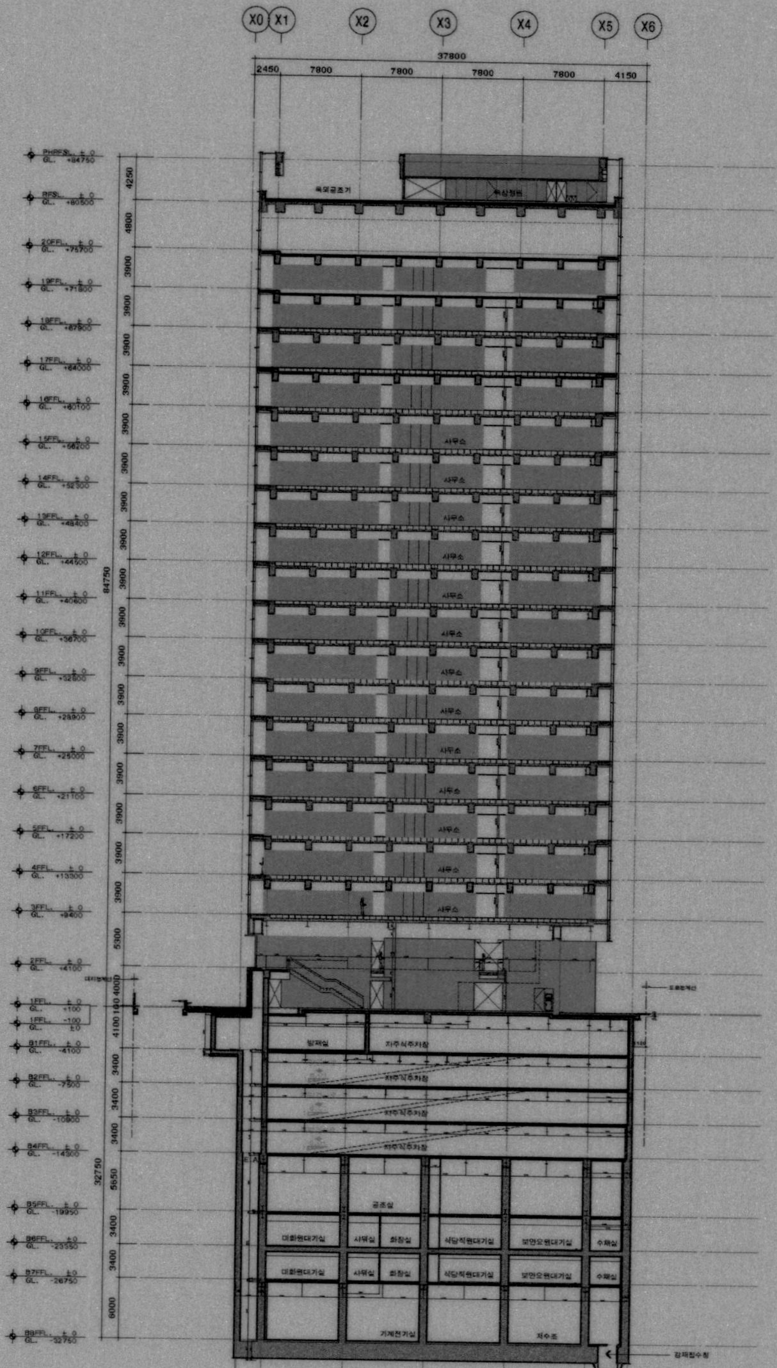

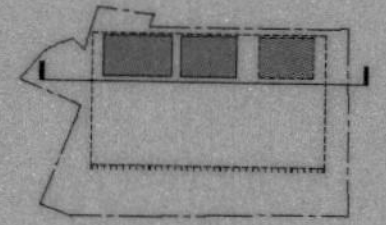

Busan Office Section

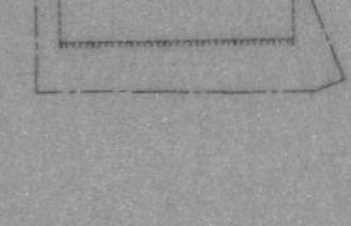

Busan Office Section

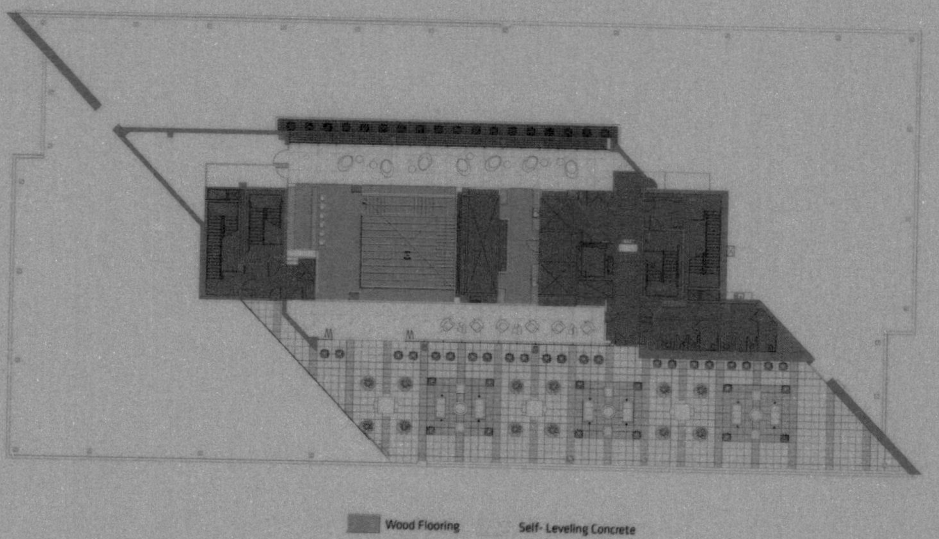

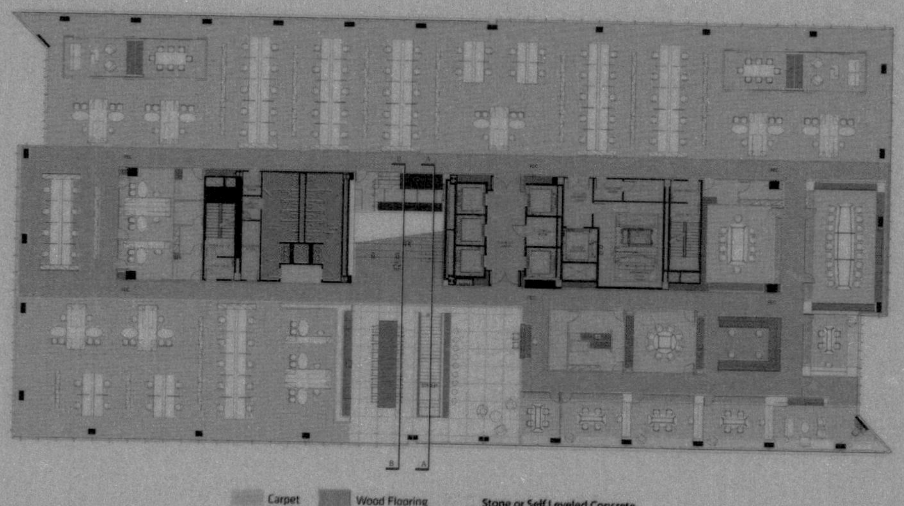

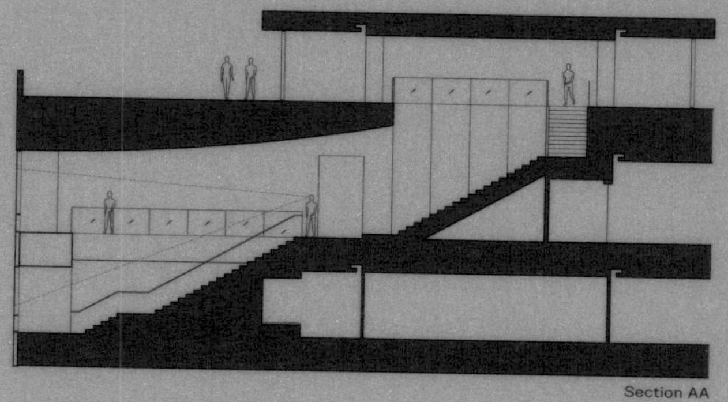
Section AA

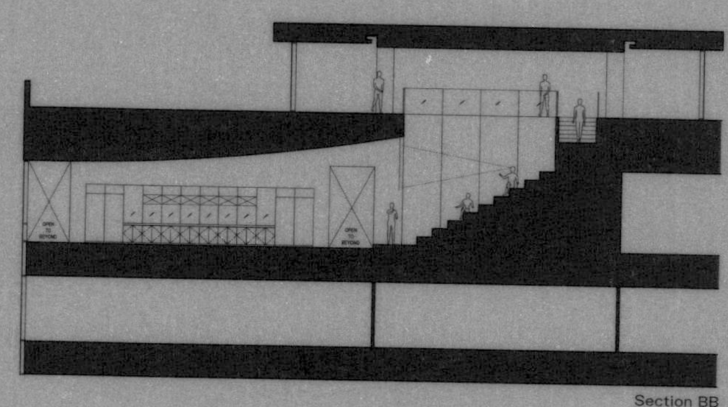
Section BB

Irvine Office 19F, 20F
Floor Plan and Section

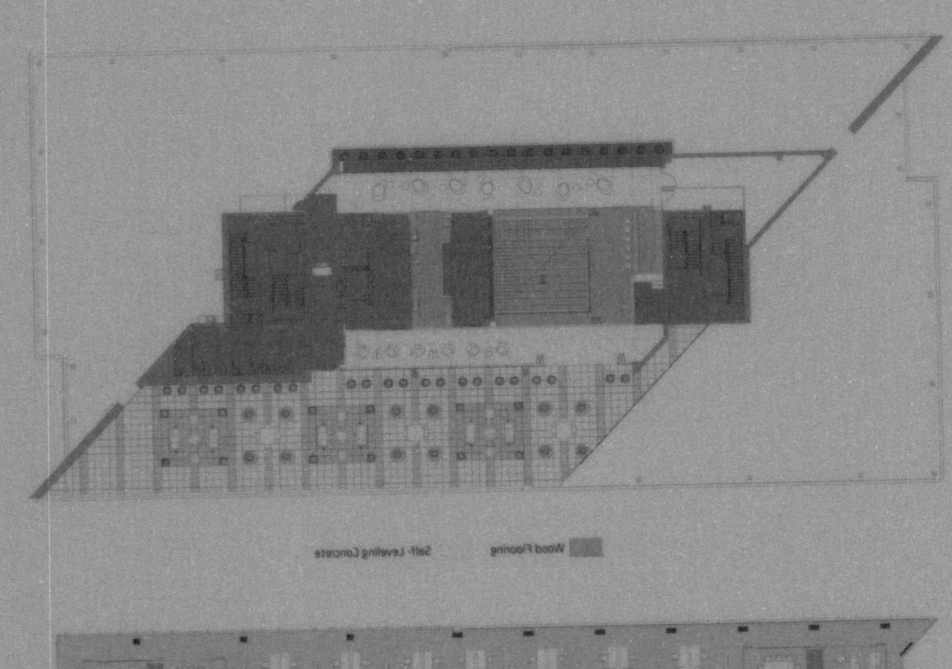

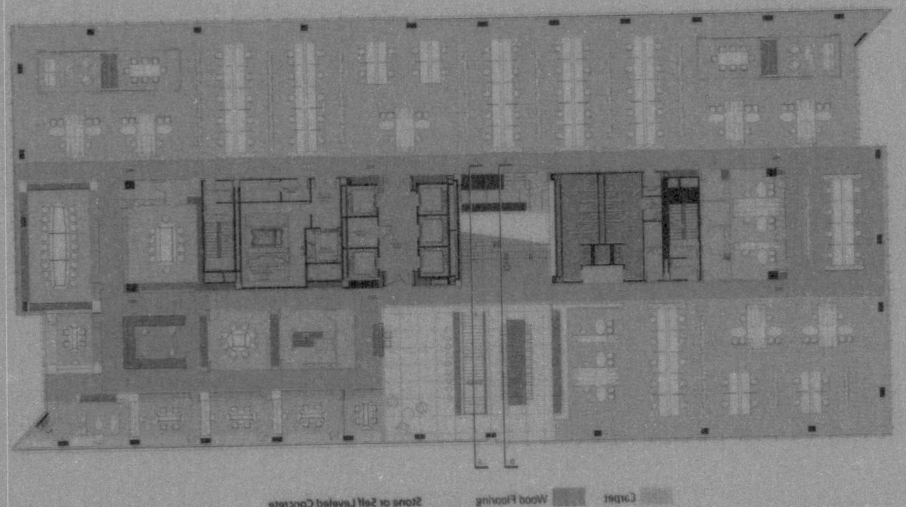

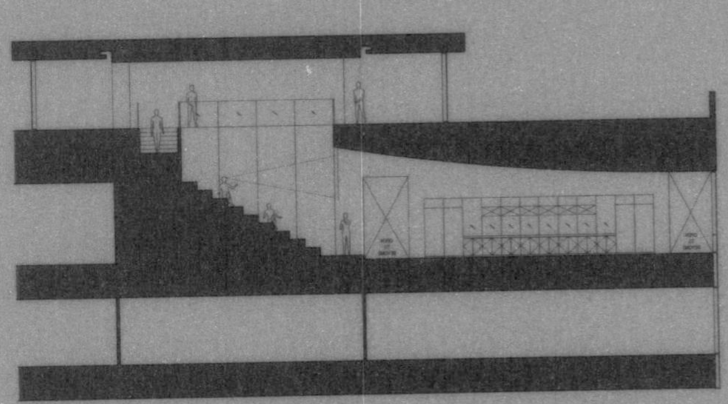
Section BB

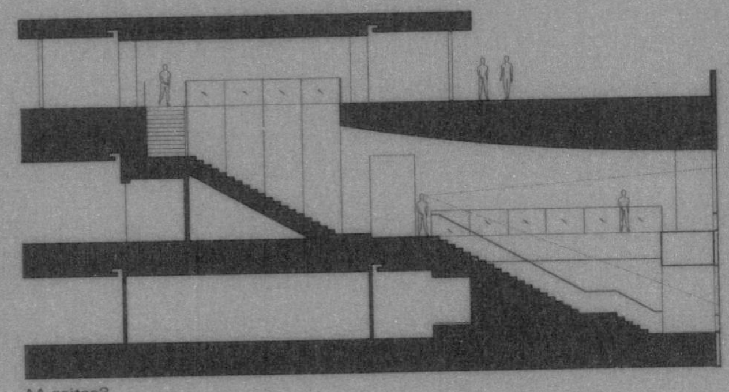
Section AA

Irvine Office 19F, 20F
Floor Plan and Section

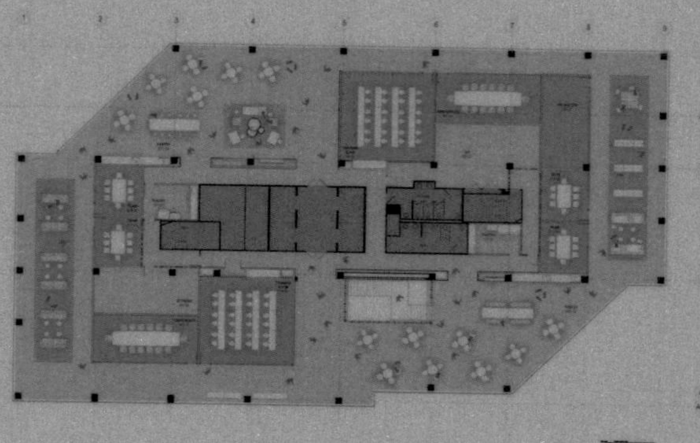

Level 10

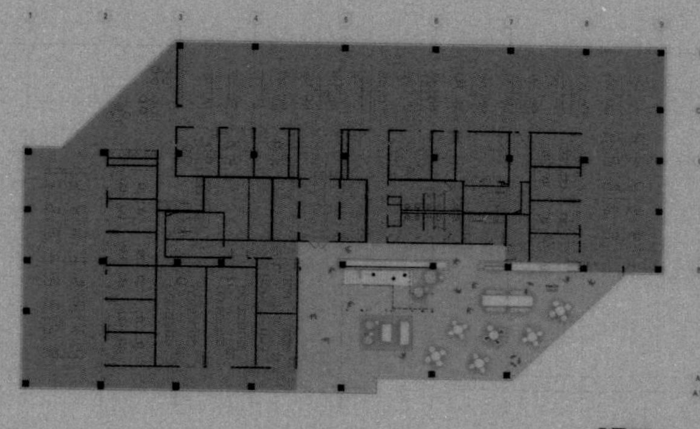

Level 9

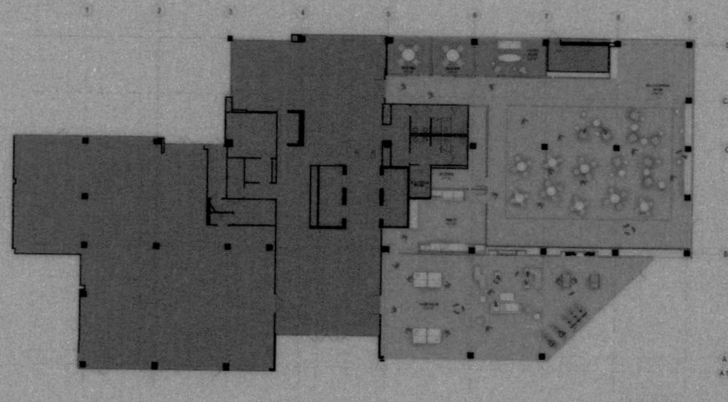

Level 1

Newport Beach Office Floor Plan

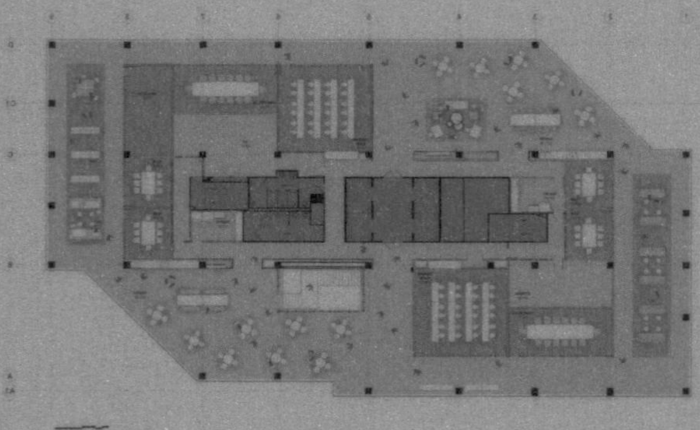

Level 10

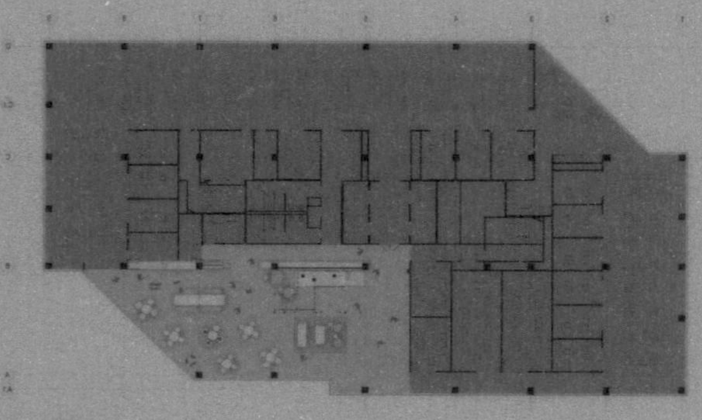

Level 9

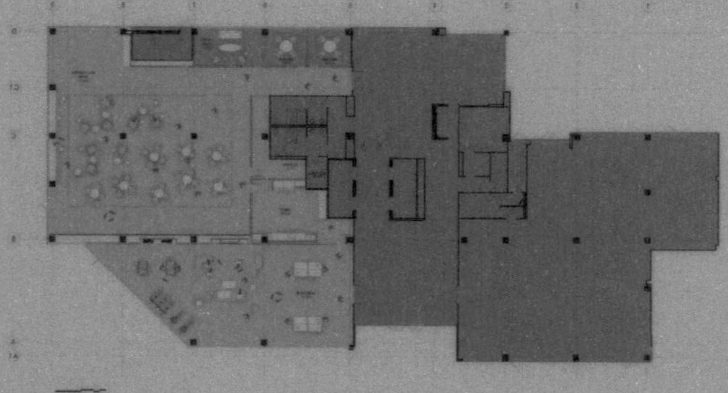

Level 1

Newport Beach Office Floor Plan

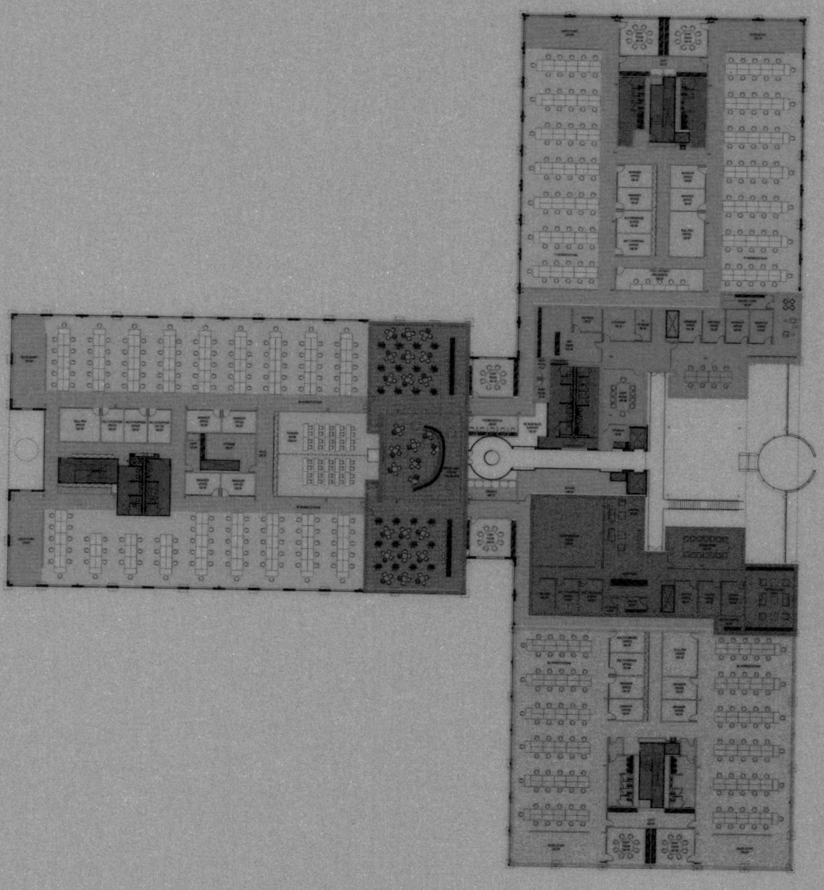

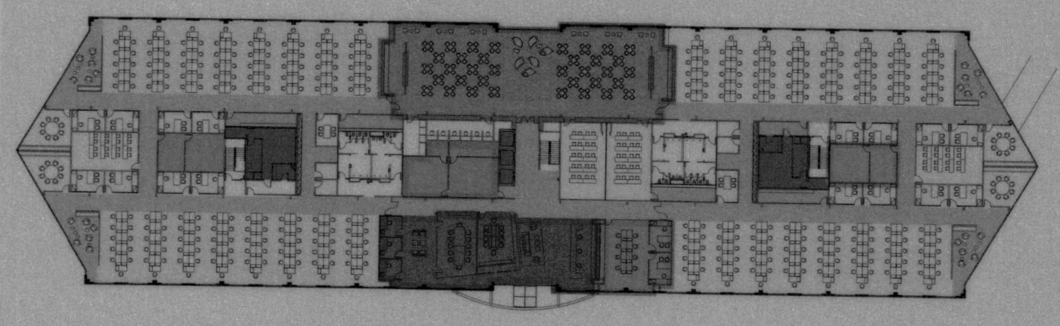

central park
reception & visitors
support services

Dallas Office Floor Plan
Atlanta Office Floor Plan

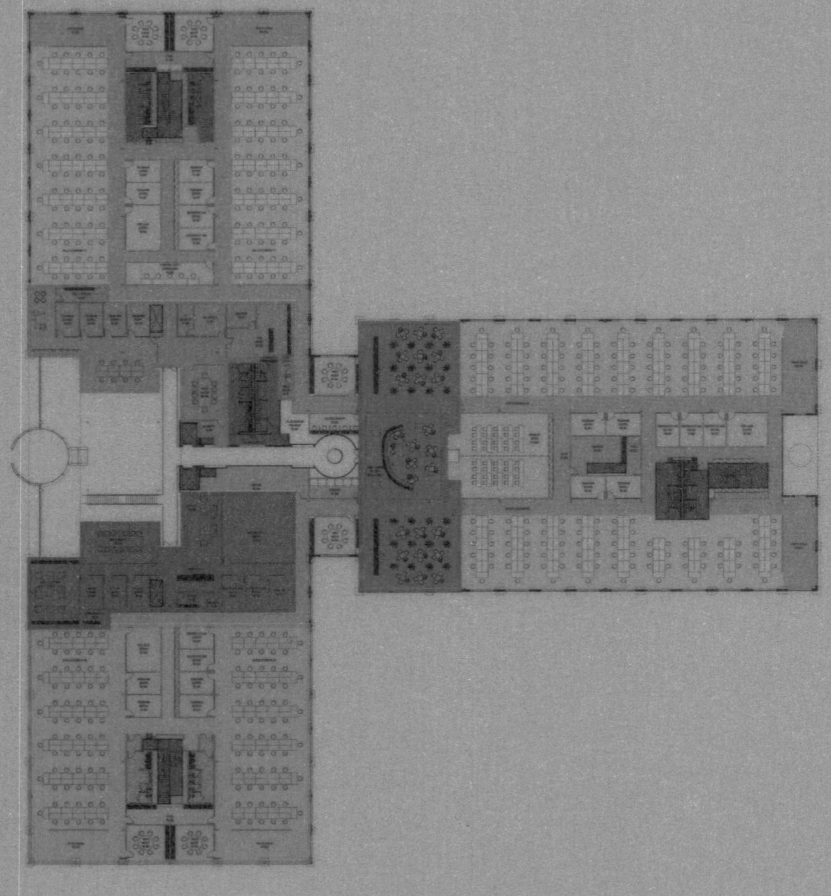

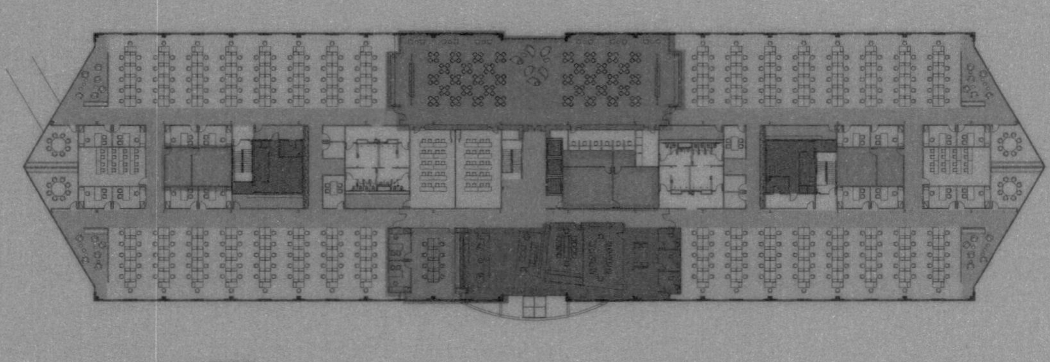

Dallas Office Floor Plan
Atlanta Office Floor Plan

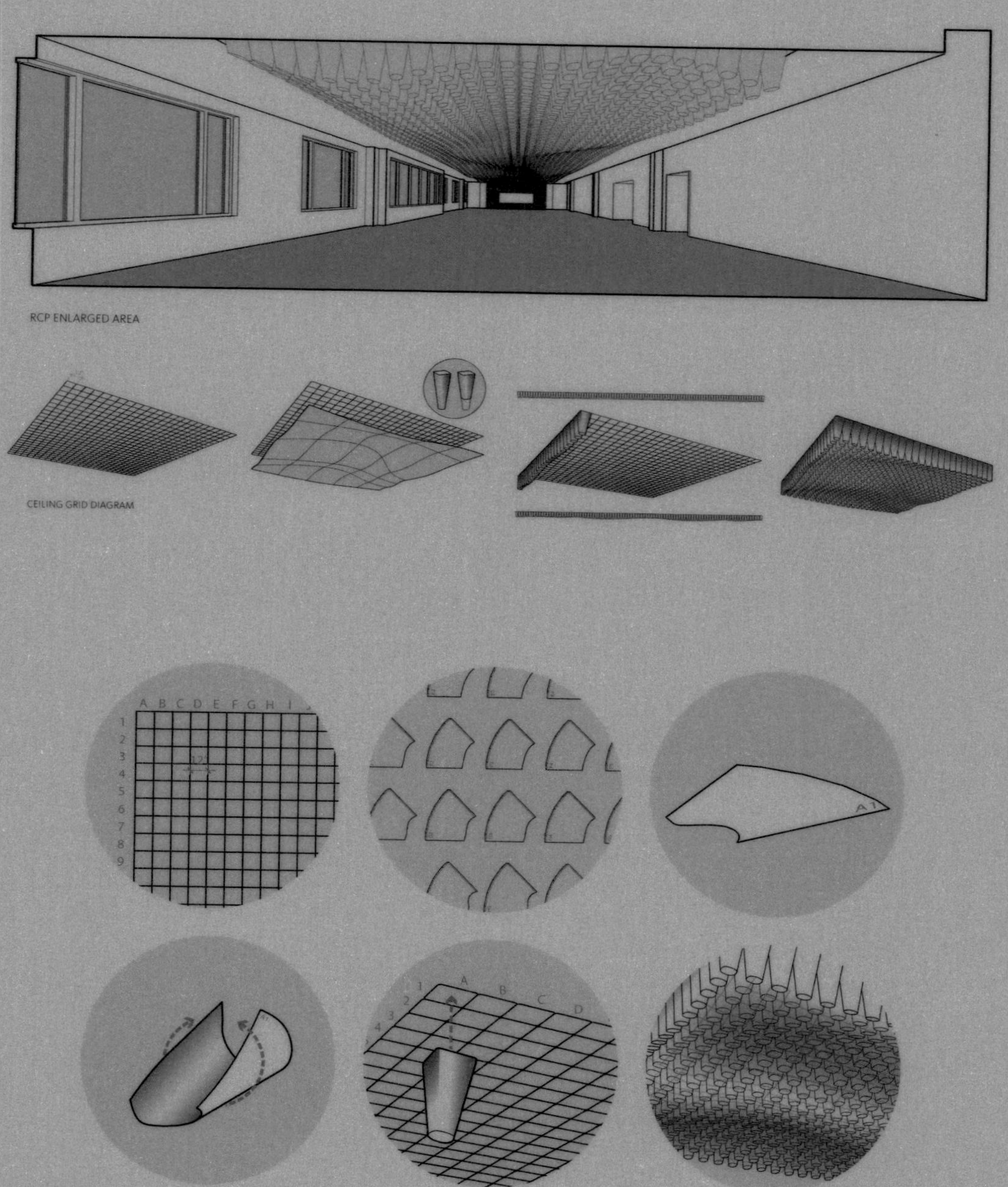

Atlanta Office Ceiling Clouds Assembly

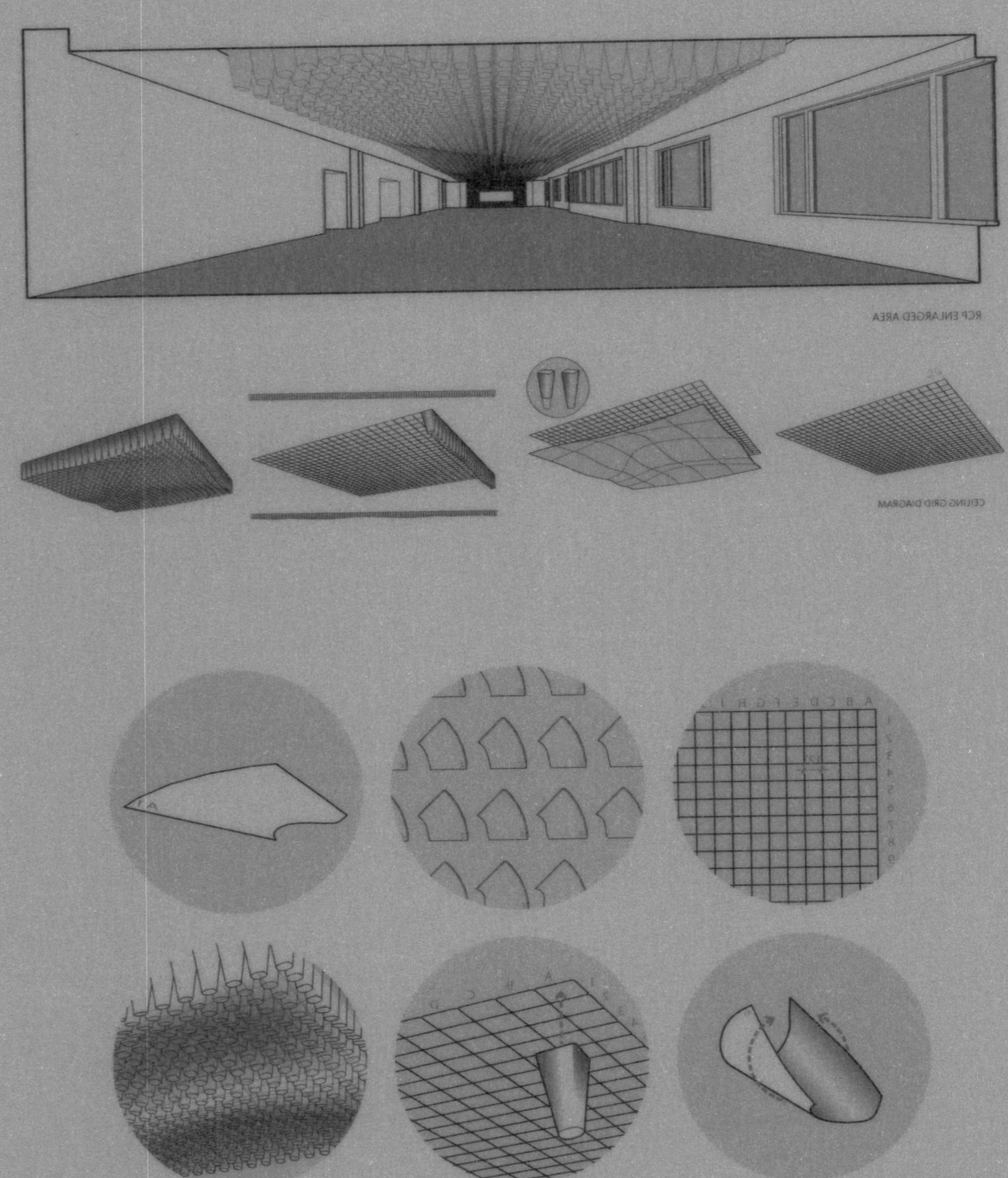

Atlanta Office Ceiling Clouds Assembly

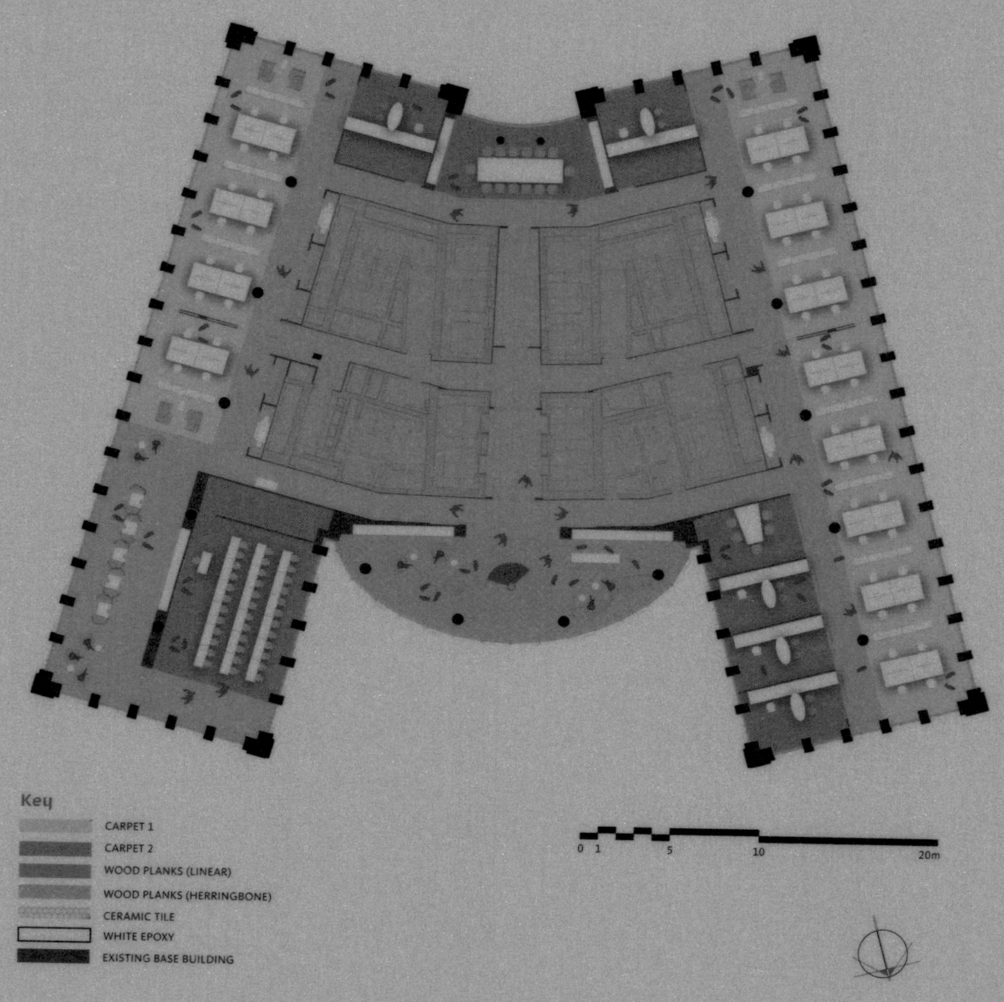

HCBE 22F Floor Plan

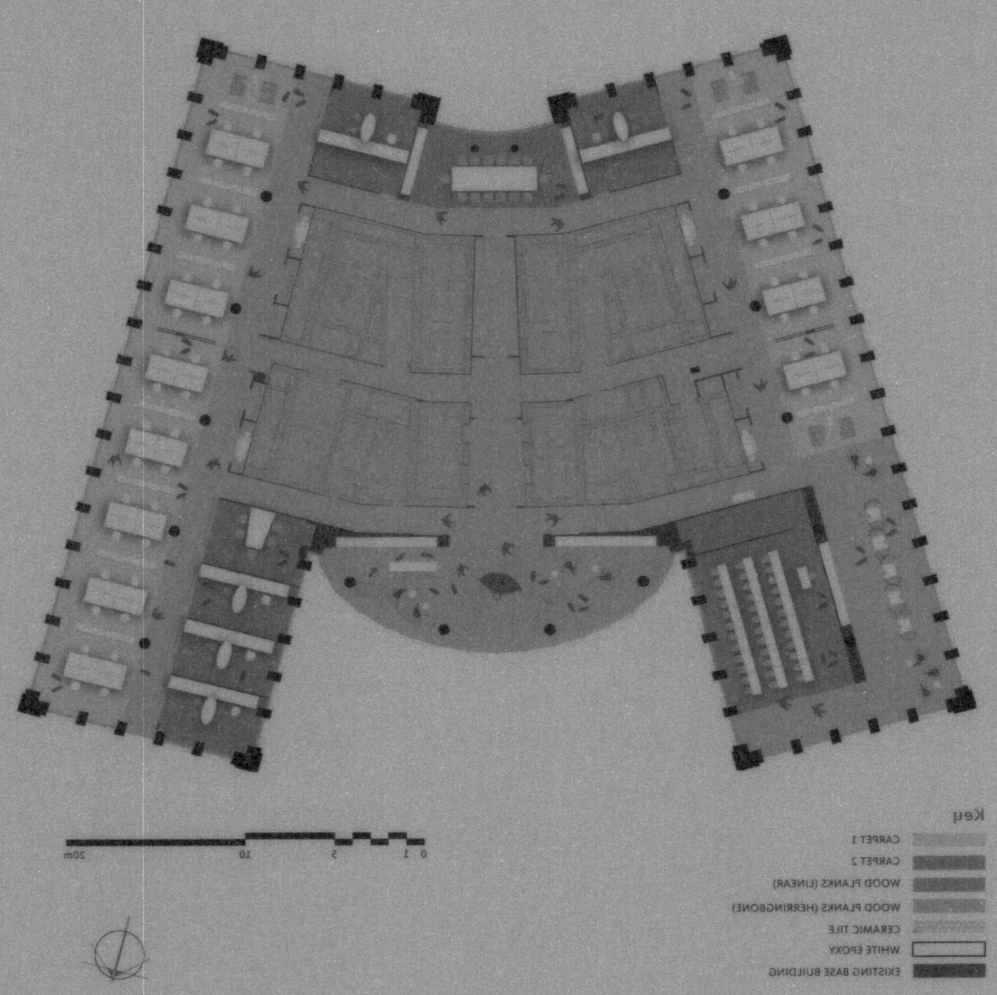

HCBE 32F Floor Plan

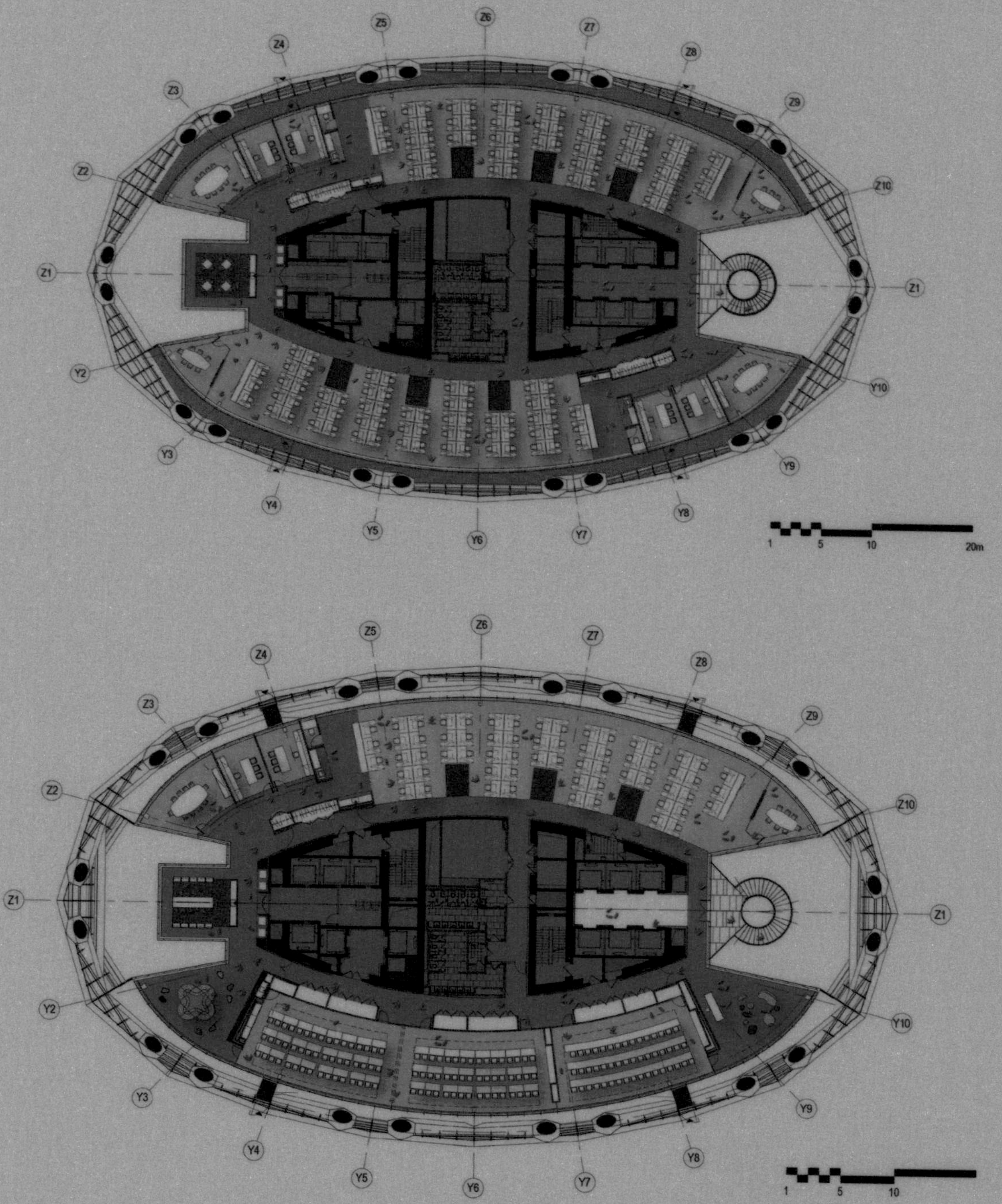

BHAF 20F, 21F Floor Plan

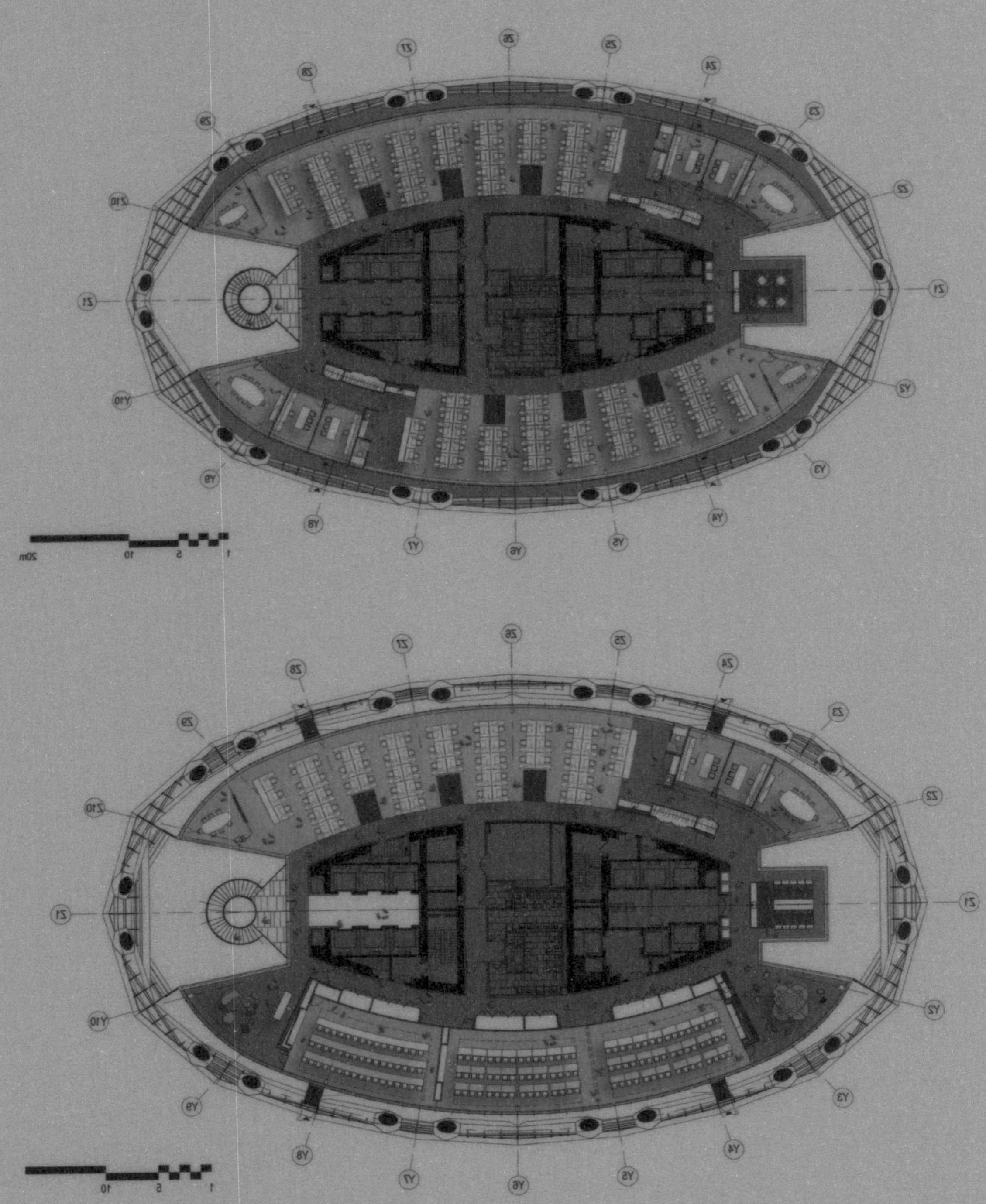

BHAF 20F, 21F Floor Plan

Concept

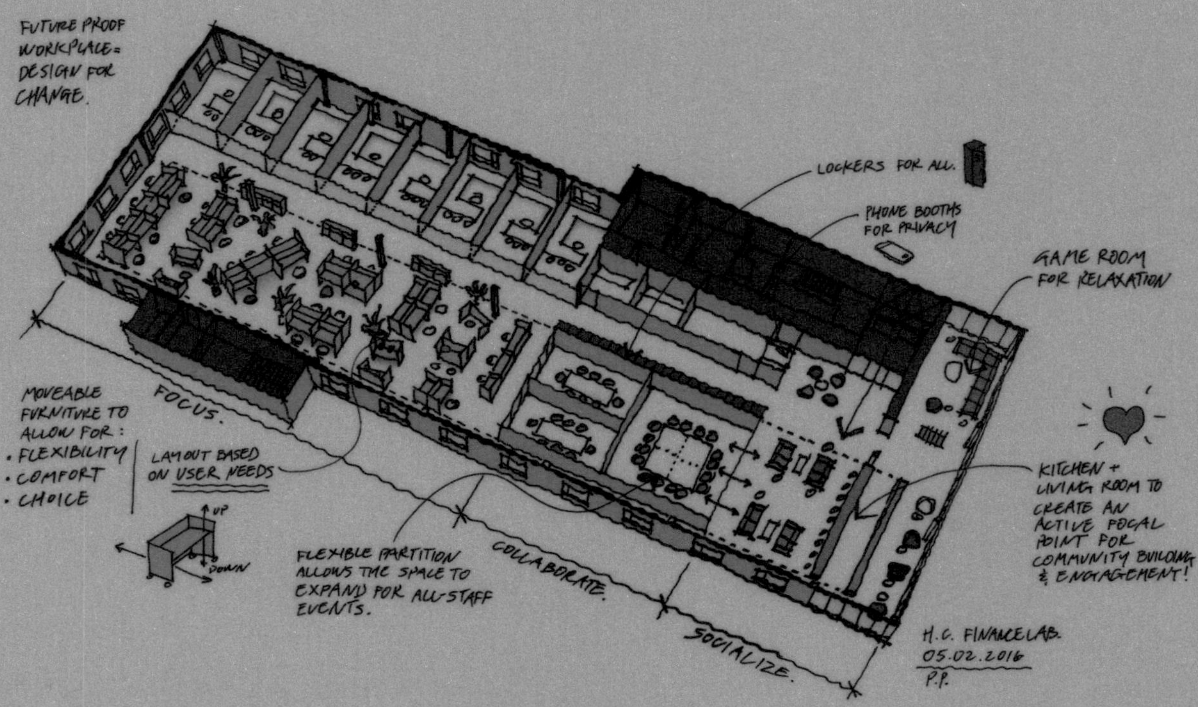
FUTURE PROOF WORKPLACE = DESIGN FOR CHANGE.
LOCKERS FOR ALL.
PHONE BOOTHS FOR PRIVACY
GAME ROOM FOR RELAXATION
MOVEABLE FURNITURE TO ALLOW FOR:
• FLEXIBILITY
• COMFORT
• CHOICE
LAYOUT BASED ON USER NEEDS
UP
DOWN
FOCUS.
FLEXIBLE PARTITION ALLOWS THE SPACE TO EXPAND FOR ALL-STAFF EVENTS.
COLLABORATE.
SOCIALIZE.
KITCHEN + LIVING ROOM TO CREATE AN ACTIVE FOCAL POINT FOR COMMUNITY BUILDING & ENGAGEMENT!
H.C. FINANCELAB.
05.02.2016
P.P.

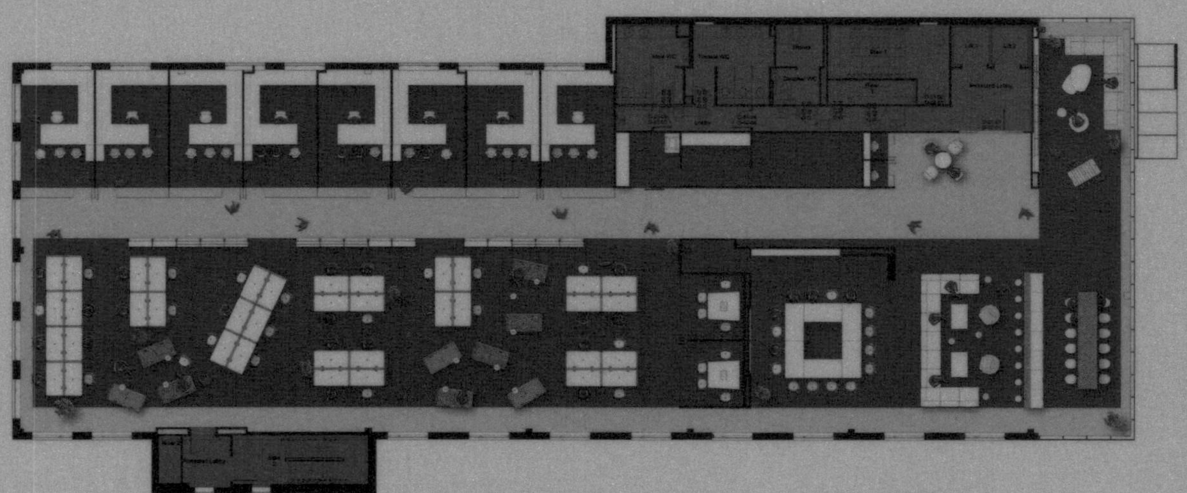

HCUK Idea Sketch
HCUK Floor Plan

Concept

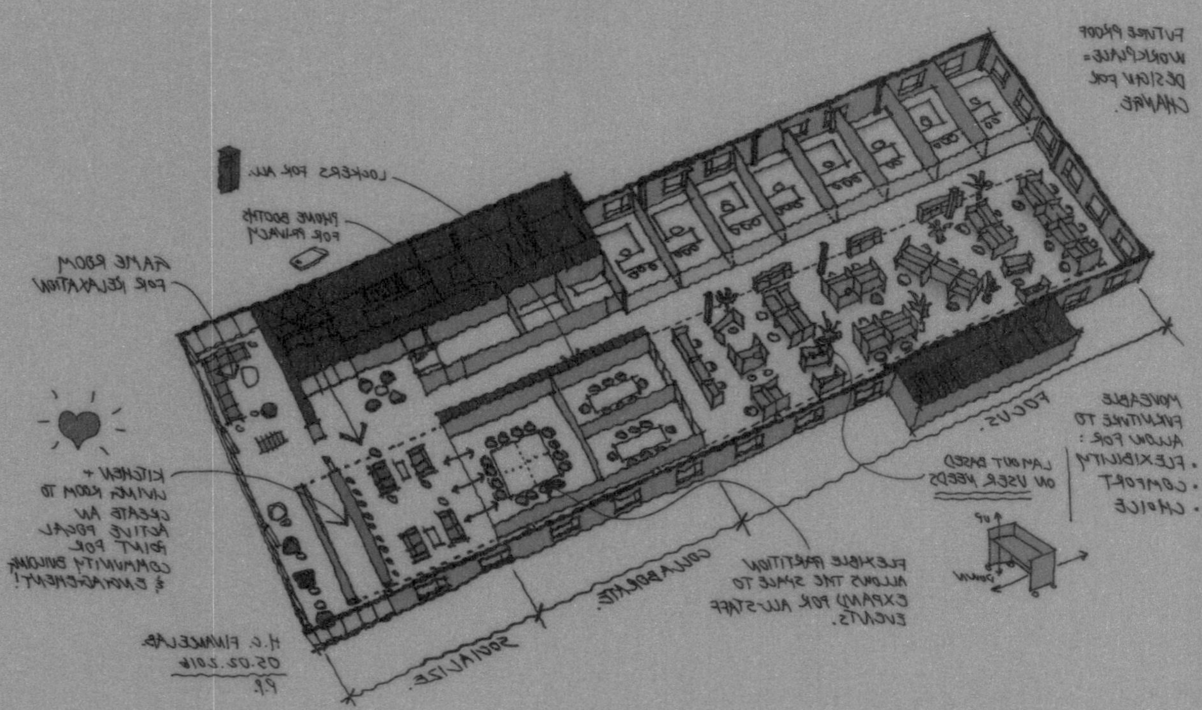

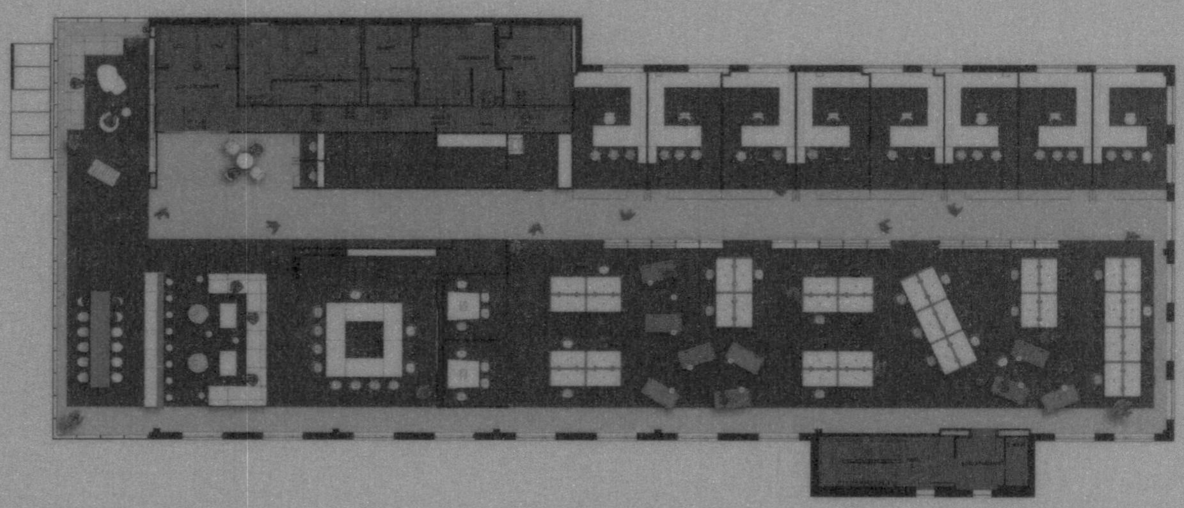

HCUK Idea Sketch

HCUK Floor Plan

Hyundai Card
Libraries

Hyundai Card's Library series realizing the brand's

ideals on Design, Travel, Music, and Cooking

Beginning with the Design Library in February 2013, and closely followed by the Travel, Music, and Cooking Libraries, the Library Series has become an important asset to the Korean modern architecture scene. A library curated through the prism of a brand with a distinctive taste and set of concerns is a rare find. The series acts as the seed of Hyundai Card's brand spirit. The three keywords running through the Library Series—Intellectual, Space, and Analogue—were ahead of their time as back in 2013 wider society was fixated on the digital realm. The decision was truly a pioneering one.

The four libraries function as a contemporary cultural hub in Seoul. There, one can experience various activities tailored to the concept governing each space, such as talk shows, concerts, exhibitions, and cooking classes, passing far beyond a simple storage facility for books and knowledge. Accordingly, they fulfill a social role beyond achievements in the corporate realm. With foundations formed of robust spatial content, the libraries exhibit a potent architectural outcome resulting from intense contemplation on the subject and expert curation.

Design Library

p. 168

The façade and courtyard of the Design Library in Gahoe-dong, display the strength and beauty of the U-shaped layout, a traditional hanok structure. Reflecting the dignity and nobility of the old town, Gahoe-dong, the design of the library preserves the architectural characteristics of the façade and courtyard in the original hanok structure.

3.1

Travel Library

p. 182

The Travel Library is not merely a place for reading, but a place with a dynamic interior, completed from the view that spatial experience is also a journey. Upon climbing the stairs, visitors appreciate various landscapes as if on an adventure. At the end of this adventure, one is met by wooden bookshelves of uneven heights, triggering visual curiosity.

3.2

Music Library

p. 194

Switching on its lights at the twilight hour, 6 pm, the Music Library's attunement to night is inimitable. The most prominent attribute is that it breaks away from the implicit rules of construction. About half of the ground floor, the area most fit for use in the whole building, has been left empty. The gently sloping floor that belongs to the original structure has been retained.

3.3

Cooking Library

p. 224

The Cooking Library, a seemingly modest building of a kind seldom found in Cheongdam-dong, is reminiscent of a mother's kitchen always poised to prepare her children a warm meal. The crossing boundaries of the floors present a space in which scents and sounds naturally blend.

3.4

3.1 Design Library

Design Library
Location: 129-1, Gahoe-dong, Jongno-gu, Seoul
Site Area: 555m²
Building Area: 289m²
Gross Floor Area: 527m²
Coverage Ratio: 53.49%
Gross Floor Ratio: 93.90%
Building Scale: 2F, Rooftop
Design: 2012.02 ~ 2012.07
Construction: 2012.07 ~ 2012.10
Architecture: 101 Architects, Wook Choi
Contractor: 101 Architects, Wook Choi

The Design Library of Gahoe-dong intimates the subtle beauty of the courtyard created by drawing upon the hanok structure's U-shaped layout.

The Design Library is the first book documenting Hyundai Card's most beloved topic: design. The Library Series is the window through which Hyundai Card makes manifest its vision, worldview, and archival collections. The Design Library advanced Hyundai Card branding beyond card design, the Super Concert, and advertisement campaigns of the early and mid-2000s.

3.1.1
The Way We Build

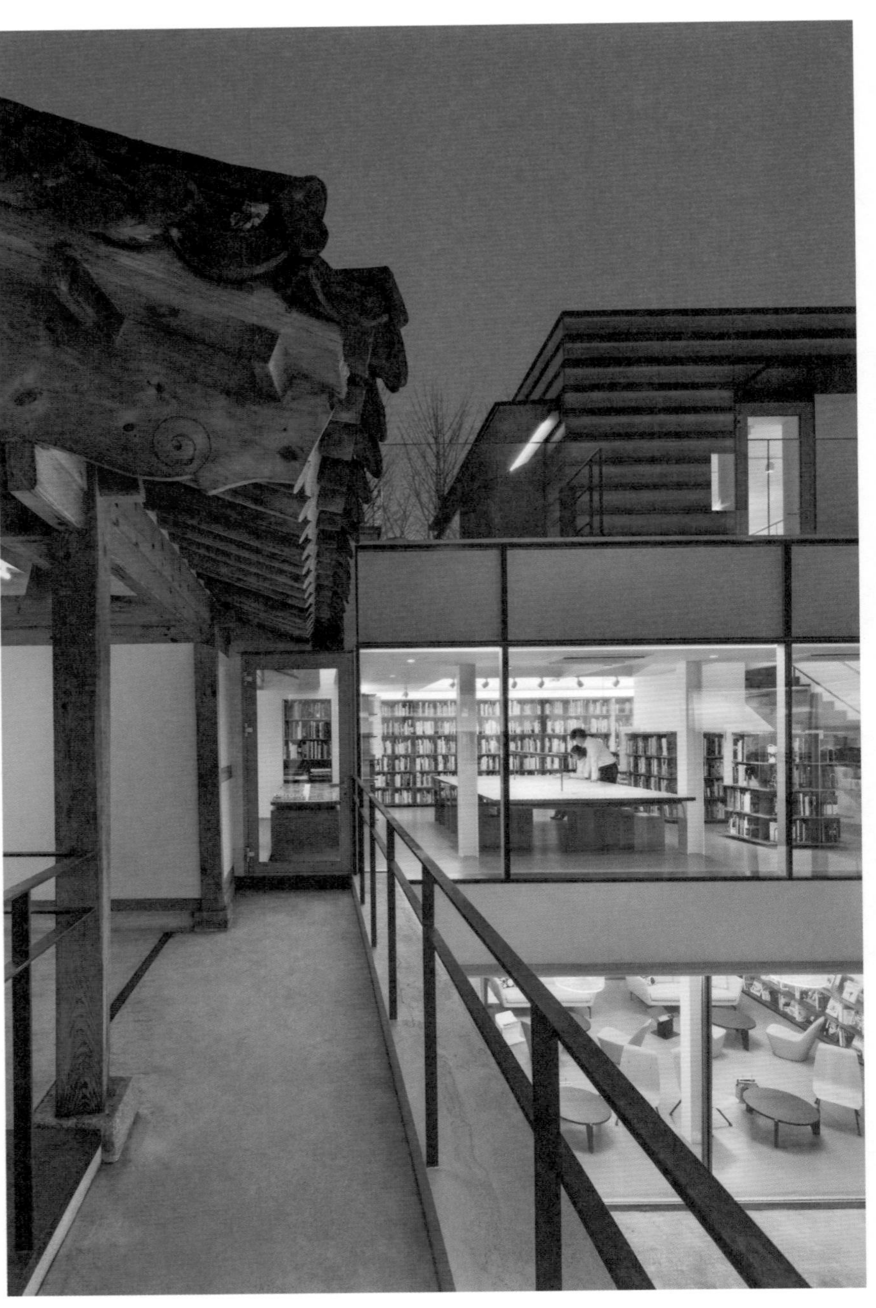

How did the Design Library become Hyundai Card's first project? Vice-Chairman Ted Chung answers: 'At first, I entertained the idea of a financial library, but I concluded that finance is knowledge, a sage spirit cared for and practiced by Hyundai Card. After much thought, the answer was in design. The fact that we didn't build a design-centered library, despite much discussion we had about the design, was a shame'. From the very beginning, Vice-Chairman Ted Chung knew the site well. Opened in 2000, the Seomi Gallery was an outstanding work of architecture worthy of the Highest Prize at the Korean Architecture Awards. Its structure, feel, location, and two-story construction were, instinctively, the perfect home for the Design Library. 101 Architects was commissioned to renovate the space to house the Design Library. However, for the said reasons, Vice-Chairman Ted Chung harnessed every opportunity to express his respect and appreciation for Taeyong Yoo, the architect who designed the original hanok structure. Once the location was fixed, the project got off the ground. The L-TFT (Library Task Force Team) was assembled, and four team members, including Team Leader Sujin Ryu (now Head of Brand Division), invested a full year in planning the library from the ground up. Vice-Chairman Ted Chung asked the L-TFT to unveil a 'Mona Lisa' that would be a testament to the unique ambition of Hyundai Card's Design Library. Against this impressive backdrop, the Design Library came to hold every volume of Domus, Life, Playboy, Visionaire, and Monthly Design, in addition to numerous rare book collections. At the heart of the Design Library is the skilfully curated collection of over 17,000 volumes. Over 43,000 books were reviewed for inclusion in the collection, and a whopping 1,670 publishers specializing in design from 36 countries were contacted, including Steidl, Taschen, Phaidon, and Hatje Cantz. The book curation began with a focus on design in the line of the Bauhaus, indisputably the origin point of modern design. The Design Library demonstrates Hyundai Card's excellent content-building abilities based on book curation. Its independent classification systems were derived from Hyundai Card's unique perspective. To secure objectivity and professionalism, the L-TFT sought advice from the Hyundai Card Design Lab and Paola Antonelli, MoMA's chief curator. Furthermore, the L-TFT invited Justin McGuirk, an architectural critic and co-winner of the 2012 Venice Architecture Biennale Golden Lion, and Alexandra Lange, a design critic and editor, to hand-pick over 5,000 books for the library. Such rigorous selection criteria meant that over 70% of the books were introduced to Korea for the first time. Indeed, about 3,000 books are now out of print or rare editions worldwide. The curation reflects the library's vital role in preserving rare collections and stocks that have rapidly depleted in recent years. The L-TFT invested unprecedented effort in locating rare or out-of-print books; in addition to scouting publishers, independent bookstores, individual collectors, and auction sites, the team visited many international book fairs worldwide, in Frankfurt, New York, Japan, and Rome, hunting for coveted out-of-print or limited-edition titles. However, the Design Library is not merely a book museum dedicated to the preservation of rare titles. A wide range of book exhibitions, launches, and talk shows are held every month at which one can meet new books and designs. Representative events include a collaborative exhibition with MoMA of New York and the Visionaire series of exhibitions. Few design-focused libraries in the world pay such careful attention to their management, curation, and access to their collections. When the doors of the Design Library were thrown open to the public, it immediately grabbed the imagination of the Korean public as well as of people across the global design scene, putting it firmly on the map of must-visit places for opinion leaders in arts and culture, such as Nicholas Serota, the director of Tate Modern, and deans of major art colleges in the UK.

The Design Library is a structure renovated by 101 Architects, redeveloped from what was once Seomi Gallery. Seomi Gallery was an outstanding architectural structure worthy of the Highest Prize in the Korean Architecture Awards.

3.1.2 Point of Space

The Design Library, nestled in Gahoe-dong, articulates the natural beauty of the U-shaped layout and courtyard of a traditional hanok. During the renovation phase, 101 Architects decided to preserve the old architectural uniqueness of the hanok in the shape of the façade and courtyard. As 101 Architects was in charge of renovating Seomi Gallery, it held a notable understanding of the existing structure even before embarking on the design of the Design Library. The courtyard is located right past the façade, made of stacked dark clay bricks. Here, the first thing one does is to raise her head and look at the vast expanse of the sky. One's gaze naturally moves to the sky as the grass-covered emptiness of this space accentuates the square framing of the sky. At first, placing an artwork in the space was considered, but they concluded that this celebration of space is the finest of artworks. The void of this small courtyard allows one to sense every small transition between the four seasons. Gazing out at the snow in the courtyard and on the roof of the hanok from the second-floor bookshelf, catching fleeting moments of light scattering across the courtyard when immersed in reading long into the evening—these are truly profound moments to be prized. The second to third floors are bona fide bookshelves. The second floor has been largely divided into three: an area for librarians, including search desks, an area with a large desk and a hanok space, and an area for the 'house within a house'. Thanks to the U-shaped spatial form surrounding the courtyard, one can use the bookshelves without interference from others. One's eyes naturally rest on the opposite side through the transparent windows. Here, people meet accounts in books and exchange glances with the space and with others. For the ease of visual engagement in this space, the key was to employ low-iron glass windows. The scenes displayed through the transparent glass also match the direction of the Design Library—that of the inner quest. In line with the context of 'looking inside from the inside', a small structure was also erected as a 'space within space' as so often implemented by Hyundai Card. The 'House Within a House', the pavilion-style reading space, is the most attractive structure on the second floor. A wide iron plate folded in its entirety, as if a work of origami, constitutes the space itself. While sitting on the bench inside the 'House Within a House', visitors become gradually immersed in their reading. The third floor includes a separate small space, Gioheon, converted from a water tank room. It was named after the Crown Prince's small study room in the Changdeokgung Palace. Meanwhile, the use of steel plates—unprocessed and pure—is a striking interior feature, accompanied by tables, stairs, and bookcases. The positioning of the bookshelves was decided after measuring daylight through-

The use of unprocessed steel plates is noticeable in the bookshelves on the second floor. A wide iron plate folded in its entirety, as if a work of origami, constitutes the space itself.

43,246: design books reviewed at the curation phase
1,678: publishers worldwide contacted at the curation phase
36: countries frequented at the curation phase
17,609: specialist books focusing on post-Bauhaus architecture, design, and art
405: rare editions valued for their existence
1,049: the entire volume of <Domus>, the world's most celebrated architecture & design magazine
2,166: the entire volume of Life magazine, the foundational voice in photojournalism
68: the entire volume of Visionaire, a publication crossing the boundary between a book and art
1,297: the entire volume of Playboy, the aggregation of 20th-century experimental designs
505: the entire volume of <Monthly Design>, the first monthly design magazine in Korea

Design Library in Numbers

out the construction period so to best protect books vulnerable to natural light. The principles behind creating the bookshelves, the beating heart of the library, were: first, books must be the protagonists; second, the designers' unique approach to reading must be understood and accounted for. Books on design and art are typically large and heavy, often employing various materials and binding techniques. For the books to be the main protagonists of these bookshelves, the shelves had to be as narrowly conceived as possible to mute their presence yet be sturdy. This is why thinly overlapped stainless steel plates were used in place of more commonly used wooden bookshelves. Moreover, considering the reading habits of designers moving across numerous books, a thin and wide iron table was deemed most appropriate. Another special element has been tucked into the walls and stairs leading to the second and third floors, one which may be easily overlooked. As the last of the traces of Seomi Gallery were stripped away, the original concrete structure was brought to light. 101 Architects added a steel plate to the structure as a new finishing material yet exposed the rough concrete. Even the thin iron stairs made of rolled steel have been spaced at slight intervals so that they do not touch the concrete. This method is common in the repair of old buildings in Italy or Germany, recording the location and time of the work by marking the differences between the structural materials.

"Beginning with the Design Library, I led the longest-running Task Force team project in the history of the company, opening four libraries in total. The request by the Vice-Chairman that 'we must create our own Mona Lisa as the public visit the Louvre to see the Mona Lisa' made me shiver. Another request was that the collection could amount to as many as 1,000 or 10,000 books—what mattered most was that the choices behind their selection had to be meaningful. That was why we worked so carefully on 'book curation', an unfamiliar concept back in 2012. However, book curation required favorable conditions. The seven principles of this phase were that the titles had to be 'Inspiring, Useful, Thorough, Influential, Wide-ranging, Aesthetic, and Timeless'. Next, we reclassified the realm of design, analyzing and recreating the existing categorization system of other libraries to manage the collections of the Design Library, which showcase a more distinctive curatorial perspective."
Sujin Ryu, Head of Hyundai Card Brand Division – Design Library project management

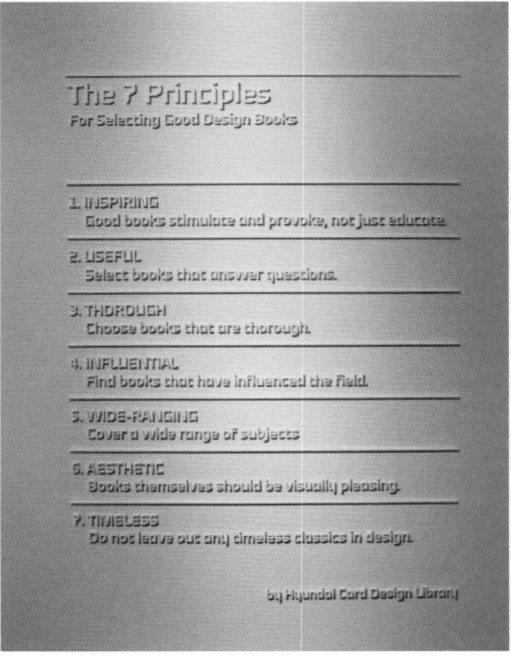
The 7 Principles
For Selecting Good Design Books
1. INSPIRING
Good books stimulate and provoke, not just educate.
2. USEFUL
Select books that answer questions.
3. THOROUGH
Choose books that are thorough.
4. INFLUENTIAL
Find books that have influenced the field.
5. WIDE-RANGING
Cover a wide range of subjects
6. AESTHETIC
Books themselves should be visually pleasing.
7. TIMELESS
Do not leave out any timeless classics in design.
by Hyundai Card Design Library

MAGNUM
MAGNUM CONTACT SHEETS
MAGNUM

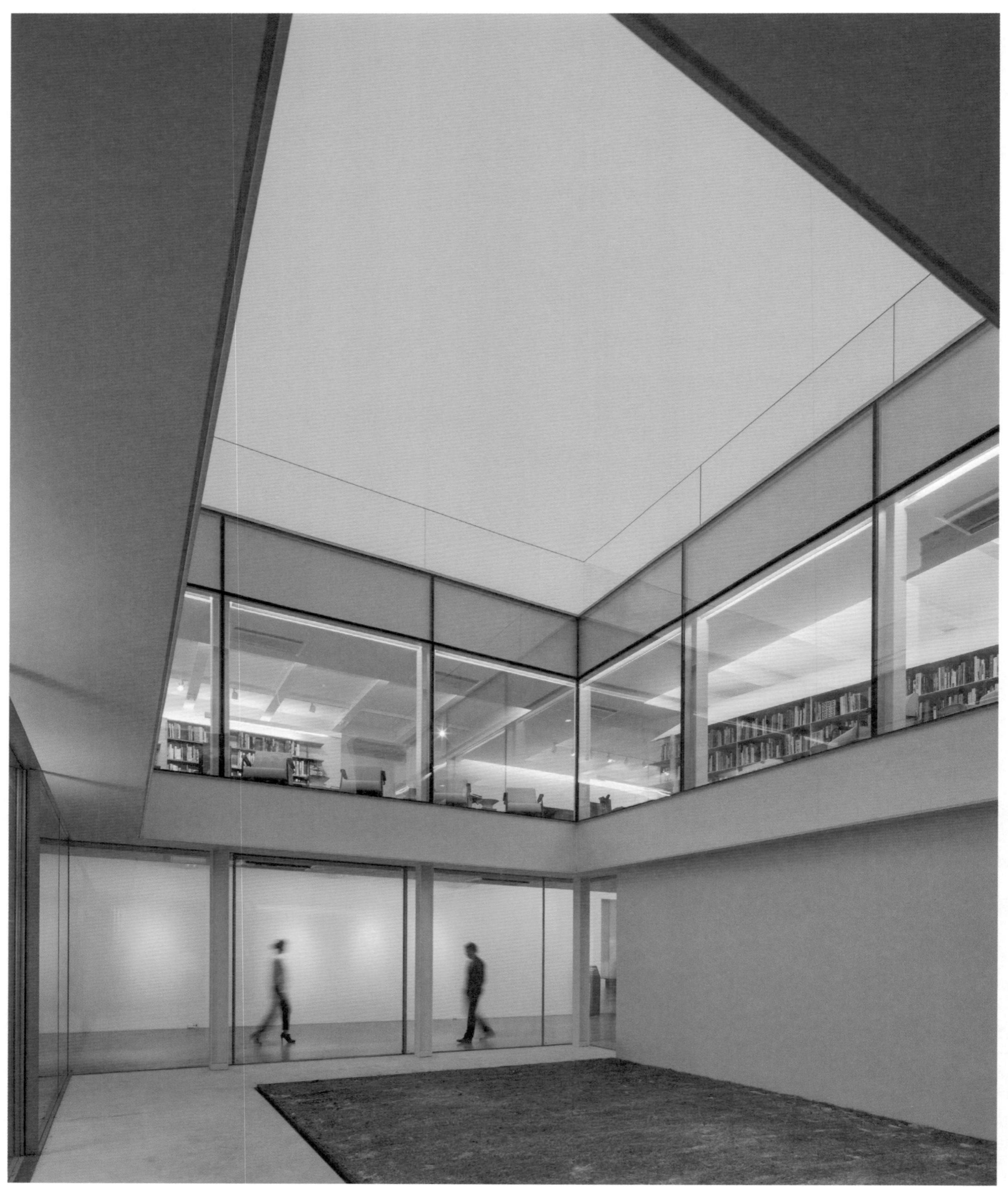

"Fundamentally, we wanted to retain the characteristics of the façade, structure, and courtyard of the old hanok. I was impressed by how beautiful the sky appeared when viewed from within the hanok's courtyard. The color of the sky pops with a more vivid life when beheld from within the courtyard. The contrast between the ground and the sky was rendered more visible thanks to the absence of the gap in height between the ground floor and the ground level itself. Above all, we thought that the courtyard ought to be emptied so that one's gaze could be directed aimlessly to views outside. When remodeling Seomi Gallery, which was used as both a gallery and a house, we stripped out a great deal of the old features. Furthermore, when turning it into the Design Library, we removed the old elements with greater confidence, uncovering the structure's original beauty as much as possible. All of the elements that could have interfered with reading and concentration were the targets of elimination. We wanted it to prioritize the creation of an analog space in which time would flow slowly."

Wook Choi, Head Architect of 101 Architects – Design Library Architecture and renovation

3.2 Travel Library

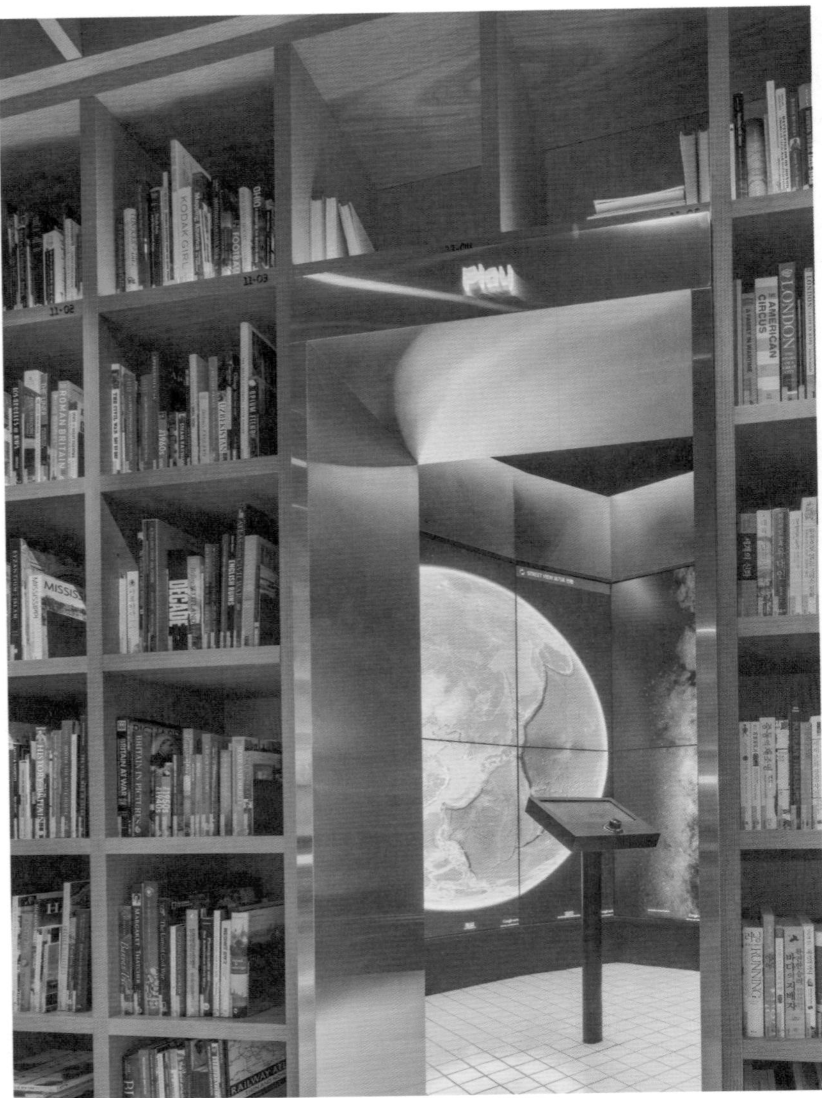

Travel Library
Location: 87-9, Cheongdam-dong, Gangnam-gu, Seoul
Area: 372m²
Design: 2012.07 ~ 2013.06
Construction: 2013.05 ~ 2014.01
Interior: Wonderwall
Contractor: Kesson

After the Design Library, Hyundai Card set upon a more popular theme, 'travel'. After Vice-Chairman Ted Chung encountered Masamichi Katayama of Wonderwall, who was on a visit to Korea, the Travel Library was born. There, wooden bookshelves of uneven heights draw into view as one climbs the stairs, prompting visitors to take in its diverse landscapes as if on an adventure.

Of all the themes explored by the Hyundai Card Libraries, 'travel' is the most popular and approachable. Its location was, inevitably, Cheongdam-dong, as the most sophisticated and trendsetting area in all of Seoul.

3.2.1
The Way We Build

Following the opening of the Design Library in Gahoe-dong, Hyundai Card fixed upon 'travel' as its second library theme. If the Design Library circles the origin of Hyundai Card, Hyundai Card decided to pursue a more accessible and animated collection through the lens of travel. The appointed location was Cheongdam-dong, the trendiest and most sophisticated district in Seoul. Around that time, Vice-Chairman Ted Chung encountered a Japanese interior designer from the (now defunct) Purple Card membership space, House of the Purple. It was Masamichi Katayama, the founder and creative director of Wonderwall. As the two conversed in Japanese, the portfolio book Katayama was holding caught their attention: 'To be honest, I had no idea who Katayama was until we met by chance at a bar that day and struck up a conversation. I was intrigued by his perspective on architecture and interiors. So, taking advantage of this chance meeting, I opened the archive book he had brought with him. Looking through it, I came to see that he was the designer most perfectly suited to the design of the Travel Library, which was under the planning stage at the time. Without hesitation, I entrusted him with the project, and a space of excellence was swiftly achieved.' The sketch was completed and delivered within three months of this first encounter. The following steps were effortless. While the Design Library focused on the 'internal' realm, the Travel Library would pursue the 'external' realm—chasing the opposite direction in light of the theme. Well-known for his dynamic presentation of central design themes, Katayama was perfect for the role. The Travel Library is not a vast space, and as such, the spatial

In designing the Travel Library, Masamichi Katayama of Wonderwall separated the first and second floors into two hemispheres with a boldly twisted central staircase that is clad in white tiles.

Located on the second floor of the Travel Library, the Plan Room is a place in which spontaneous ideas, routes, or travel plans can be freely scribbled and erased on the white wall.

impact upon visitors depended on elements placed in the structure's center. Katayama solved this problem through two patterns. First, the first and second floors were divided into two hemispheres with the central staircase twisted into a bold shape and finished with white tiles. When climbing the stairs, visitors are encouraged to appreciate various scenes of landscapes as if they are on an adventure. At the end of this adventure, one is met by wooden bookshelves of uneven heights. These bookshelves bend into angles and heights of increasing variegation as they reach the ceiling, generating intriguing movements and moments. The Travel Library possesses a collection of over 15,000 books, selected from the unique perspective on travel held by Hyundai Card. The introduction of unparalleled new categories is inspiring, worlds apart from the conventional notion of destination-oriented travel. The collections were created in collaboration with Kevin Rushby, a travel columnist for the British daily newspaper The Guardian, Shawn Low, an Asia-region editor for Lonely Planet, Carolina Miranda, a columnist specializing in architecture, travel, and art, and Yoshitaka Haba, a renowned Japanese book consultant, among others. The Travel Library divided the space into regions and themes and applied detailed categories accordingly. A total of 13 themes were established including 'Arts·Architecture', 'History·Heritage', and 'Adventure', with subcategories including 'Travel for Inspiration (Arts/Heritage)', 'Travel for Adventure (Adventure/Activities)', and 'Travel for Recharge (Lifestyle/Landscape)'. Separate sections designated by region feature guidebooks for 196 countries for visitors unfamiliar with the dynamic new approach. The Travel Library is the world's only place to meet every volume of Imago Mundi, the world's first and only travel-geography magazine, and Transaction, the Korean branch academic journal of the Royal Asian Society of England. Also included are every title of National Geographic.

3.2.2 Point of Space

Of note upon first entering the Travel Library is an old-fashioned analog airport schedule board. The board displays flight departure times with soft clicking sounds recording changes every 10 minutes, linked in real-time with Incheon International Airport Terminal 1. The interior is a polyhedral space where diagonal lines extend and meet in intricate formulations, creating a new appearance on every visit. Katayama stressed the presence of the stairs in his design, connecting the first and second floors as a focal point of attention. 'When we think of stairs, what usually comes to mind is the simple act of going up and down. Stairs, however, can be part of an exploration on their own terms. I thought the stairs themselves ought to be symbolic in a library on the theme of travel'. Finished with white tiles reminiscent of airplane hangers and high-gloss paint, the stairs display a wide range of landscapes from a bold, twisted angle. The stairs go beyond the simple function of connecting floors. The elements scattered along the stairs encourage visitors to be inspired by the possibilities of their next trip. Ascending the stairs from the first floor, visitors are greeted by the 1.5-floor space dedicated to National Geographic and Imago Mundi, leading them to the bridge on the second floor that connects the bookshelves by theme and by region. When the floor area is minimal, a lack of distinction between the wall and the ceiling may further congest the space. This shortcoming was overcome as the staircase skyrockets vertically in combination with the see-through floor. Except for a small passageway on the second floor that barely fits two people, the floor was made transparent through

Finished with white tiles reminiscent of airplane hangers and high-gloss paint, the stairs display a wide range of landscapes from a bold, twisted angle.

"The Travel Library is themed under the guiding concept, a box of curiosities. Visiting the space is the same as embarking on a journey. That is why every corner of the space is replete with information, experiences, and objects connected to travel. We always try to approach every project with the mind of a novice. Before setting out on a design, we secretly visit our clients to better understand them. We also produce comprehensive research on the company's history. Rarely is a company as conscious of design as Hyundai Card is—they are an ideal client for designers, as they put up no obstacles to design ambitions and accept demanding proposals."

Masamichi Katayama, Founder of Wonderwall - Travel Library Space Design

to the first floor. The walls are made of wooden shelves, and their frames extend to the ceiling. This removes a perceivable border between the wall and the ceiling. A ceiling of large and small triangles reminds one of a sheer cliff or a cave concealing a secret. The irregular pattern reflects the concept, 'travel is a heading to the unknown'. This complex ceiling structure was a challenge to create. Lines expressed simply on the drawings carried a certain thickness when materialized within the space. As such, a three-dimensional complex shape had to be devised to force the lines emerging from various spheres to meet seamlessly at one point. Kesson, the company in charge of the construction, attempted to use 3D drawings, which was challenging. In the end, the shape was completed by cutting it little by little on-site owing to the level of craftsmanship and expertise. This is why the ceiling construction took 45 days when the usual timeline for such work is 3 weeks or less. The highlight of the space on the second floor is the Play Room where visitors can control the street view on the large screen with the kiosk's joystick. The display is beyond the familiar, simple visualization of Google Maps on a large screen. The system was available only at the Google headquarters and combined two software programs: Google Earth and Google Galaxy. Vice-Chairman Ted Chung specifically requested it for use in the Travel Library after visiting and experiencing the technology at the Google headquarters. It was a result of significant and persuasive labor, and Google even sent a technician to create the new design. The chairs in the Café and around the bookshelves create an ambiance that transcends generations and regions. The Wagner Chair, the high point in Nordic design, the 1950s chairs restored by British fashion designer Margaret Howell with Ercol, the Shaker Chair that represents the functionalist attitude of American Shakers, and animal-shaped stools from Africa, have all been scattered around the space. The process of installing the stuffed dear head on the Café's wall was somewhat difficult. Katayama, the designer, was so fond of taxidermy that he displays a stuffed polar bear, moose, and bull in his office. However, Vice-Chairman Ted Chung's taste is the exact opposite. At first, the company looked for something that could stand in for this idea but could not find a match. After a long discussion over whether to respect the designer's choice, the decision made was to go to Paris to purchase one.

94,324: travel books reviewed at the curation phase
14,504: books focusing on regions and travel
69: rare books that are of high value
1,545: the entire volume of National Geographic, the earth's diary
89: the entire volume of Imago Mundi, the world's first travel geography books
1,910: guidebooks introducing travel destinations around the world
196: countries covered by the Travel Library

Travel Library in Numbers

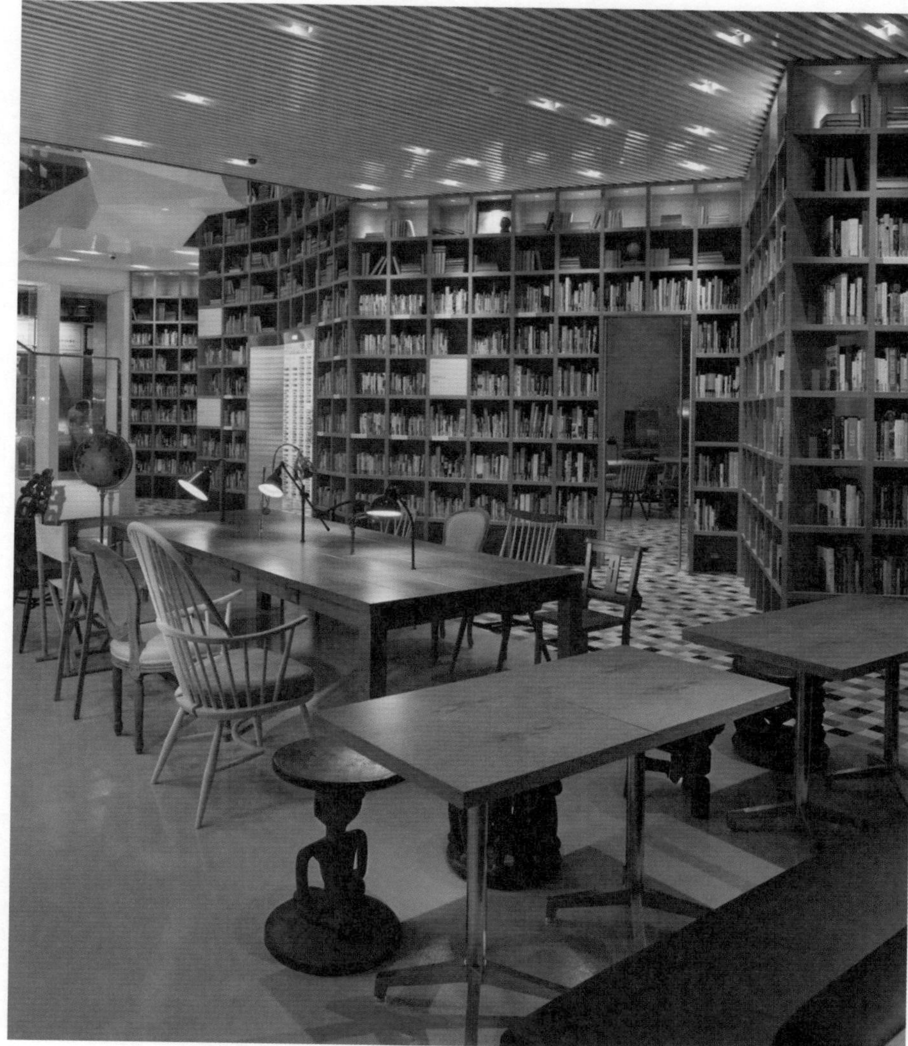

The Play Room, where visitors can control the street view on the large screen using the kiosk's joystick, employs two software programs, Google Earth and Google Galaxy (a combination that was only available at the Google headquarters), to achieve its animated space. Classic features, such as the old-fashioned airport schedule board with its nostalgic clicks, a vintage globe, and vintage furniture, all evoke memories of different times and places. Here, visitors can encounter a second-hand experience of the lives lived in numerous regions and of diverse peoples while sitting on the Wagner Chair, the Shaker Chair, or the animal-shaped stools.

3.3 Music Library

Music Library
Location: 683-132, Hannam-dong, Yongsan-gu, Seou
Site Area: 738m²
Building Area: 409m²
Gross Floor Area: 2,963m²
Coverage Ratio: 52.65%
Gross Floor Ratio: 82.12%
Building Scale: B5~2F
Design: 2012.11 ~ 2014.04
Construction: 2011.10 ~ 2015.05
Architecture: Ga.A Architects, Moongyu Choi
Interior: Gensler
Contractor: Hyundai E&C

The two key projects that ennobled Hyundai Card's branding were the card design and the Super Concert. Along with the design, music has become a central axis in Hyundai Card's brand expression. The location of the Music Library had to be Itaewon, as it holds a unique significance in the history of Korean popular music. The contents therein were filled with vinyl, unlike those of other libraries. The Music Library, where new musical experiences are always on offer, is also the most popular place among visitors.

The gently sloped floor and the slanted U-frame are elements that add vitality to the Music Library. The intention is to create a place for people to gather instead of filling the space with monumental structures.

3.3.1
The Way We Build

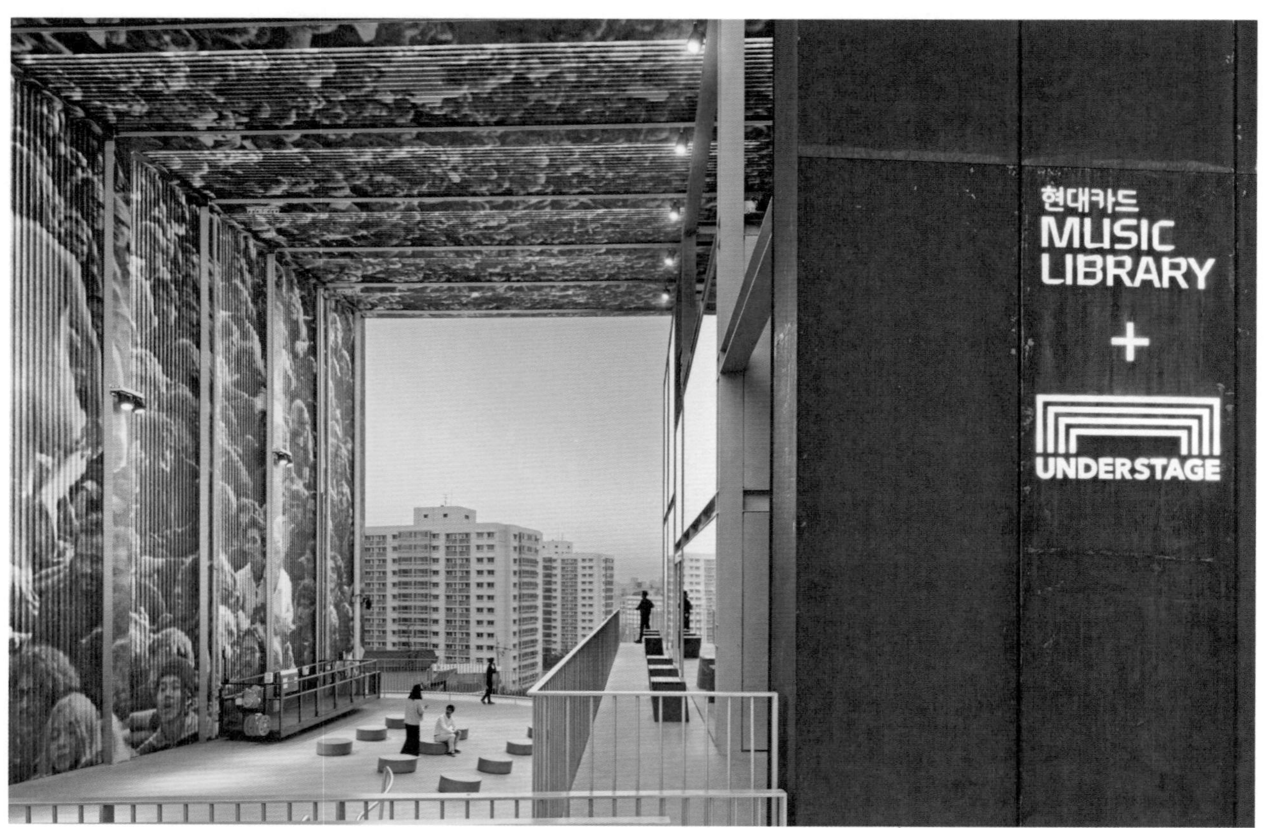

Hyundai Card initially planned a small-scale classical concert hall. The first architect to take on the project was Kazuyo Sejima, the winner of the 2010 Pritzker Architecture Prize. Kazuyo Sejima is known as an architect with a formative language of 'controlled beauty', which would be well-suited for Hyundai Card. The structure Sejima proposed was a four-story glass building completely transparent on all sides. The transparency reflected Hyundai Card's wish to let the Music Library function as a porthole—a passage through which Nam Mountain's energy could pass. Before the Music Library was built, the street of Itaewon was crowded with buildings. Hyundai Card wanted to transform the landscape to give Nam Mountain greater breathing space. Although the project did not proceed under Sejima's proposal, Vice-Chairman Ted Chung expressed a great appreciation for Sejima, who recognized the value of Nam Mountain even though he was a foreigner. Inspired by his proposal, Hyundai Card upheld its efforts to incorporate the scenery of Seoul and Nam Mountain with the Music Library, half of which was daring in its emptiness, even after the architect and the purpose changed. However, during a design period of one and a half years, Hyundai Card revised its strategy and took a drastic turn away from a classical concert venue to a rock-centered popular music venue. One concern prompting this shift was a significant budget overrun if the project were to proceed under Sejima's plan. The disappointment was so great that Vice-Chairman Ted Chung even personally visited Tokyo to politely ask for the architect's understanding: 'The initial proposal and concept, of a glass pavilion with a slanted floor, would have been a tremendous challenge considering the construction details, and a significant budget overrun was expected. The plan would have resulted in an outstanding work of architecture. I determined that going forward it would be unachievable for a range of reasons. However, as I was so grateful that a great architect like Sejima had proposed such a superb plan, that I had to apologize to him in person'. After the transition from classical music to rock, Moongyu Choi, a professor at Yonsei University, took charge of the architectural design, and Gensler of the interior design. The Music Library was a project that had every reason to be stranded in time as there were a number of obstacles involving modifications in the design concept, a change in the architect, and concerns over budget overruns. However, owing to the decisive decisions and will of Vice-Chairman Ted Chung, the project proceeded swiftly. The progress was so quick that a two-hour emergency meeting managed to reach decisions on the concept, key structures, and design-related issues. The bumpy path endured by the project at the beginning ultimately strengthened teamwork, architecture, and content. The Music Library is a structure best viewed at nighttime, a building that is bathed in light after 6 pm. After changing the concept to a venue for popular music performance, Hyundai Card embarked on

The blank space, like the windows of the Music Library, provides a time and space in which to peruse the landscape of Seoul, including Nam Mountain and Han River. The building is see-through as it is made of transparent low-iron glass.

extensive research of international concert halls and record shops, including the Roundhouse in London, the Brooklyn Bowl and Bowery Ballroom in New York, Amoeba Music in Los Angeles, Rough Trade in London and New York, Other Music, and sites in Oslo and Helsinki. The Music Library's focal point was to build a collection of vinyl, not books or magazines in the Design, Travel, or Cooking Libraries. Vinyl resulted from an attempt to turn musical experience into physical realization in a nostalgic space. To curate the music collection, the big data of the Music Timeline was used instead of selecting popular songs or albums by period. The Music Timeline categorizes periods and genres. A Google search of the Music Timeline retrieves graphs that archive an insurmountable number of records for each period and genre. The basic strategy was to add curators' sharp reflection on the data and process. To this end, four global curators for each area, including DJ Soulscape and Scott Mou, were selected for the task. Since then, the curators and Hyundai Card collated vinyl records after visiting individual collectors and record shops in 11 countries. The outcome of these efforts was the discovery of 10,000 vinyl records of popular music post-1950 and 3,400 books.

The process of collecting vinyl records was arduous. Hyundai Card had accumulated ample command of how best to obtain rare copies and acquired salient practices in the collection via the Design and Travel Libraries. Yet, music albums are a different animal. Albums tended to be acquired by scouring eBay and Discogs, the largest online market for pre-owned vinyl records, and visiting Amoeba in LA, Disc Union in Osaka, and Rough Trade in London. The collection was completed by crossing the borders of dozens of countries, online and offline stores, and personally meeting world-class dealers and lines in possession of rare albums. Consequently, all of the albums currently owned by the Music Library are rare and collectible. Representative examples are 250 rare vinyl collection albums, including the famous Butcher Cover of Yesterday and Today by The Beatles and A Special Radio Promotional Album in Limited Edition, a distillation of the musical world of the Rolling Stones, only 100 albums of which

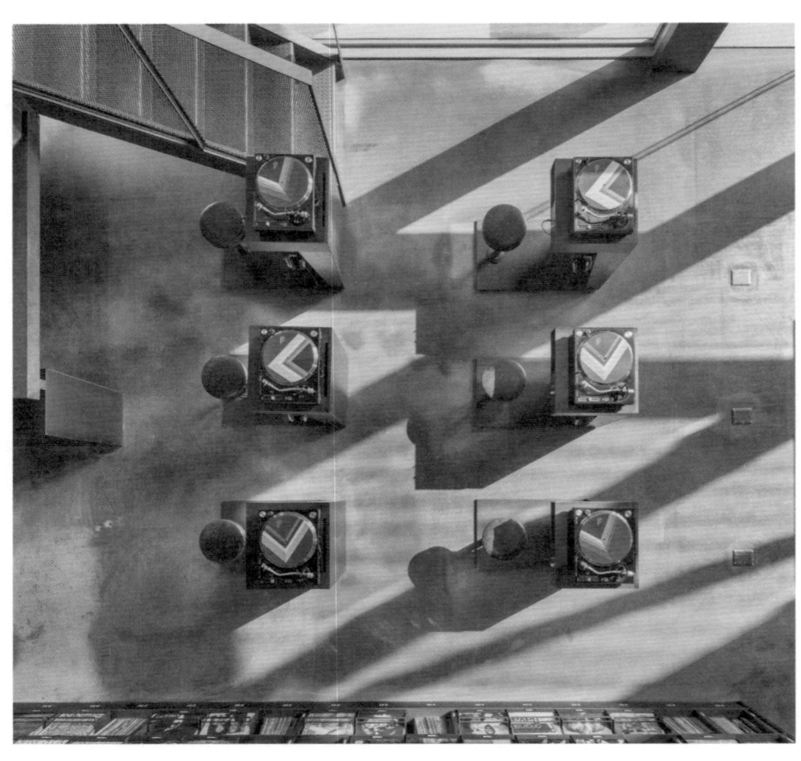

Notwithstanding the small floor area, the Music Library features a relatively high ceiling, making it spacious and suitable for appreciating a DJ's selection of music while browsing magazines or albums. The vinyl cover and the artwork on the wall transmit energy that extends out to the city through the windows and due to the apparent lack of physical boundaries.

have been released. The Music Library also holds every title of the music magazine Rolling Stone, from the first issue in 1967 to the most recent. The role of the underground music venue must not be ignored. The inspiration felt by musicians in the Music Library has melded with the performances on the Understage, made possible through the labor of highly influential curators who have superb connections and performance planning skills in the Korean pop music scene. Understage is central to Hyundai Card's experience and competency in performance and spatial marketing, knowledge gained through the Super Concert program and various cultural projects. This space even sponsors musicians experiencing difficulties in finding professional concert venues. This is why the Understage is a must-see space for famous musicians and creators visiting Korea.

MAGAZINES
1950s
1960s
1960s
1970s
1970s
ROCK
SOUL
JAZZ
ELEC-
TRONIC
Pioneer

THEORY
SUBCULTURE
GENRE
OTHERS
WORLD
KOREA
1980s
1990s
2000s~
HIP HOP
NIRVANA

3.3.2
Point of Space

"This was my first project in Korea and it offered an excellent opportunity to develop my personal project 'Unframed'. The open structure of the Music Library truly inspired me. As an architecture enthusiast, I have always dreamed of placing an artwork that connects the street and the building, and this project was a wonderful opportunity to make that dream come true. Above all, it is amazing that people can see my work as they drive by, without the need to enter the building itself."

JR, Artist – Music Library mural

The Music Library is uniquely structured. About half of the ground floor, the most serviceable section of the building, is empty. The most prominent attribute of the Music Library is that has visibly broken away from the implicit rules of construction. First, the structure has no frontality (façade) in the traditional sense. This absence contrasts with the unmistakable desire of the neighboring buildings to be noticed. Instead, an empty space—a void—exists within the frame of a grand roof. The blank space, like the windows of the Music Library, provides a time and space from which to survey the landscape of Seoul, including Nam Mountain and Han River. In a high-density city like Seoul, multi-use facilities like the Music Library with a total floor area of 5,000m² or higher must open at least 10% of its total floor space to the public. This rule, requiring the provision of land for public use, is often in vain, and yet Hyundai Card voluntarily expanded the legally required 10% of the land area to half of its area. The empty space of so great a magnitude encourages the activities of passersby and houses the energy of the city. Furthermore, the doorstep has been removed to enhance accessibility while expanding a line of sight into the building. By resolving the sloping land with the slope as is, and not with the introduction of stairs, a natural flow extends from the road to the depths of the Café on the first floor. After transforming the intention behind the project from a classical music-oriented venue to a pop music venue, the new building turned to iron as its main material suited to a trendier setting. Adhering to the construction budget was of utmost concern but bold investments were made as and when necessary. The Music Library and the Understage are not extravagant in their

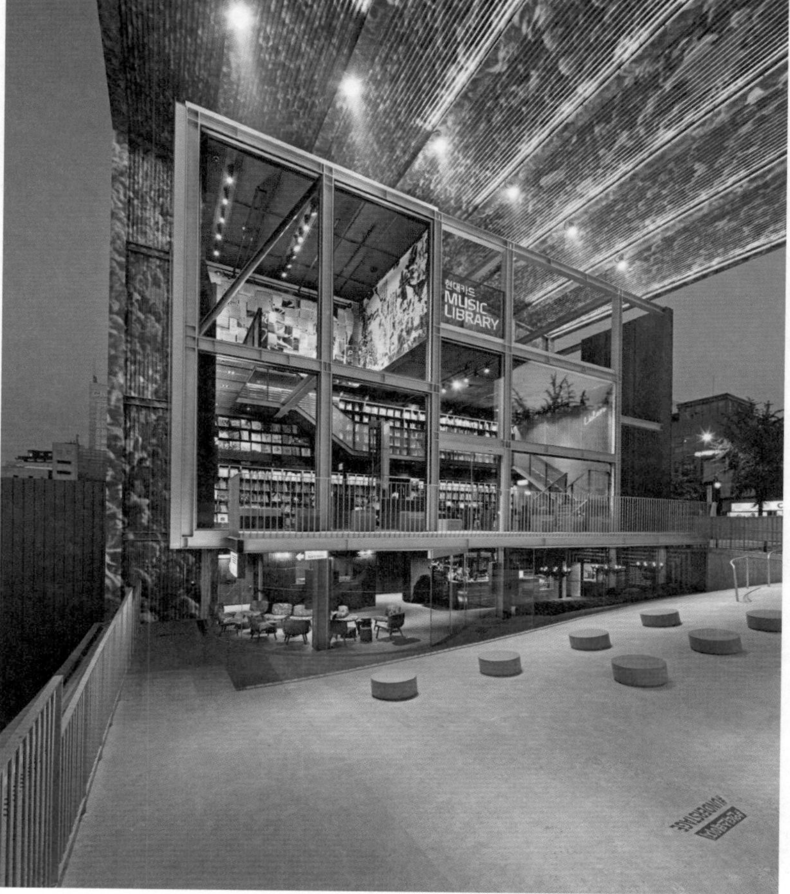

Metal used inside and outside is generally regarded as a low-quality material. However, here, it serves as a background that accentuates the murals of world-renowned artists and valuable collections.

surface appearance but are more in tune with street culture. However, no expense was spared on the floor, ceiling, restrooms, air conditioning systems, waiting rooms, and convenience facilities. Another aspect that was insisted upon was the expensive low-iron glass to ensure total transparency. The Music Library has combined seemingly inharmonious elements, marking a departure from dichotomous thinking around performance venues that equates a warehouse with an independent cultural space and opulence with an opera house. And yet this reinforces the library's presence. The second floor of the library has two intermediate floors connecting the high ceiling height of 12 m. Thanks to these floors, the area does not feel small despite its actual size. The thick steel structure is the secret to the long span of the space, achieving a height of 12 m without columns. The ground floor, which was simply finished with 11 plates of 5 m x 3 m glass, reinforces the path between the interior and the exterior thanks to the vast scale and transparency of a single glass plate. Contemplation was invested, to the extent that the completed construction of a 1/3-sized plate of glass due to budgeting concerns was later replaced with wider glass plates to usher in a visual expansiveness. The Music Library is to be celebrated for the contrast it strikes between its ground floor's cold and hard properties, based on its steel structure and use of glass, and its basement floor's reinforced concrete and light-blocked structure, used as the Understage. Aiming to integrate a space divided between the Music Library and the Understage, Gensler proposed consistency in the materials used and general ambiance through the industrial styling. Moreover, since modern design can be restrained and even dull, the concept was revised at the suggestion of Vice-Chairman Ted Chung so that the space on the first floor would include Victorian decorative elements. In the case of the Café, the black ceiling

The interior space is finished with industrial elements, such as the use of black iron and galvanized steel. Here, the Café on the first floor adds a particular charm to such a contemporary climate. Taking Victorian-era spaces as the central motif, a chandelier made by Murano artisans and fabric sofas have been chosen. The mural, which became a signature of the Music Library, has been replaced with a work by Los Angeles photographer Alex Prager.

51,586: vinyl records and books reviewed at the curation phase
35: countries traveled for vinyl and book collection
10,404: vinyl records documenting the history of popular music since the 1950s
3,433: music books to help understanding history of popular music
389: rare albums valued for their existence
1,256: the entire titles of Rolling Stone, the representative magazine on popular culture and music
6,100: musicians and artists introduced by the Music Library
115,911: songs from albums housed in the Music Library

Music Library in Numbers

"The Understage, when viewed from a musicians' perspective, functions perfectly as a concert hall. All operations, including its sound systems, are close to their ideal realization within this composition. Not long ago, I collaborated with female artists, who told me that they had never experienced musical creativeness that could expand as such owing to the height of the ceiling until that moment. Almost all of the artists who have worked or performed here will experience some kind of leveling up."

DJ Soulscape, Musician - Music Library Curator

tiles, glass chandeliers, exposed concrete of pine plywood, and fabric sofas in primary colors are all drawn into juxtaposition. The sofa made of various fabrics rises like a mound, making it perfect for leaning into from any angle. Those standing on the top half of the floor can exchange gazes. The chandelier is from La Murrina and made by artisan makers on the island of Murano, Venice. The Understage on the first and second basement floors is an incubator for artists. It is equipped with features for performance such as a practice room, a recording studio, a lounge, and a performance hall. The practice rooms, which look like containers, have been fitted with Pinta Acoustic sound-absorbent material for sound insulation and a Bentley Mills carpet on the floor. The performance hall has interchangeable stages between performers and audiences depending on the type of performance, enabling varying types of performances. The two basement floors are partially open for the more efficient use of space. The fact that the ceiling of the Understage is punctuated with a mezzanine on the first basement floor has occasionally proven controversial among musicians. Some say it weakens the quality of the sound, while others say it purifies the sound. Meanwhile, the key element in the Music Library is a collaboration among aspiring young artists. Most particularly, the mural by JR is his only work completed in Korea. JR finished the mural with a picture taken by Bill Owens from a 1969 Rolling Stones performance in Altamont. This work—free-spirited, passionate, and animated—serves as an expression of and entrance to the spirit of the Music Library. Due to the nature of the material, the mural had to be replaced after six years. It was switched for a work by the L.A.-based photographer Alex Prager in early 2021. Also, an enormous graffiti work by Vhils was installed on the wall, spanning the two basement floors of the Understage. Vhils usually sculpts graffiti by carving and drilling holes into a wall instead of drawing a picture across its surface. His unique way of working is reminiscent of the historical context innate to the space.

The basement floors (the Understage) are equipped with features for performances such as a practice room, a recording studio, a lounge, and a performance hall. A uniform industrial concept has been implemented throughout with its reinforced concrete structure and interiors.

ON
20:25

KURZWEIL

NIGELOLSSON

"I host and curate shows on the Understage under three conditions: first, whether the space is truly suited to this musician; second, what it means for the musician to perform here in terms of his or her future endeavors; and lastly, whether the musician's music fits well within this space."
Heeyeol Yoo, Musician/CEO of Antenna Music - Understage Curator

"Similar to hip-hop and rock music imitating various genres to create a new sound, I am inspired by a variety of things and project them into my work. I am fascinated by work that connects the lines from various roots as one. When creating the graffiti work for the Understage, I wanted to demonstrate how art is connected to music despite its obvious differences."

Vhils, Artist - Understage Wall Artwork

The spatial concept of the 'house within a house' frequently employed by Hyundai Card is reinforced by the container box-like practice rooms. The containers can be moved after disassembly, and a space for musicians to take a break is also free and flexible rather than fixed and formal. In the two basement floors of the Understage, the void in the middle extends the space vertically—Vhils has taken advantage of this to create his gigantic mural.

"People visit Hyundai Card's space to immerse themselves in music and make new discoveries. I hope that my work stimulates the curiosity of visitors and gives them a moment to escape from their daily life, further enriching their experience."

Aakash Nihalani, Artist - Music Library, Vinyl & Plastic Exterior Wall Graffiti

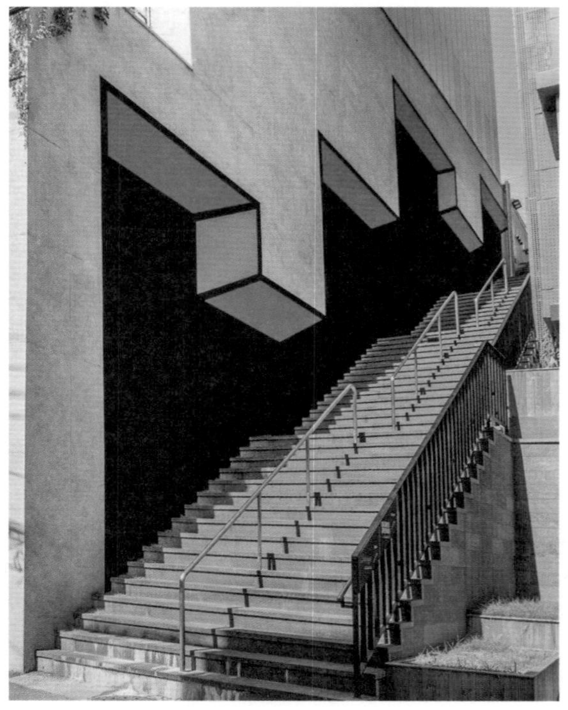

Between the Music Library and Vinyl & Plastic, there was an old, narrow, and long staircase. It was a sloping staircase with stairs of irregular depths and widths. However, without it, visitors had to go a few hundred meters. Hyundai Card renovated this staircase and connected it to the alleyways of Itaewon, drawing a new topographical map in the neighborhood. The graffiti on the wall is the work of Aakash Nihalani.

현대카드
MUSIC
LIBRARY

54
RD28
53

RD32

3.4 Cooking Library

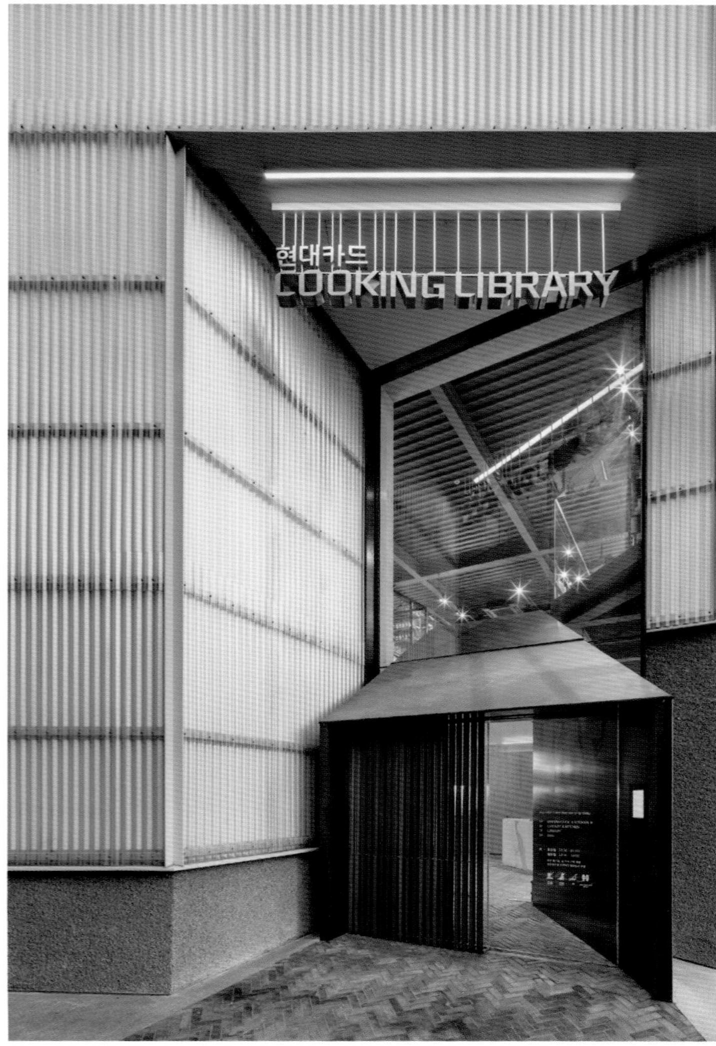

Cooking Library
Location: 645-22, Sinsa-dong, Gangnam-gu, Seoul
Site Area: 405m²
Building Area: 240m²
Gross Floor Area: 879m²
Coverage Ratio: 59.30%
Gross Floor Ratio: 149.78%
Building Scale: B1~4F
Design: 2015.01 ~ 2015.09
Construction: 2015.10 ~ 2016.12
Architecture: 101 Architects, Wook Choi
Interior: Black sheep
Contractor: Hyundai E&C

The Cooking Library, opened in 2017, is the most vibrant of the four libraries. It is a library for exploring new recipes, featuring a restaurant for enjoying gourmet dishes and a kitchen for experiencing cooking firsthand. That is, the Cooking Library is a space in which a wide range of activities connected to food takes place. With the completion of the library, Hyundai Card finalized its content & architecture series to encompass the hallmarks of the modern lifestyle; design, travel, music, and cooking.

The core concept behind the Cooking Library is reminiscent of a cozy family kitchen.

3.4.1
The Way We Build

The architectural concept guiding the Cooking Library was quickly decided. Initially, Vice-Chairman Ted Chung thought, 'I wanted the space inside to offer a wide range of experiences through books, albeit with an ordinary appearance from the outside. On the roof of the building, the finest aspect must unfold. This is for those who desire to experience the next level in the Michelin Guide. I equated this with enjoying a meal in a greenhouse of a farmhouse. I assertively my wish for the structure to exude the scent of the baking bread'. Vice-Chairman Ted Chung sketched out these thoughts and delivered this to Wook Choi, Head Architect of 101 Architects, who immediately grasped the essence of the space Vice-Chairman Ted Chung envisaged. A successful result was achieved, thanks to the synergy between Vice-Chairman Ted Chung's concept proposal based on his fluency in creating new paradigms founded upon his intuitive understanding of architecture, and the architect's interpretation derived from his understanding of Hyundai Card's corporate culture. The Cooking Library, as with the other libraries, features books that equip visitors with an intellectual experience beyond recipes. By defining cooking as a creative pursuit that expands the knowledge and experience of gastronomy that fills our daily lives, an interdisciplinary approach to the library's curation has endeavored to cross the humanities, sociology, and art, and to accept global food culture. Moreover, a pet food shelf, which explores animal food as an unexplored area of cooking, has been arranged to heighten the comprehensiveness of book curation. To this end, Leisa Tyler, a food columnist for Time and National Geographic, Celia Sack, the owner of Omnivore Books on Food, a bookstore specializing in cookbooks in San Francisco, and Catherine Phipps, a food

Inside the Cooking Library, the light pouring in from above reaches the basement evenly, and the mouth-watering scents and sounds of cooking utensils making their pleasant clink spread throughout the space. While enjoying a meal in the dining area on the first floor, one is immersed in such a pleasant ambiance.

writer and cookbook reviewer participated as the curators. The five principles behind the curation were inspiring, practical, reliable, flavorful, and timeless. That is, each book selected must be 'inspiring, practical, or delivering reliable information worthy of the subject, food, flavorful by triggering one's appetite with visual representations, and timeless'. As in other Hyundai Card libraries, the Cooking Library has its own complete collection of renowned books and magazines. The Cooking Library owns the cooking magazines The Art of Eating, Art Culinaire, and Cook's Illustrated. Another exceptional collection is the award collection that holds all of the winning works from the James Beard Foundation Book Awards, a.k.a. the Oscars of the culinary world, and the internationally-recognized Cookbook Awards of the International Association of Culinary Professionals. The two awards constitute the backbone of cookbook awards worldwide. Skimming this collection is sufficient to identify the history of cookbooks and cooking trends. The most unique section of the Cooking Library is the Ingredients House, which stocks over 150 real spices and herbs, as well as 20 kinds of salt and 20 kinds of oil, and a variety of food seasonings. In the form of the 'house within a house', the space is like a small laboratory where one can enjoy five-sense experiences. Inside the library, there are hanging exhibits including 'Untitled 2013' (How to Cook Wolf) by Rirkrit Tiravanija, who preaches the aesthetics of an improved relationship through food, and 'Mind over Matter' by Elmgreen & Dragset, who employ black humor using the balcony as their stage. The Cooking Library allows visitors to experience the process behind the production and eating of food, reading books, and cooking as a healthy and luxurious pastime, from the deli and restaurant on the first floor, the library on the second and third floors, the kitchen extending over the third and fourth floors, and private dining room in the greenhouse with a mini garden on the top floor. As it is the fourth library to be built in the series, Hyundai Card's expertise, detailing, and services have been implemented to the highest degree of perfection.

The Cooking Library was constructed as an open space without clear inter-floor divisions. The dining area on the first floor has one wall that can be opened or closed, giving it the feeling of an open-air café.

MARC FORGIONE
1001
LEON
Ingredients
MICHELIN
donna hay
CHINESE

The rooftop greenhouse of the Cooking Library houses a small garden, in which one can enjoy a private dining experience.

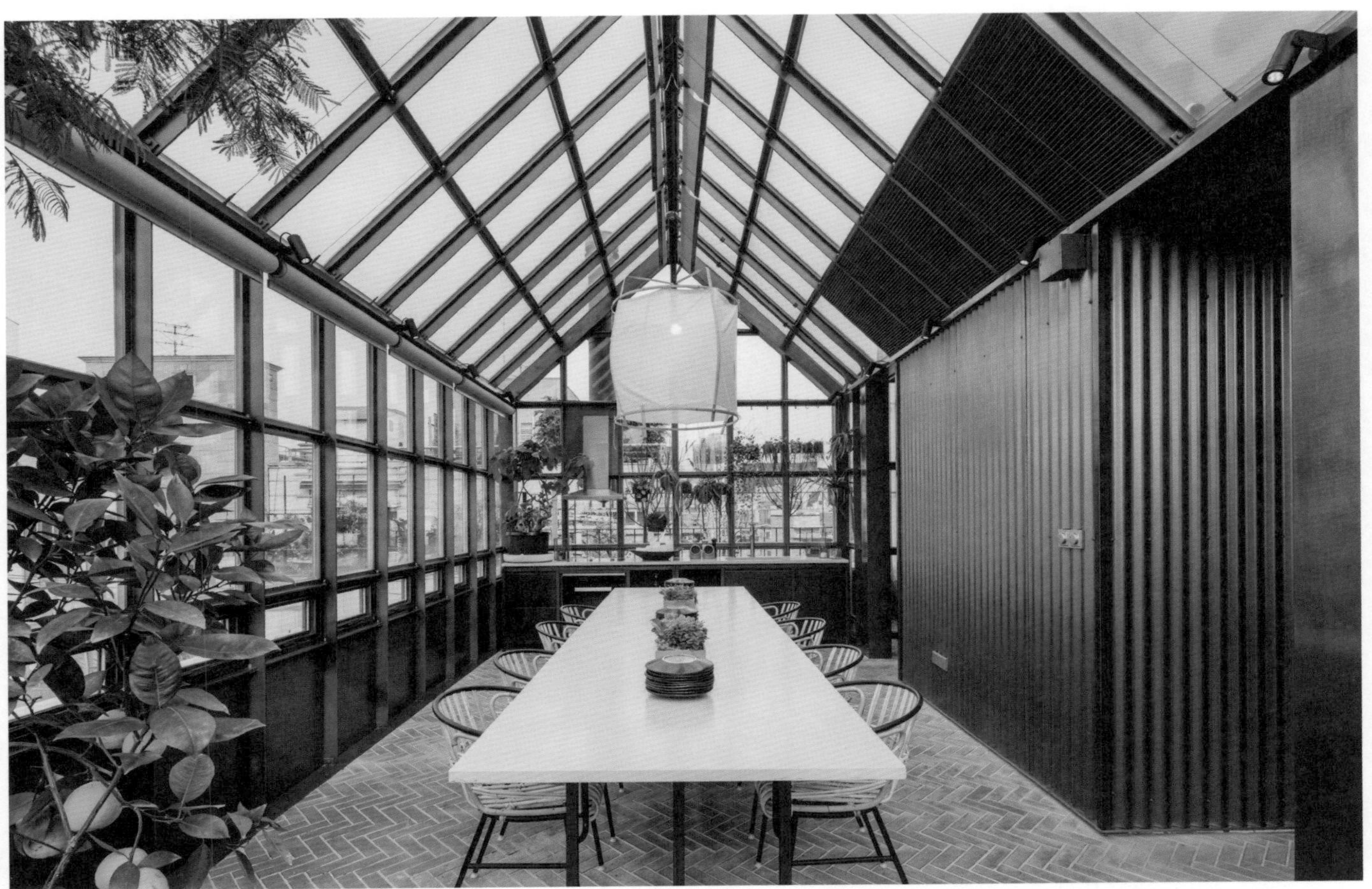

3.4.2 Point of Space

At first glance, the Cooking Library is a modest building, rare for in Cheongdam-dong filled with extravagant buildings. The exterior is finished with dark gray concrete and a milky polycarbonate corrugated panel, a material used for makeshift roofs or warehouses, which gives a gentle and rustic impression by softening the reflection of light. The galvalume steel plates that support the exterior finishing materials are invisible yet were chosen as a high-end specification to maintain a sense of architectural elegance. The entrance is in the form of an independent annex, similar to the concept of Gioheon displayed on the third floor of the Design Library. The aim was to express something of the drama upon meeting the grander space after passing through the compact entrance. The interior is an open structure without clear separations between floors, thereby naturally connecting the floors through visual expansiveness. Thanks to this open structure, the light pouring in from above reaches the basement with an even temper, and the mouth-watering scents and sounds of cooking utensils making their pleasant clink spread throughout the space. The architecture and basic interior concept of the Cooking Library were handled by 101 Architects, and the interior design and selection of faucets and props were carried out by Blacksheep, an interior design studio well-known for its work on Jamie Oliver's restaurants. Here, however, the architects overlooked interior design as Vice-Chairman Ted Chung felt that the direct involvement of the architects in the interior design was not necessary. As such, he paid attention to the delineation of the firms' responsibilities so that each would not encroach on the other's realm. Considering the nature of the Cooking Library, its overriding spatial concept was crucial. Thus, the architects conceived of a comfy and cozy space like a family kitchen in which one can spend any amount of time. The kitchen was defined as a place that could reflect the passage of time, the human touch, and appetizing scents. Above all, a 'kitchen' reminds Korean people of rising steam. Realizing this idea within a building's design required a section of the building to remain open. Inspired by the traditional hanok, where the smells and sounds of the kitchen permeate the house, 101 Architects created

The inside and outside of the Cooking Library employed materials that mature over time along the theme of 'aging'.

40,212: cookbooks reviewed at the curation phase
1,049: books from the James Beard Foundation Book Awards and IACP Cookbook Awards
12,232: cookbooks covering local ingredients and recipes from around the world.
6,354: chefs and food writers introduced by the Cooking Library
396,570: recipes introduced in the books of the Cooking Library

Cooking Library in Numbers

the concept of connected floors instead of clearly defined floors. The mezzanine, which forms a natural vertical structure, is of a small area but is the best place in which to grasp the concept of the building as it houses the Ingredients House and bookshelves. The Cooking Library is not large in size, with one floor at about 132m^2. For efficient spatial division, thus, dozens of architectural scale models were developed. The narrow stairs were laid to be reminiscent of alleyways, and there are windows viewable from every angle, preventing the feeling of tight spaces. Overall, the space is open without clear partitions of floors. As the kitchen, bookshelves, and restaurant are visually connected, the visual pathways of visitors intersect in various directions. As people can look at each other while eating food, reading books, or participating in cooking classes, a lively sense of collectivity organically arises. Meanwhile, there are maze-like spaces through the diverse range of divisions. Wook Choi, Head Architect of 101 Architects, explains, 'The fact that one feels like in a maze means that eyes travel in various directions'. If the composition of the above-ground spaces signifies luxurious gourmet experiences in the top floor's greenhouse, the high-end spaces downstairs are none other than restrooms. The restrooms are in fact as cozy and pleasant as the living room and can be even interpreted as a gallery as visitors can appreciate interesting artworks through the restroom's expansive glass that draws natural light down into the basement. The artwork is from the 'Fruits 1, 2, 3, 4, 5' series, which the British artist David Shrigley created especially for the Cooking Library. This work, which combines objects such as golf balls, nails, and coins with common ingredients, can only be seen from the restroom on the first basement floor. The consistency of the interior and exterior finishing materials is achieved in the materials that mature with time, such as wood, metal, and concrete, based on the concept of 'aging'. The polycarbonate corrugated panel, which is the exterior material, has to be replaced every six years as the colors fade. The cooking equipment was also made of steel so that fingerprints can be left behind. As such, the traces and memories leave behind their imprint organically with use. Small items reminiscent of those from a kitchen are scattered throughout the space, such as a pump made of pulleys, a large forklift wheel, and straws. The idea is that the building itself is a dynamic kitchen as well as a maturing vessel for creativity.

Tucked in every corner of the Cooking Library are witty works from the entrance on the first floor to the walls of the bookshelves on the second floor, as well as the restrooms on the basement floor. On the first floor, you will meet Rirkrit Tiravanija's 'Untitled 2013 (How to Cook a Wolf)'. Inspired by American culinary essayist M. F. K Fisher's book <How to Cook a Wolf>, this work archives photos showing the author personally cooking the dishes introduced in the book and objects. On the wall of the second-floor bookshelf, 'Mind over Matter', a work by Elmgreen & Dragset, is installed, which poses questions on everyday space through the balcony.

"We wanted the Cooking Library to be a space reminiscent of a family kitchen where one can stay at any time and rest. To this end, the exterior had to be simple, so I thought it would be nice to create a building with an extremely plain and tranquil look in this trendy and sophisticated part of Cheongdam-dong. As one steps inside, the building's section opens up like a kitchen engulfed in rising steam, allowing the smells of cooking food to travel upward. We eliminated the clear divisions between floors so that people could more easily mingle and revitalize the overall structure. Hyundai Card wanted the Cooking Library to be an open space and a place in which those who cook are visible to others."

Wook Choi, Head Architect of 101 Architects - Cooking Library Architecture

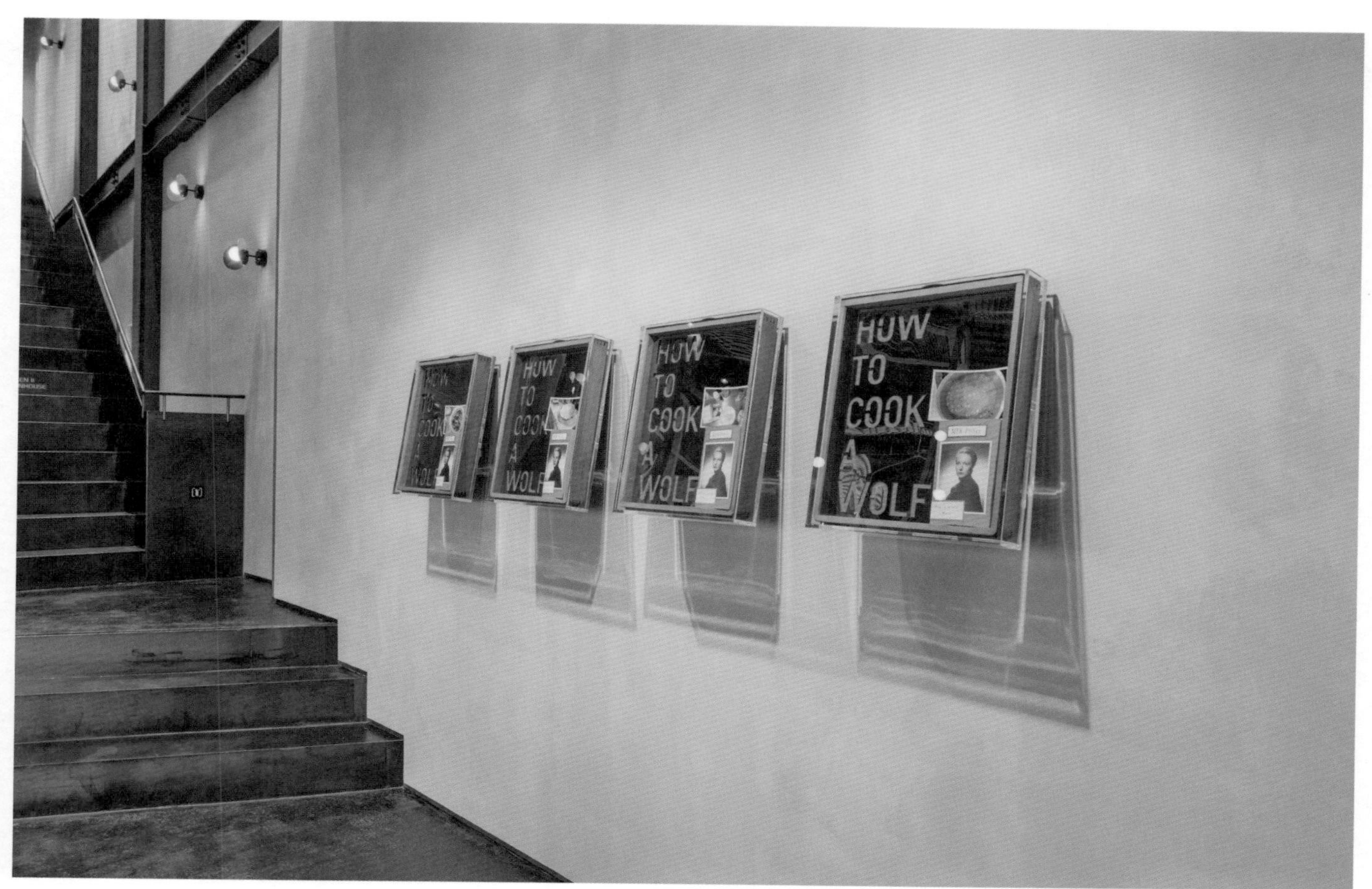

“The Cooking Library is a complex space. Here, one can read a range of books and interact more directly with chefs from diverse fields. It is a place capable of satisfying a desire for more detailed and complex tastes and culinary experiences while also offering diverse experiences to a general public interested in cooking.”

Joonwoo Park, Chef and Food Columnist

In addition to the high ceiling height, the Cooking Library presents an expansiveness and open spatial sense with ambiguous divisions between floors. In the kitchen, you can join various cooking classes.

Periodicals
정기간행물

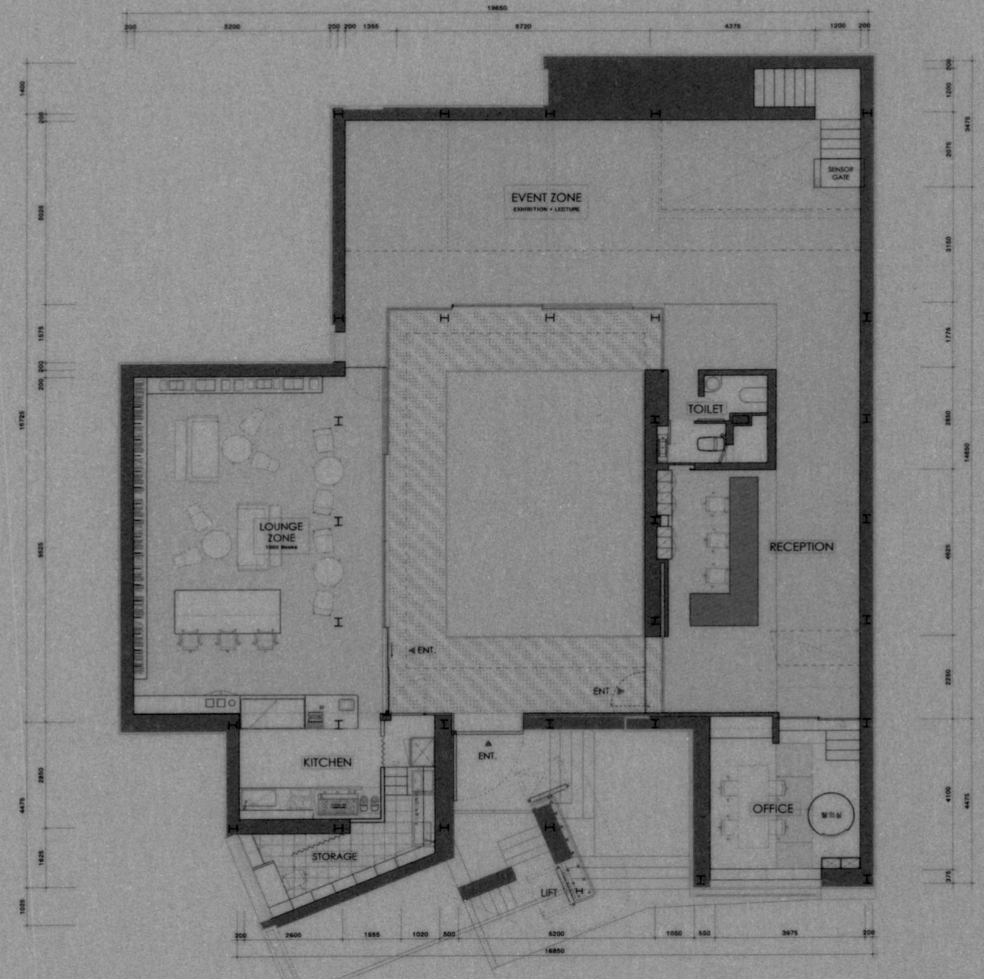

Design Library 1F Floor Plan

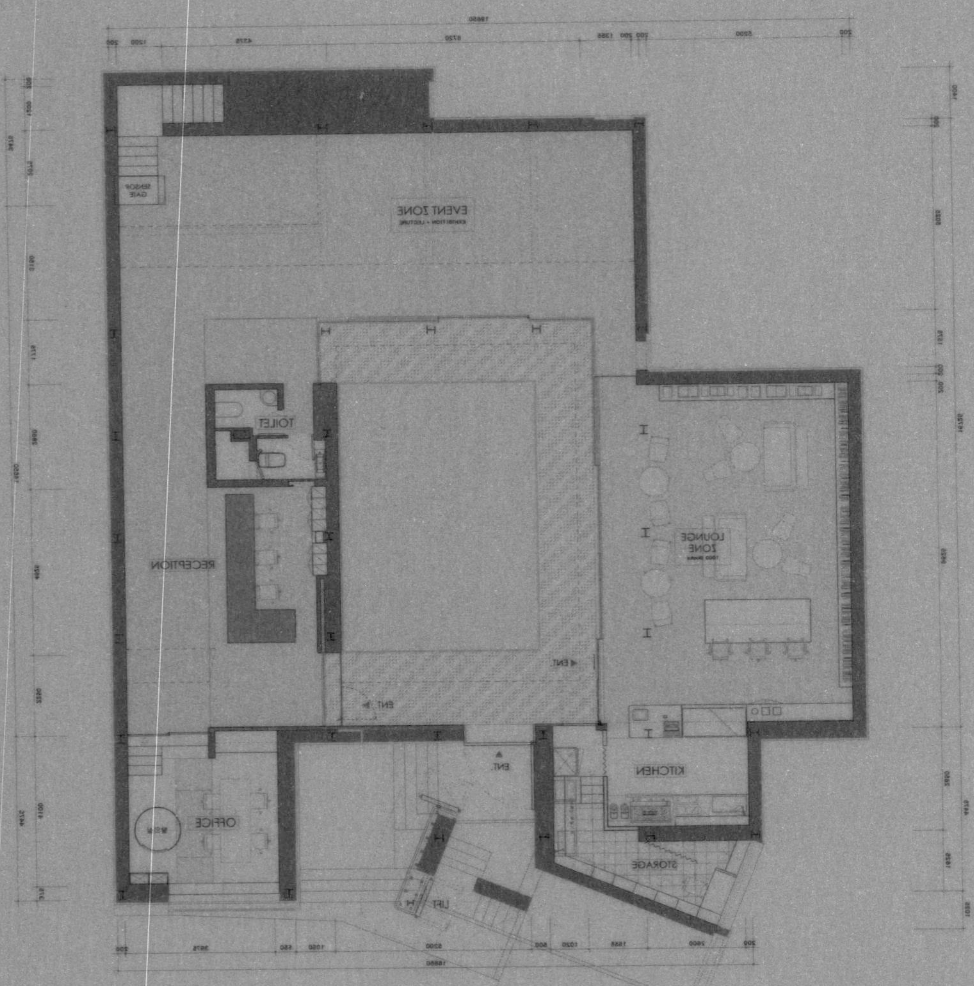

Design Library 1F Floor Plan

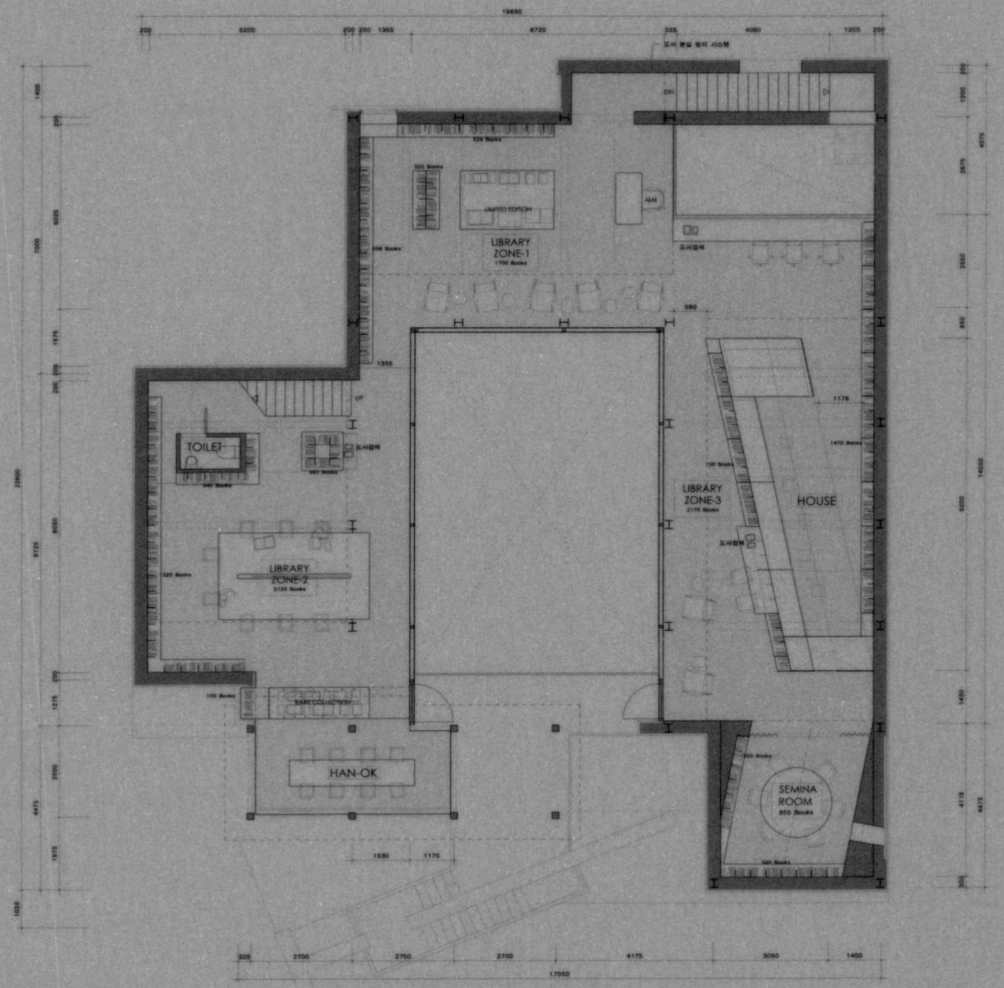

Design Library 2F Floor Plan

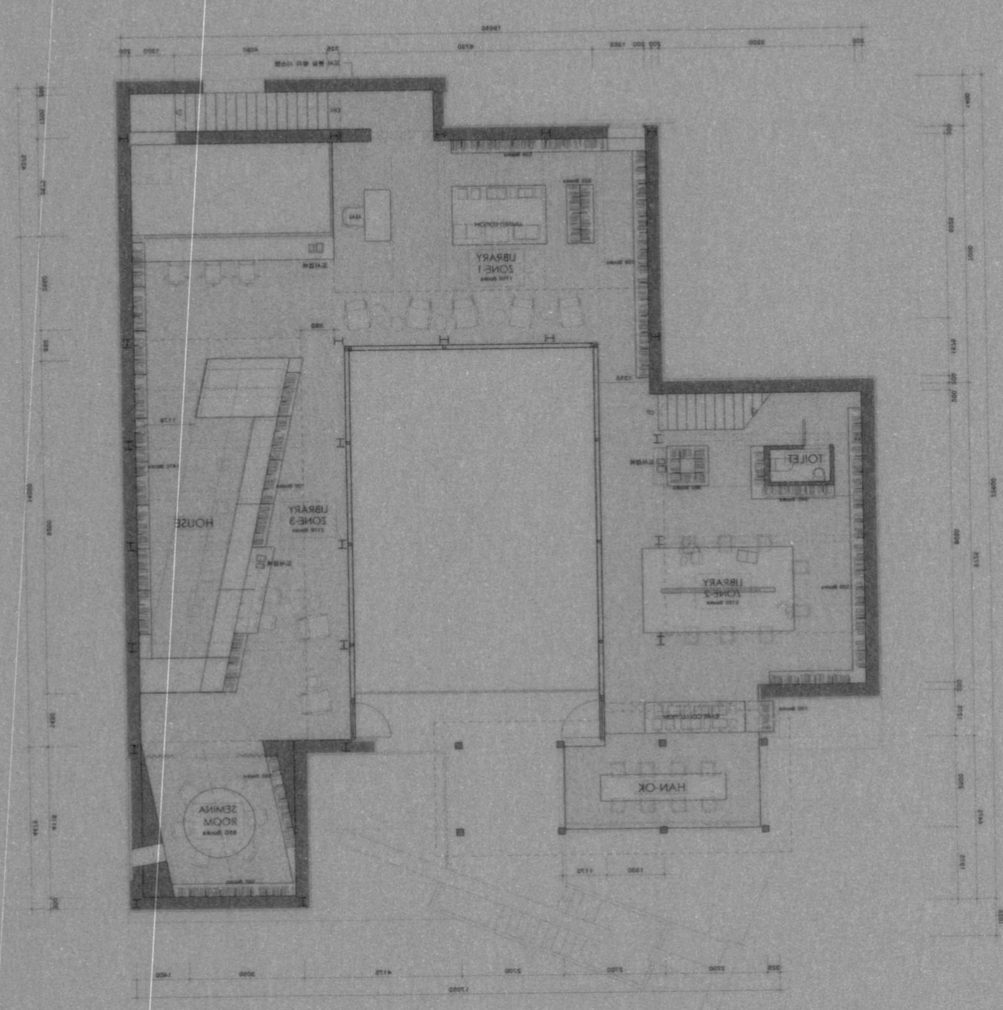

Design Library 2F Floor Plan

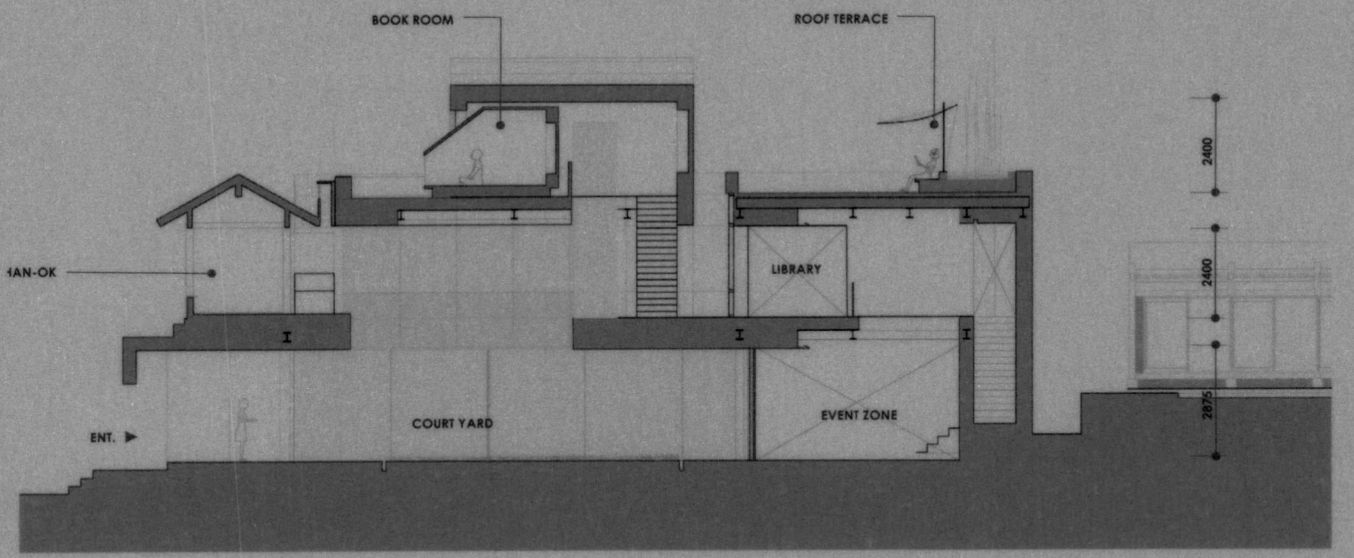

Design Library Section

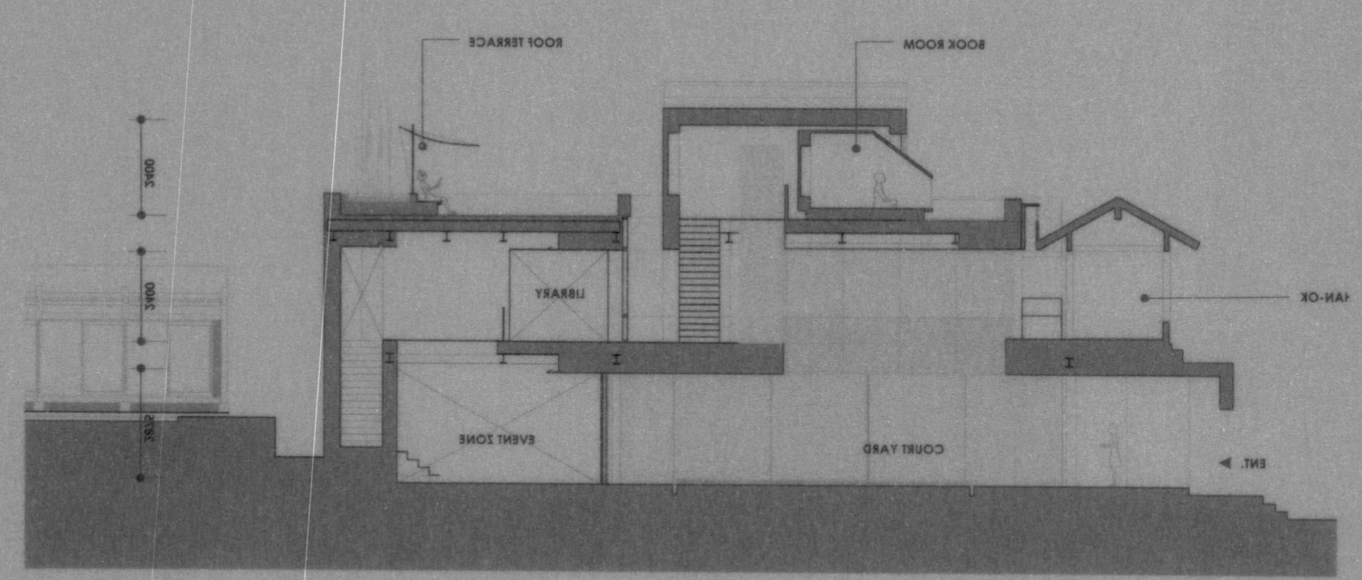

Design Library Section

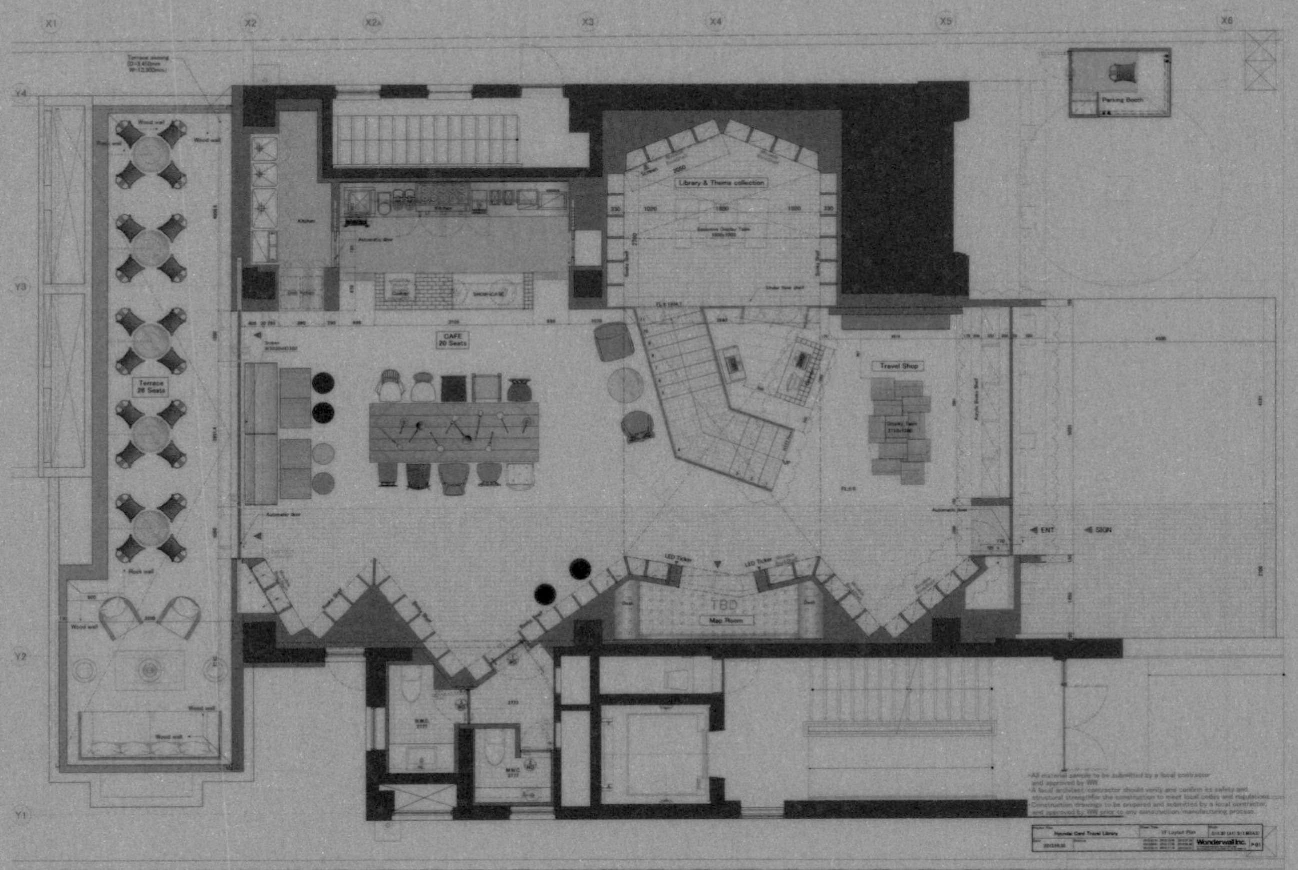

Travel Library 1F Floor Plan

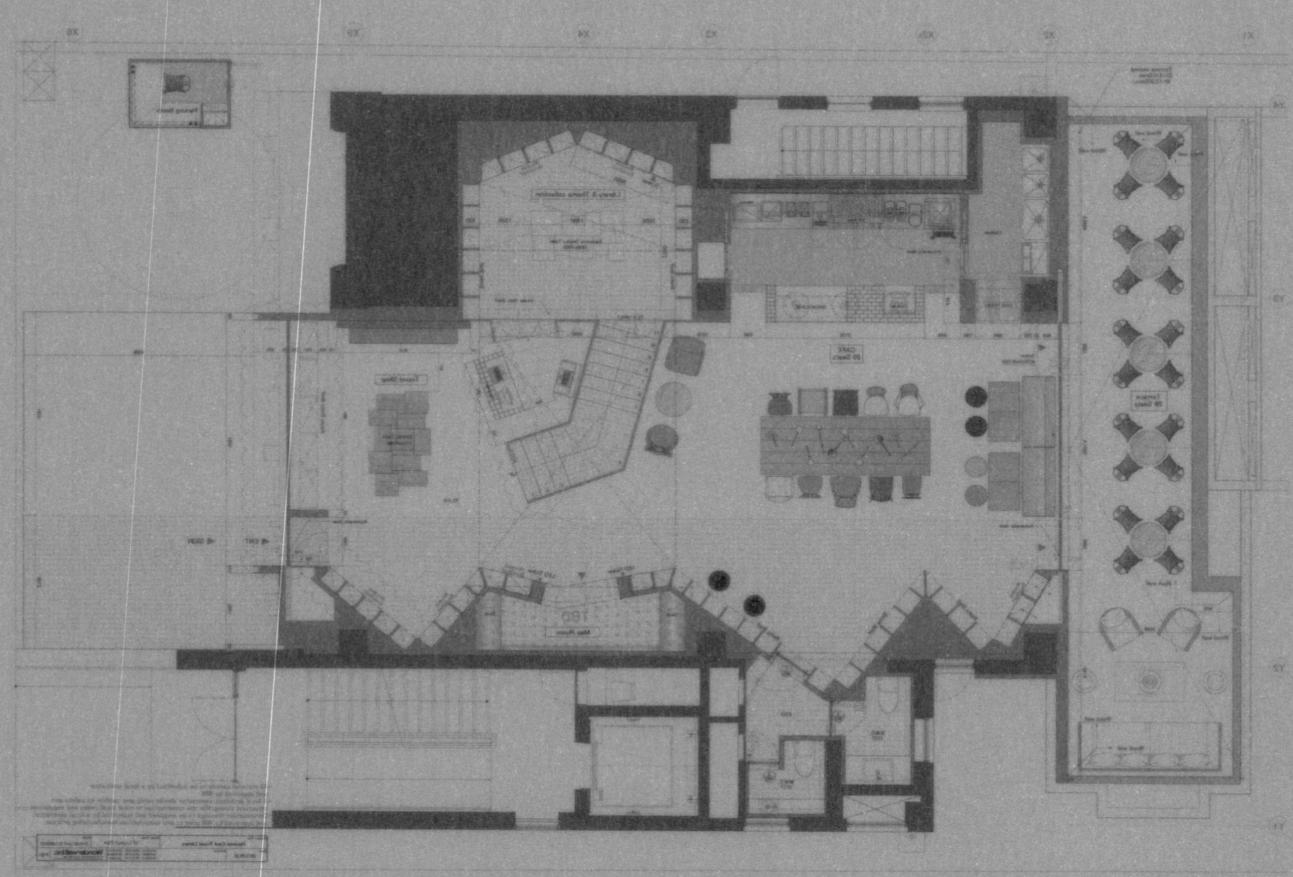

Travel Library 1F Floor Plan

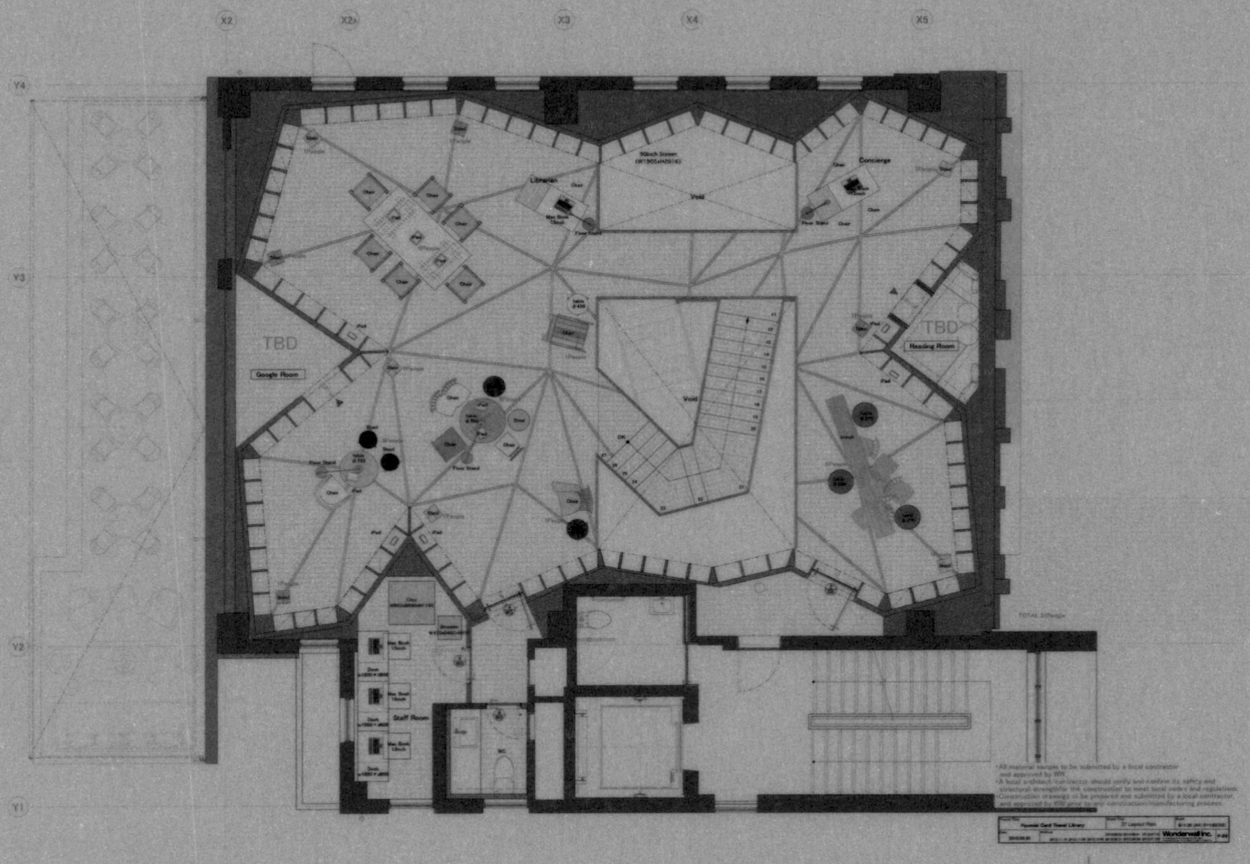

Travel Library 2F Floor Plan

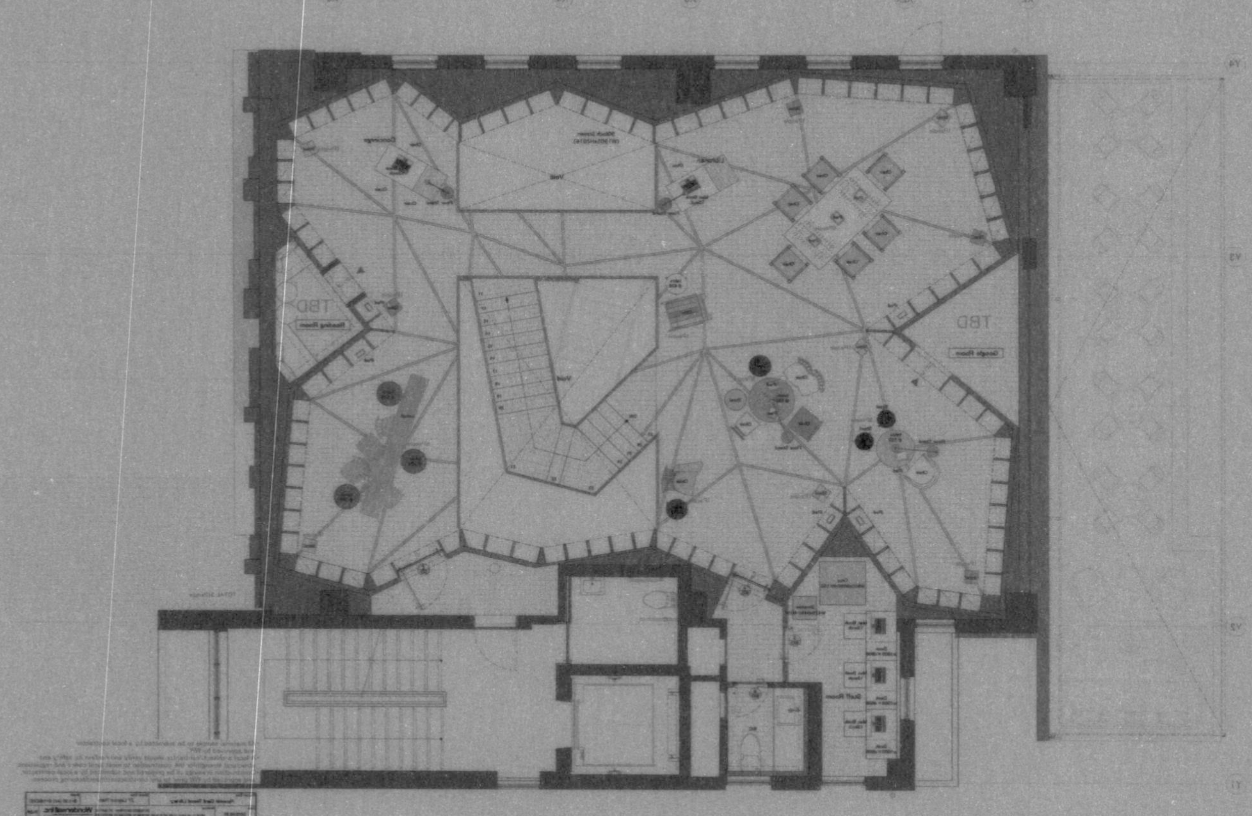

Travel Library 2F Floor Plan

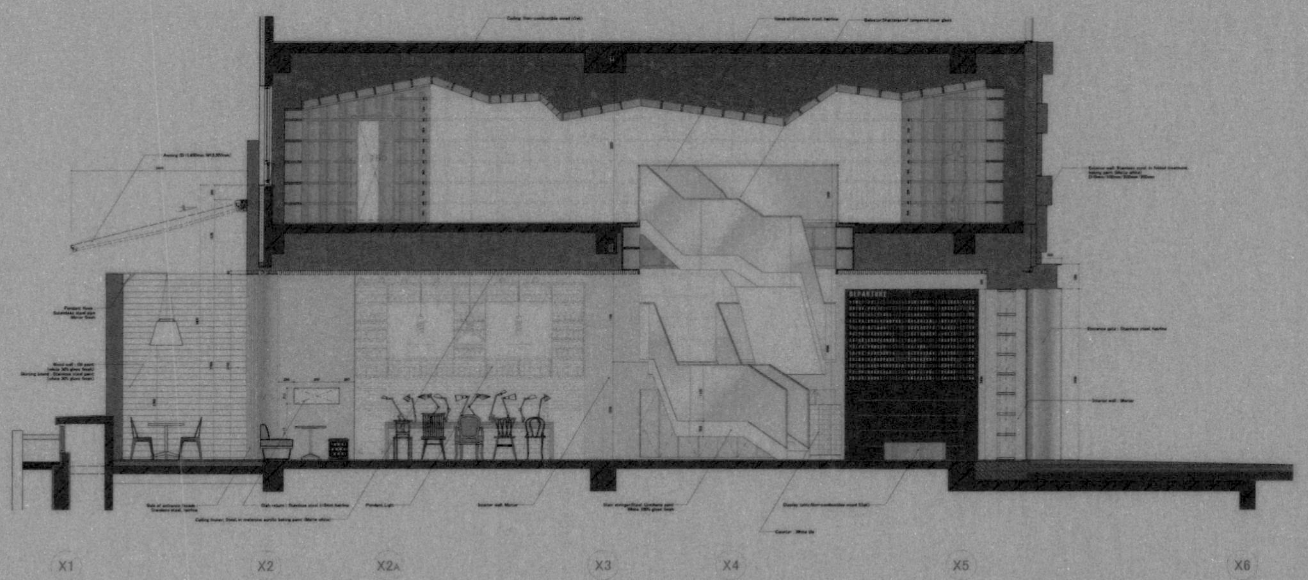

Travel Library Section

Travel Library Section

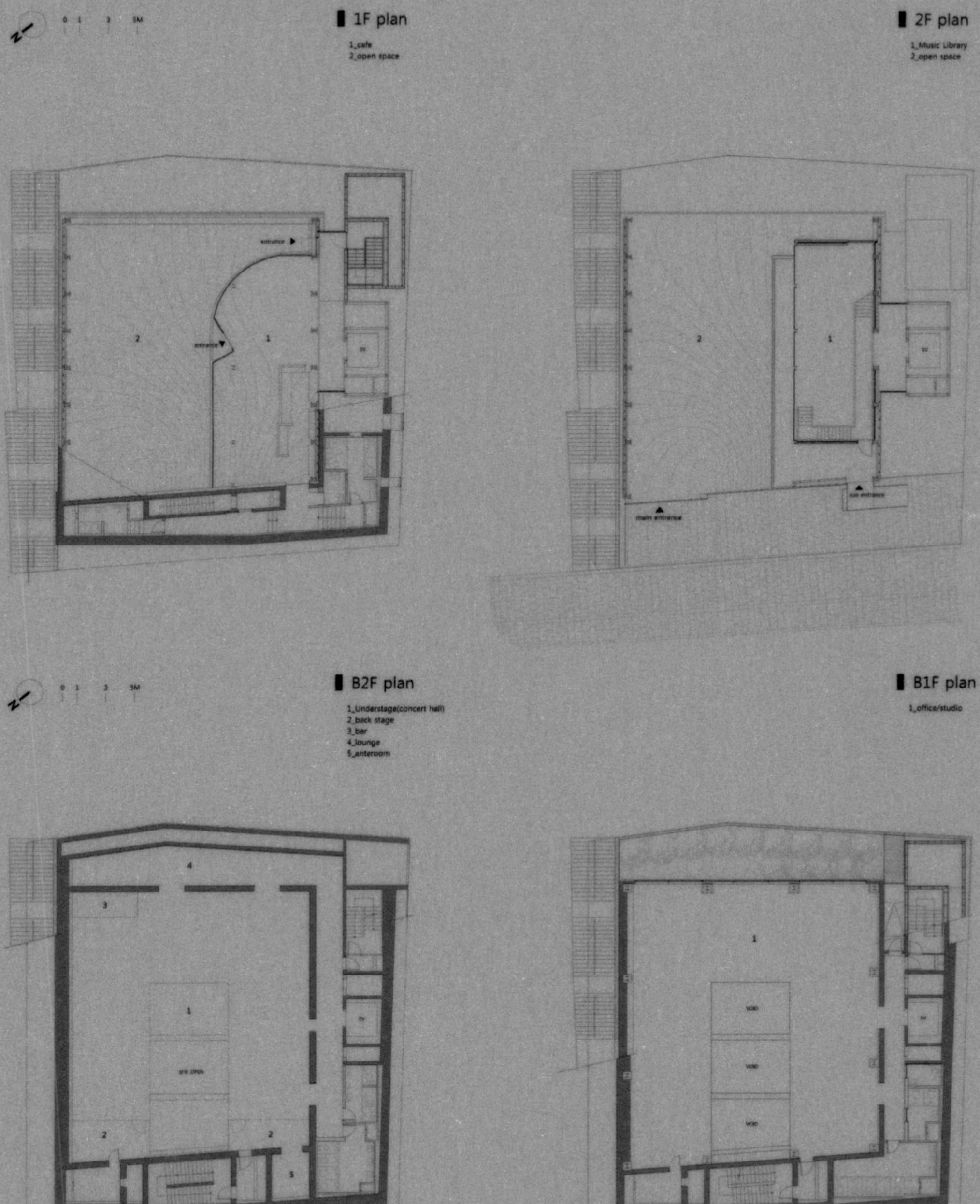

Music Library Floor Plan
Understage Floor Plan

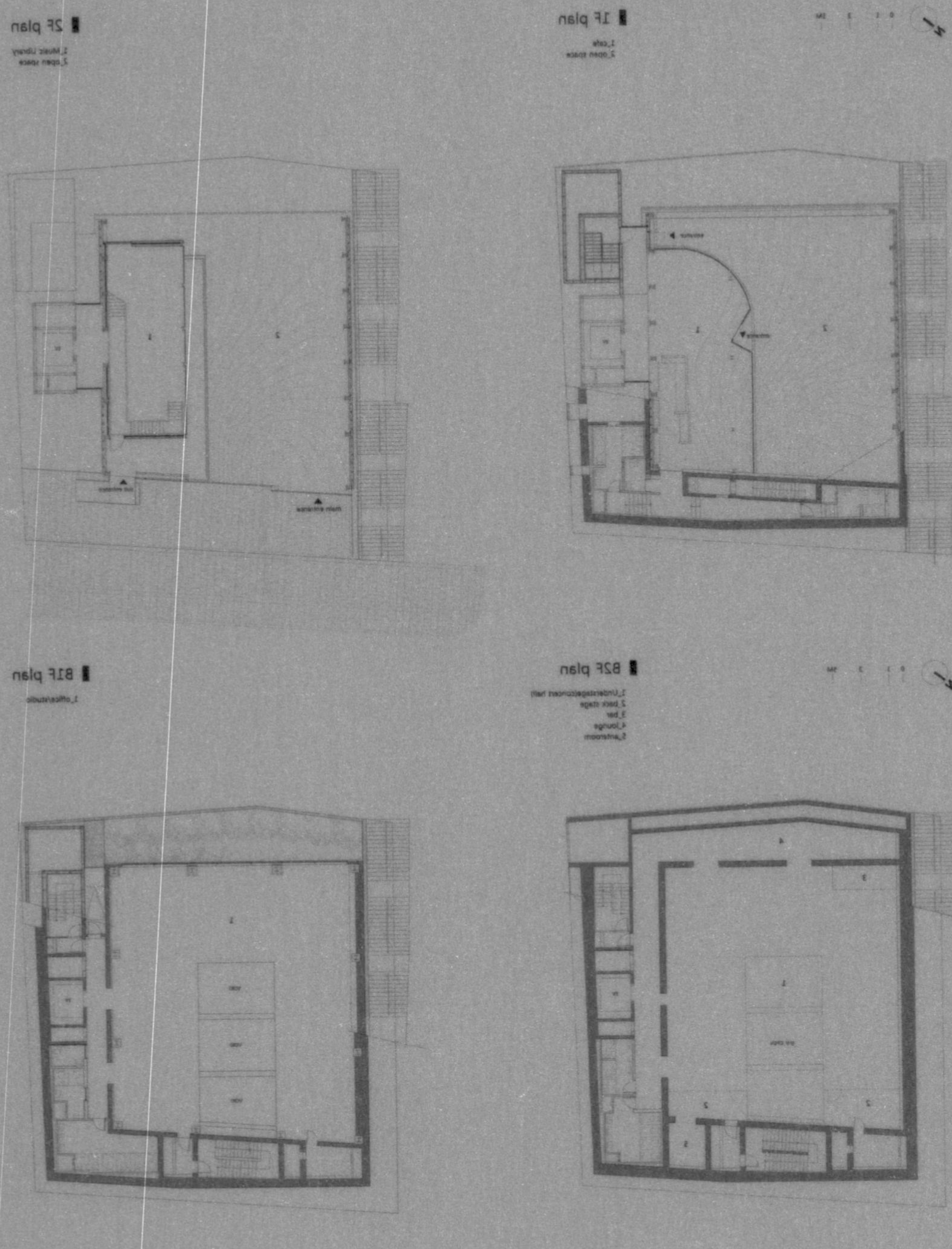

Music Library Floor Plan

Understage Floor Plan

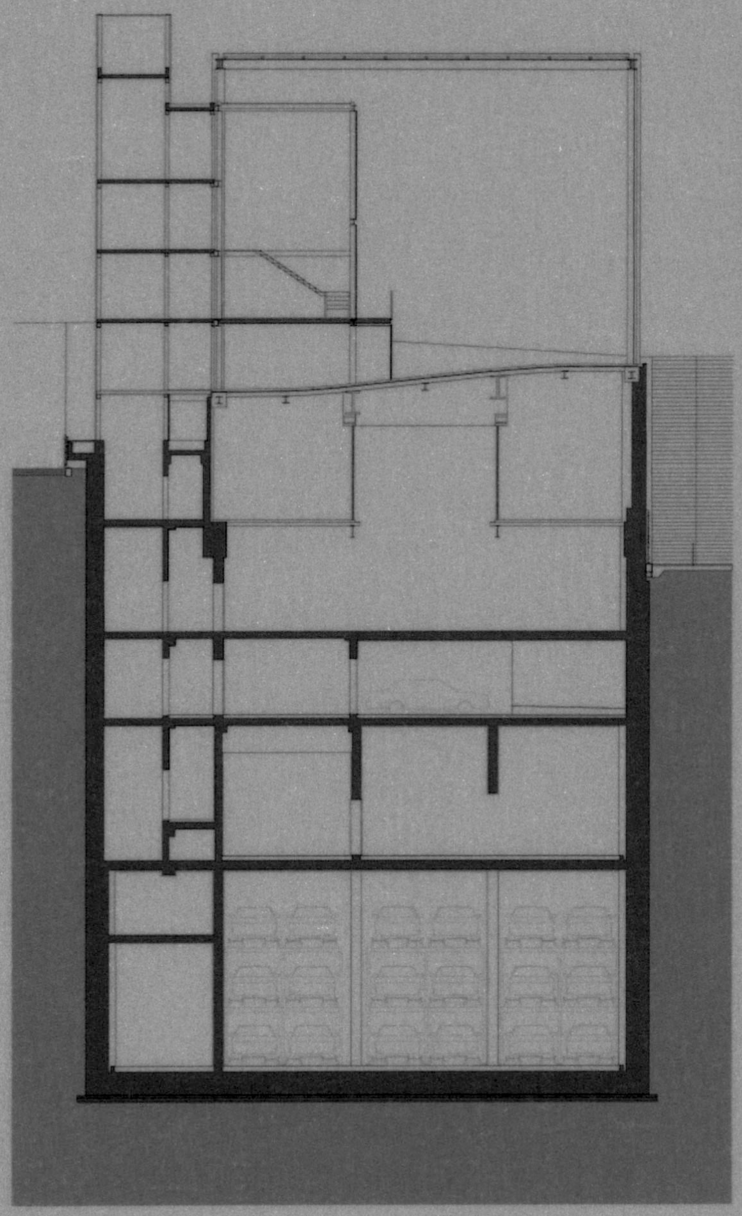

Music Library & Understage Section

Music Library & Understage Section

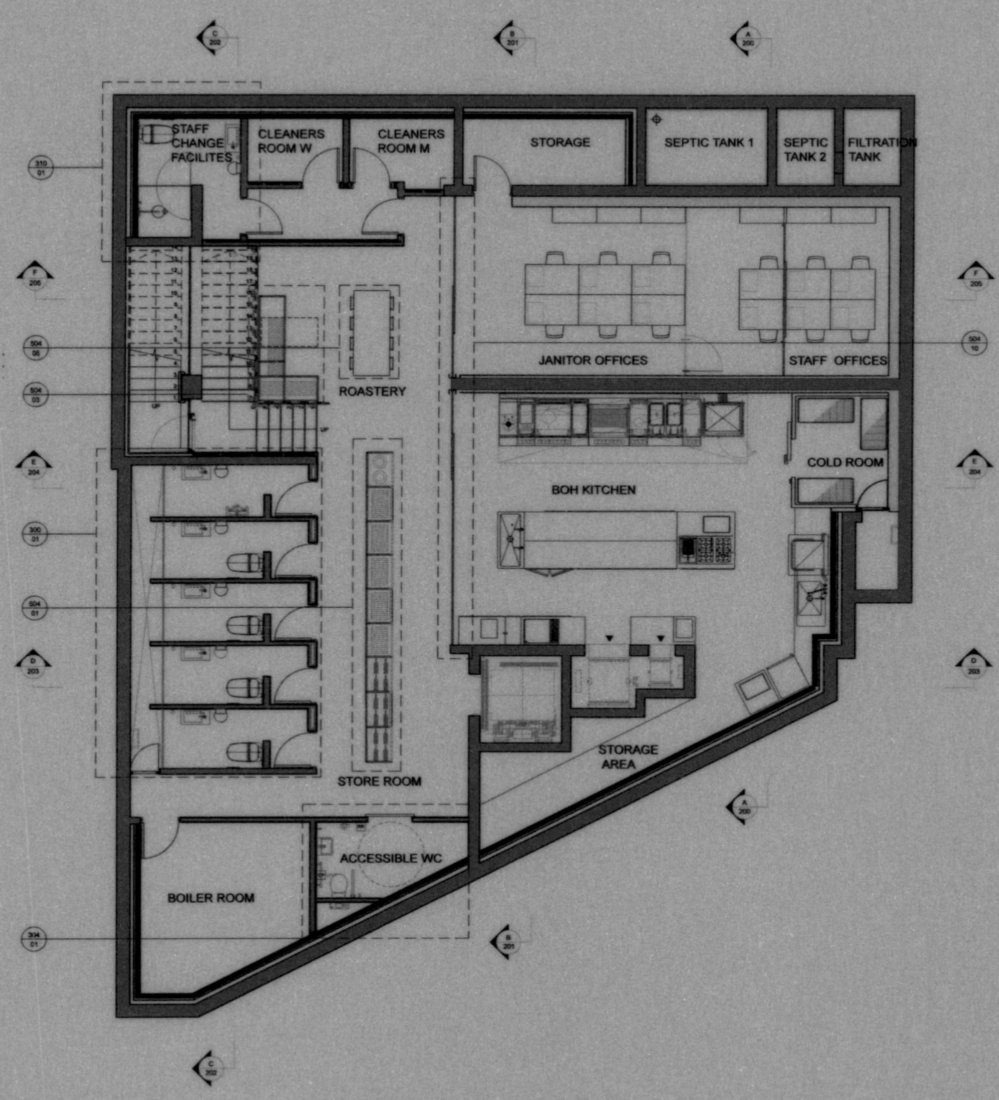

Cooking Library B1 Floor Plan

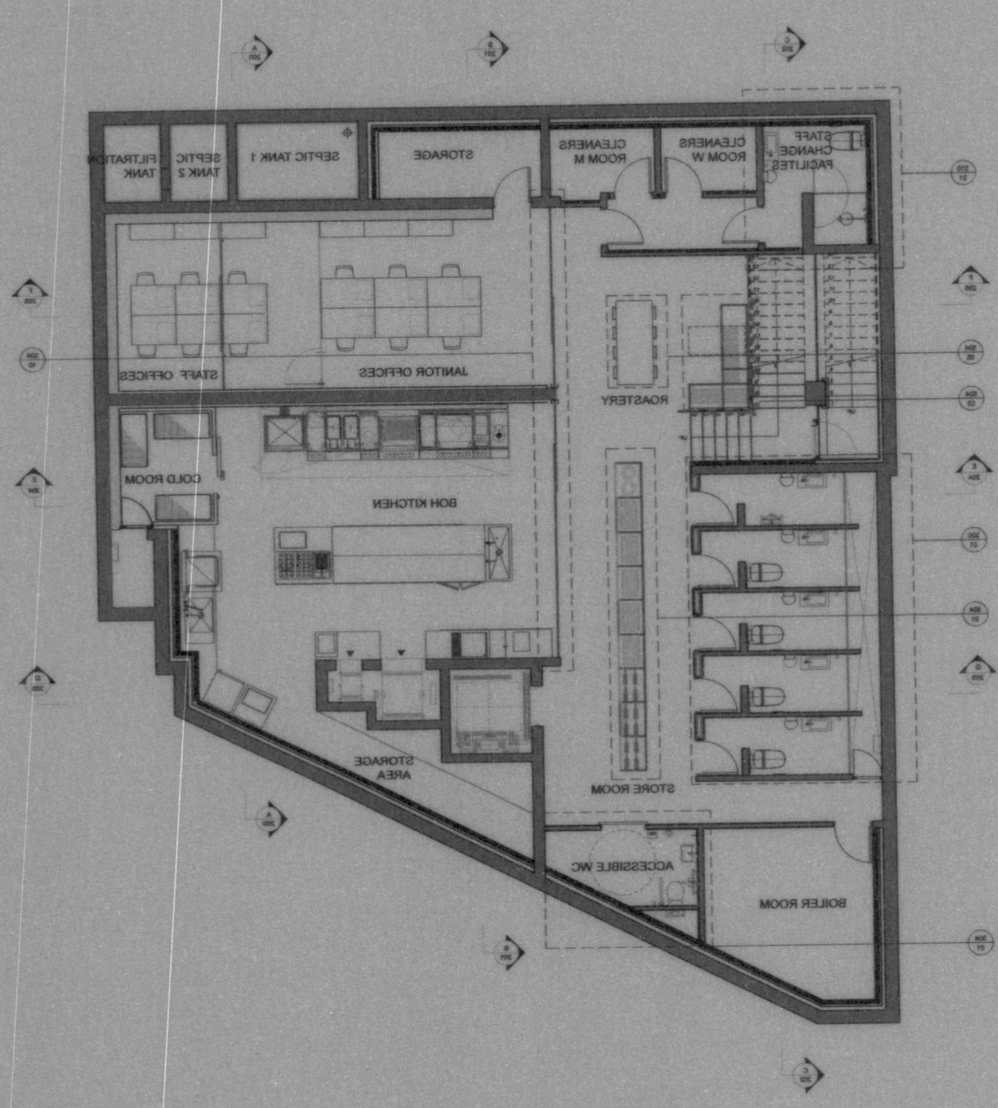

Cooking Library B1 Floor Plan

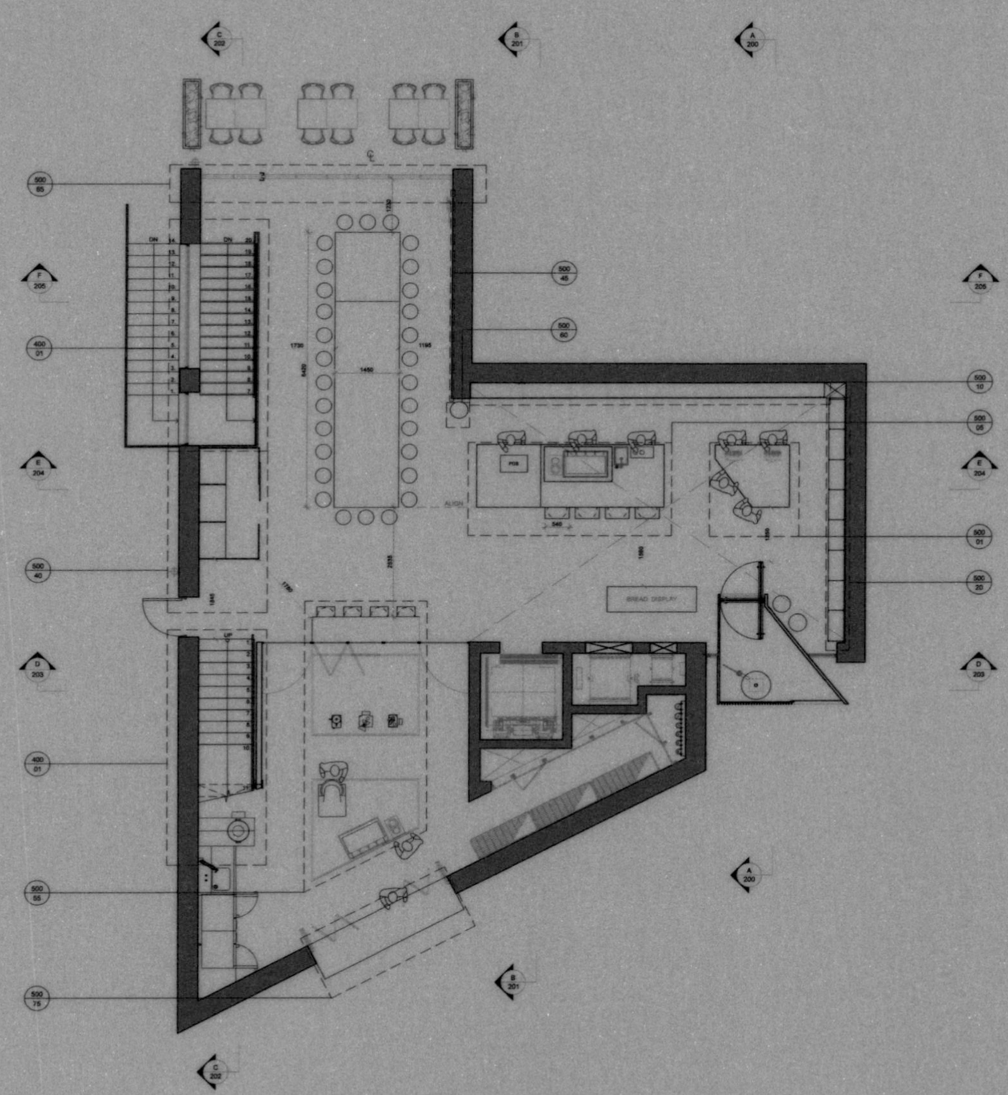

Cooking Library 1F Floor Plan

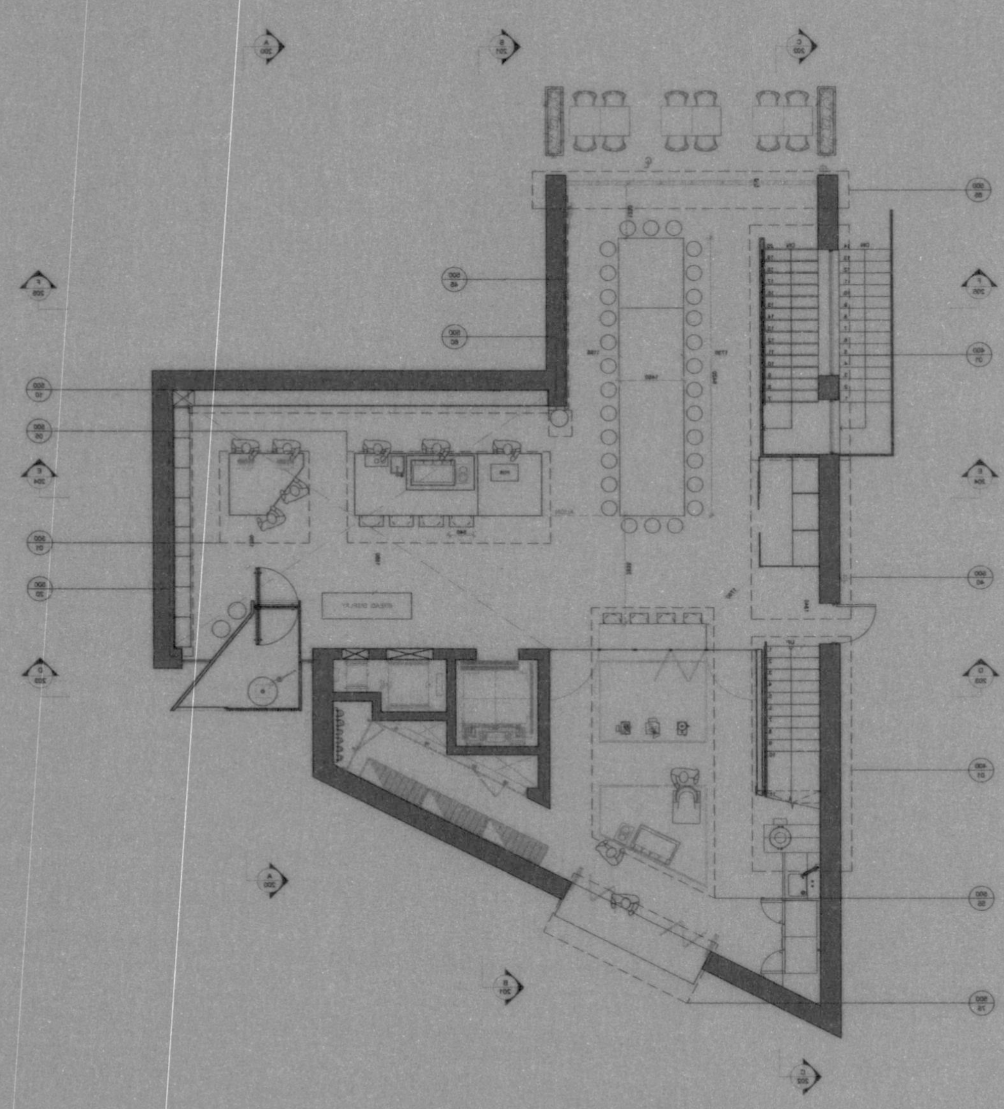

Cooking Library 1F Floor Plan

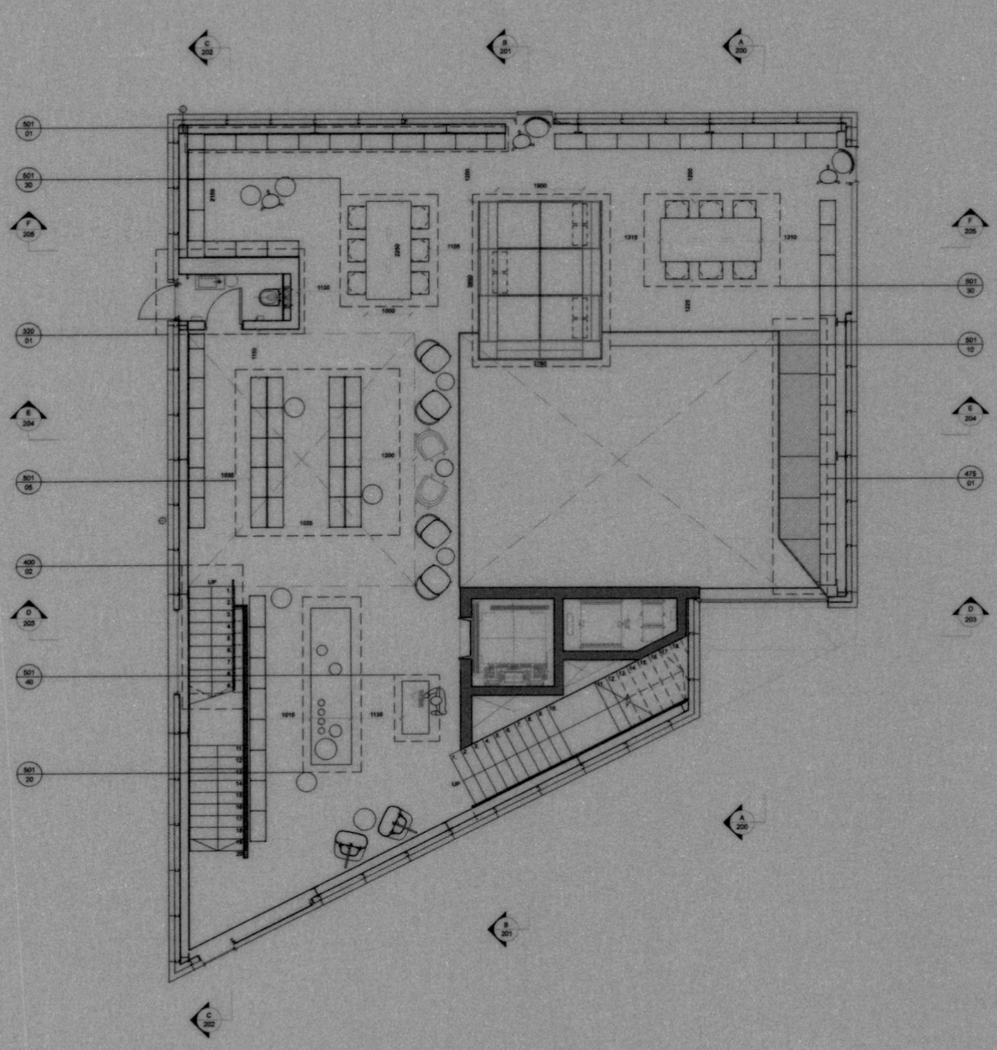

Cooking Library 2F Floor Plan

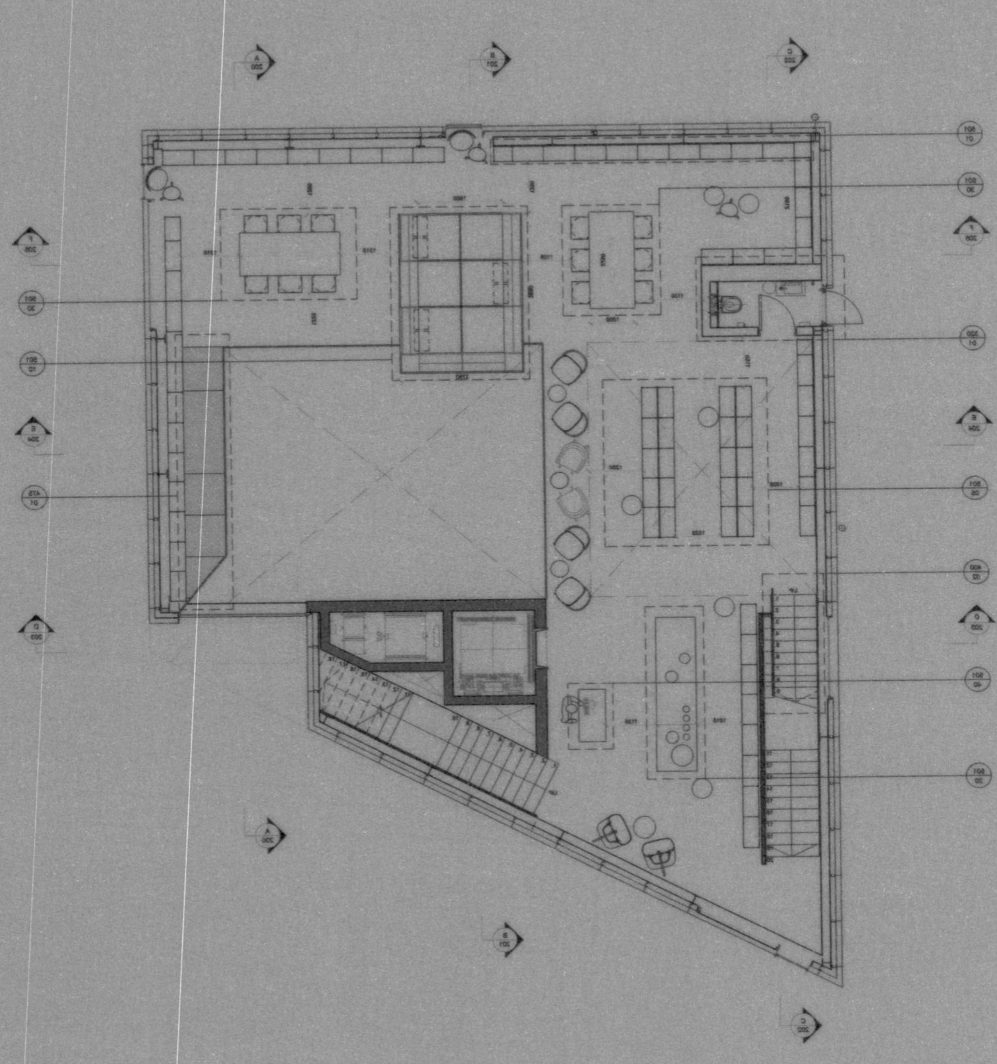

Cooking Library 2F Floor Plan

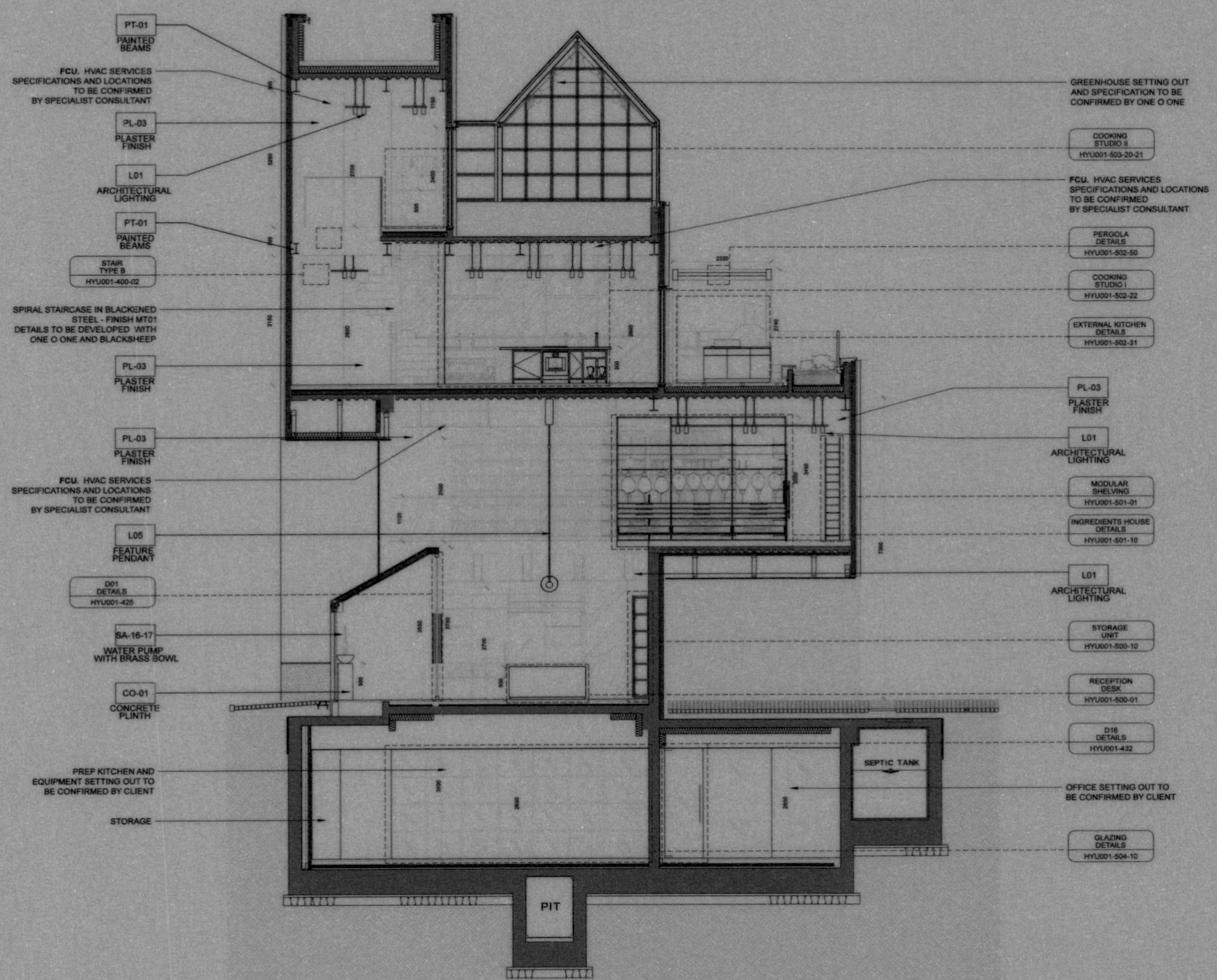
PT-01
PAINTED BEAMS
FCU. HVAC SERVICES SPECIFICATIONS AND LOCATIONS TO BE CONFIRMED BY SPECIALIST CONSULTANT
PL-03
PLASTER FINISH
L01
ARCHITECTURAL LIGHTING
PT-01
PAINTED BEAMS
STAIR TYPE B
HYU001-400-02
SPIRAL STAIRCASE IN BLACKENED STEEL - FINISH MT01 DETAILS TO BE DEVELOPED WITH ONE O ONE AND BLACKSHEEP
PL-03
PLASTER FINISH
PL-03
PLASTER FINISH
FCU. HVAC SERVICES SPECIFICATIONS AND LOCATIONS TO BE CONFIRMED BY SPECIALIST CONSULTANT
L05
FEATURE PENDANT
D01 DETAILS
HYU001-425
SA-16-17
WATER PUMP WITH BRASS BOWL
CO-01
CONCRETE PLINTH
PREP KITCHEN AND EQUIPMENT SETTING OUT TO BE CONFIRMED BY CLIENT
STORAGE
GREENHOUSE SETTING OUT AND SPECIFICATION TO BE CONFIRMED BY ONE O ONE
COOKING STUDIO II
HYU001-503-20-21
FCU. HVAC SERVICES SPECIFICATIONS AND LOCATIONS TO BE CONFIRMED BY SPECIALIST CONSULTANT
PERGOLA DETAILS
HYU001-502-50
COOKING STUDIO I
HYU001-502-22
EXTERNAL KITCHEN DETAILS
HYU001-502-31
PL-03
PLASTER FINISH
L01
ARCHITECTURAL LIGHTING
MODULAR SHELVING
HYU001-501-01
INGREDIENTS HOUSE DETAILS
HYU001-501-10
L01
ARCHITECTURAL LIGHTING
STORAGE UNIT
HYU001-500-10
RECEPTION DESK
HYU001-500-01
D16 DETAILS
HYU001-432
SEPTIC TANK
OFFICE SETTING OUT TO BE CONFIRMED BY CLIENT
GLAZING DETAILS
HYU001-504-10
PIT

Cooking Library Section

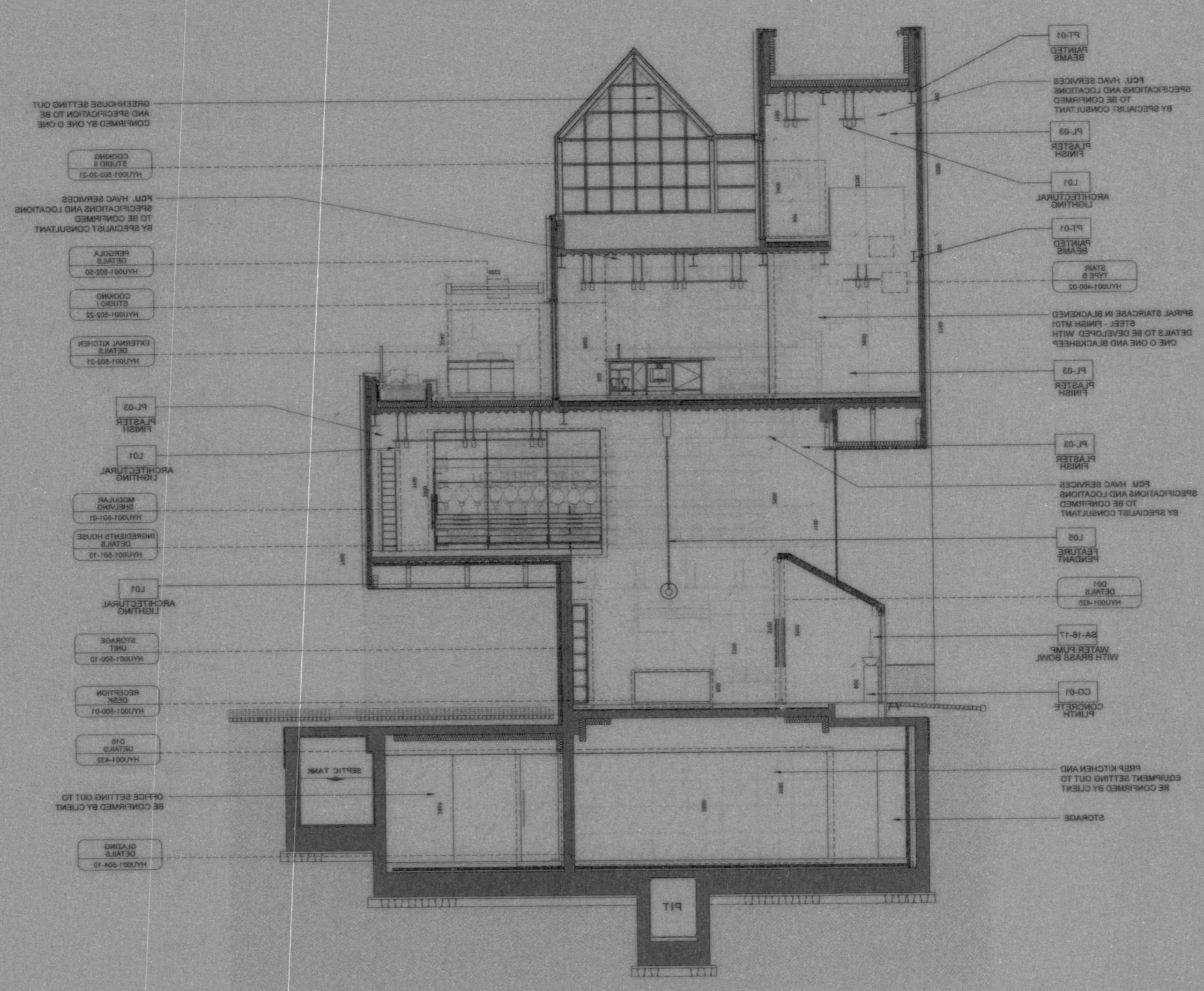
GREENHOUSE SETTING OUT AND SPECIFICATION TO BE CONFIRMED BY ONE O ONE
PT-01 PAINTED BEAMS
FCU, HVAC SERVICES SPECIFICATIONS AND LOCATIONS TO BE CONFIRMED BY SPECIALIST CONSULTANT
PL-03 PLASTER FINISH
L01 ARCHITECTURAL LIGHTING
SPIRAL STAIRCASE IN BLACKENED STEEL - FINISH MT01 DETAILS TO BE DEVELOPED WITH ONE O ONE AND BLACKSHEEP
L05 FEATURE PENDANT
SA-16-17 WATER PUMP WITH BRASS BOWL
CO-01 CONCRETE PLINTH
PREP KITCHEN AND EQUIPMENT SETTING OUT TO BE CONFIRMED BY CLIENT
STORAGE
OFFICE SETTING OUT TO BE CONFIRMED BY CLIENT
SEPTIC TANK
PIT

Brand Spaces

Spaces representing brand identity and cultural

initiatives led by Hyundai Card · Hyundai Capital

Vinyl & Plastic and the Storage, Studio Black, House of the Purple, Air Lounge, and Castle of Skywalkers are the emblematic branded spaces of Hyundai Card. Vinyl & Plastic, Korea's largest music store and music playground, is central to Itaewon youth culture, attracting diverse creators and brands. Meanwhile, the Storage is an experimental exhibition space. Initially, the spaces were set out to supplement the Music Library. However, as both have developed into spaces with unique personalities, their dynamism has been greatly augmented. Studio Black is a shared office; Hyundai Card embarked on this project to share management expertise garnered from its experimentation with workspaces, such as the lecture and meeting rooms tested at the Yeouido headquarters. Providing optimized facilities, information technology, and programs to build a startup eco-

system tailored to creative groups and initiatives, Studio Black functions as a shared office symbolic of a collaborative spirit. House of the Purple, a now-defunct bar opened exclusively for the Purple members of Hyundai Card, drew nationwide attention during its period of operation and became the envy of members of competitor credit card schemes. The Air Lounge opened at the Incheon International Airport also enjoyed great popularity with various innovative design touches that far surpass the conventional airline lounge. The Castle of Skywalkers is equipped with the finest grade of sports facilities, unprecedented among the training sites for world-leaders in volleyball. As a touchstone in design, visitors from around the world flock to the building day and night.

Vinyl & Plastic

p. 272

The main concept behind Vinyl & Plastic was a breed of vintage industrial design. The free-spirited atmosphere is multiplied in its exposed concrete ceilings and floors, load-bearing columns, rough wooden tables, the 1950s and 1960s-style objects, and neon-colored details juxtaposed with more subdued lighting.

4.1

Storage

p. 288

Under the aegis of the neon signage facing the entrance, the Storage has a slightly artificial edge applied to its design as implied by the name, i.e., storage place and warehouse. The original wall's rougher elements were revivified to amplify a cruder impression. Even the traces of old graffiti found by chance during construction have remained in the space.

4.2

Studio Black

p. 298

The lounge floor, equipped with a concierge desk at the entrance, serves as a hub like a hotel lobby. Meanwhile, the studio floor adopts a sophisticated modular structure. Composed of the same module, all of the spaces can be enlarged or shrunk as necessary.

4.3

Castle of Skywalkers

p. 304

This structure serves as accommodation and the training facility for the Skywalkers volleyball team of Hyundai Capital. It functions as a training center and a residence, a combination unparalleled among sports facilities. The space is also full of geometrical play and cutting-edge construction techniques.

4.4

House of the Purple

p. 316

House of the Purple is an exclusive membership space for Hyundai Card the Purple members, fashioned in the form of a speakeasy bar, a concept that was considered suspect even in large cities like New York at the time. This space even served as a 'design salon' filled with over 150 artworks of world-class designers and artists.

4.5

Air Lounge

p. 322

The Hyundai Card Air Lounge was planned as 'a preparatory space for travel'. Whereas existing airport lounges are merely a space to pass the time, the Air Lounge differentiated itself as a space to tie up any loose ends before traveling while also browsing various facilities as if one was in an exhibition hall.

4.6

4.1 Vinyl & Plastic

Vinyl & Plastic
Location: 1F~2F, 248, Itaewon-ro, Yongsan-gu, Seoul
Site Area: 729m²
Building Area: 409m²
Gross Floor Area: 2,160m²
Coverage Ratio: 56.02%
Gross Floor Ratio: 231.47%
Building Scale: B1~6F
Design: 2015.06 ~ 2016.02
Construction: 2016.02 ~ 2016.06
Interior: Spackman Associates
Façade: Samuso Hyojadong Architects
Contractor: Kesson

A wall graffiti by New York-based artist Aakash Nihalani. The work also functions as the signboard for Vinyl & Plastic that seizes the attention of passersby in Itaewon.

While the Music Library serves as a museum for rare vinyls, Vinyl & Plastic is a space for experiencing and possessing music in the form of vinyl records and CDs. Coupled with the Storage, it is a space that reflects Hyundai Card's prodigious spatial ambitions, a prospect evidenced by the fact that Vice-Chairman Ted Chung personally named the place.

4.1.1
The Way We Build

From the outset, Vinyl & Plastic aimed to serve as a vinyl (LP) & plastic (CD) record store that would satisfy post-show exhilaration and fill the empty hands of an audience leaving the venue after performances in the Music Library and the Understage. As such, Vinyl & Plastic complements the Music Library and Understage, forming an enriching cycle of activity between these spaces. In a similar fashion to the Music Library, Vinyl & Plastic is tailored to events that take place in the evenings and at night, forgoing daylight. On the nighttime streets of Itaewon, it is impossible to miss Vinyl & Plastic with its hot pink façade. The façade illuminates the interior of the first and second floors and immediately catches the eyes of passersby, realizing the two points of contemplation in its design: drawing in an impression of the surrounding Itaewon neighborhood and keeping pace with the Music Library. The façade's design, 22.7m in width and 3.4m in depth, allows natural light to permeate the interior throughout the day. Thanks to this design, the geometric block artwork by Nihalani fills an entire wall on the second floor, standing out in vivid animation. The process behind Nihalani's work drawing to completion is one of the focal points of Vinyl & Plastic architecture. Vice-Chairman Ted Chung, who led this project, did not agree with the artist's original suggestion. As the artist's integrity was also at stake, Vice-Chairman Ted Chung opened a dialogue on the nature of the work after much careful consideration. He frankly pointed out that the initial proposal did not resemble work in the artist's portfolio, offering words of encouragement and understanding that the artist might have felt somewhat apprehensive as this was to be his first permanent installation. This was an accurate estimation. Indeed, Nihalani has mainly practiced as a street graffiti artist and has never completed a commission for permanent display. In response, Gensler's Philippe Paré, who was in L.A., and Mary Spackman, the Principal Designer of Spackman Associates, joined the discussion from Seoul to

As the largest record store in Korea, Vinyl & Plastic features all-embracing album corners.

continue to encourage Nihalani and support his work. Special paints were brought in from abroad in consultation with the artist, and, finally, after three weeks of agonizing and frustrations, the present artwork was brought to life. When opening the door and passing by Nihalani's work, an intersection resembling Abbey Road, across which the Beatles so famously walked, has been drawn on the floor. Exposed concrete ceilings and floors in different shades of grey, thinly revealed load-bearing columns, rough wooden tables, 1950s and 1960s-style objects, and neon-colored details, all in juxtaposition with subdued lighting, amplify this free-spirited atmosphere. The main concept behind the space of vintage industrial design brings to life the peak of LPs' popularity. The first floor houses 4,000 types of music and 9000 records of vinyl (LP), covering contemporary albums from the 1950s to the present day and classical music of all periods. Situated on the second floor are 16,000 plastics (CDs) cataloged by genre. Furthermore, the space has been fully equipped with devices for listening to music and gadgetry or accessories such as cassette tapes, turntables, headphones, snapbacks, and T-shirts loved by musicians. As various brands including Louis Vuitton, Vans, Converse Jack Purcell, and Bottega Veneta have collaborated with Vinyl & Plastic on special pop-up stores, this space has rapidly emerged as the center of 'Itaewon youth culture'.

electronic

STAIRS TO 1ST FLOOR
CAFE BAR

“During the design phase, we noted that the space would be open to the general public, so to those who are not Hyundai Card members unlike in the library series. Therefore, the desired spatial ambiance was ‘comfort’ so that everyone can enjoy a moment’s pause and have fun, as in a local pub or regular hangout. As the area is frequented by large crowds, we ensured the space would draw people in by privileging the raw joy of music experienced inside. Stores should attract visitors by offering a new atmosphere on every visit. For instance, one may notice the red ladder on the first visit, while on a subsequent visit, a wooden pallet might catch the eye. As such, multiple layers within the space paint a range of spatial impressions.”
Mary Spackman, Principal Designer of Spackman Associates – Vinyl & Plastic Space Design

Contrary to the prevailing approach of a vintage industrial feel, which overpowers the album corners on the first and second floors, a more casual attitude is affected within the café on the second floor through its daring use of color.

"I believed that the immediacy of browsing and listening to albums, the opportunity to buy bread, and to dive deep into periods of contemplation while enjoying a cup of tea would galvanize the roads of Itaewon. The fish, meat, and fruit sold at the open market radiate unique energy while exuding aromas and a lively atmosphere. Conversely, the packaged fish, meat, and fruit at a supermarket may look regimented and sterile yet lack a certain vivacity. Itaewon is more like an open market than a supermarket. For the workings of Vinyl & Plastic to be considered unrefined, the presence of the façade had to be erased."

Seungmo Seo, Founder of Samuso Hyojadong Architects Office - Vinyl & Plastic Façade Renovation

4.1.2 Point of Space

Vinyl & Plastic is a renovated, not a new, building. The façade renovation was commissioned from Samuso Hyojadong Architects Office and the interior design from Spackman Associates. The façade, capturing the imagination of the passersby and illuminating the first and second floors of Vinyl & Plastic, could only be achieved by remaining true to its supporting role to the Music Library while reflecting Itaewon's regional personality. The architect decided that the façade of traditional design had to be eliminated to better deliver the energy of the musical acts performed inside Vinyl & Plastic and strengthen its connection with the roads of Itaewon. To this end, the huge glass curtain wall façade, with a width of 22.7 m and a height of 8 m, plays a crucial role. The main four doors unfolds freely, and along with the sloping approach of the Music Library, this naturally encourages passersby to enter, implying the building's relationship with the Music Library. As visitors enter Vinyl & Plastic via a façade that unfolds towards the Itaewon townscape, they run into a more informal and old-fashioned industrial sensibility. While the Music Library is a space that lavishly redefines the vintage, Vinyl & Plastic focuses on comfort despite its aged and rough appearance. Creating an ambiance that prompts a certain nostalgia through the preponderance of LPs, a technology popular at midcentury, was also vital to the project. In the end, it was designed to appear rough and worn like a warehouse, while adding the ambiance of a hideout. Vinyl & Plastic exhibits a resilient look in line with its surroundings. Sincere efforts invested in the space are also evident from all of the furniture and accessories, which were made to order. As the building does not have an elevator, a mini-elevator, often used in kitchens, has been installed for employees carrying heavy items or bags. For Vinyl & Plastic, the industrial sensibility has been calculated to the finest detail. On the second floor, visitors can relax in a café overlooking Itaewon and enjoy CDs prepared for each table. A real highlight is the balcony-styled glass wall on the second floor. Here, through the clear window, people outside can observe all sorts of things taking place inside, from listening to music to enjoying drinks in the café. Another noteworthy feature is the CD Wall, in which the six-compartment wall is a gigantic CD player. The CD Wall was designed and manufactured directly by Hyundai Card Design Lab at the suggestion of Spackman Associates.

The concept is a vintage industrial design that reimagines the post-war period when LPs were extremely popular.

A
B
THE SIDEWINDER

VINYL & PLASTIC by Hyundai Card

4.2 Storage

Storage
Location: B2~B3, 248, Itaewon-ro, Yongsan-gu, Seoul
Area: 588m²
Design: 2015.06 ~ 2015.12
Construction: 2015.12 ~ 2016.06
Interior: Spackman Associates
Contractor: Kesson

Opening along with Vinyl & Plastic, the Storage began life with a proposal for an experimental exhibition space. To meet this expectation, experimental projects at home and abroad were planned for the space, spanning art, film, architecture, and design, regardless of genre. It serves as a spatial symbol of the most progressive slants in Seoul's culture today.

The Storage was set in motion as one opens the door with its pink neon signage. The interior space has been renovated to preserve something of the crudity of the original building.

4.2.1 The Way We Build

As the name suggests, both storage and warehouse, the Storage presents experimental projects in and out of the country, from film to architecture to design, crossing various genres. Operated with a self-curated program as its basis, the Storage held the exhibition, “Traces: The Origins of Hyundai Card Design”, to mark its opening, looking to design as a means of expressing Hyundai Card’s brand philosophy and identity. Since then, the exhibitions have been numerous: British artist David Shrigley’s solo exhibition “Lose Your Mind”; “Love/Love” by graphic design duo M/M(Paris); “Void” by the installation artist group Numen/For Use active in Europe; “One Minute Forever”, the solo exhibition of world-renowned contemporary artist Erwin Wurm; “Good Night: Energy Flash” which reinterpreted club culture; “RE:ECM”, an exhibition commemorating the 50th anniversary of ECM; and “OSGEMEOS: You Are My Guest”, which was the Brazilian artist Osgemeos’s first solo exhibition in Korea. In fact, the Storage is not an open space that welcomes everyone. This is because the Storage was planned as a space that would address both niche and broader tastes, reachable only after enduring the agony of descending the stairs and pushing the heavy door, unlike Vinyl & Plastic with its welcoming, wide, and transparent façade. This is why the Storage is now the flashpoint of the most progressive culture movement in Seoul. The initial plan for the Storage was not an exhibition space; it was originally planned as a speakeasy bar. Vice-Chairman Ted Chung, who hankered after a ‘music-centered complex’, thought of a red bar in light of House of the Purple. It was because a bar setting was perfect for the continuation of the excitement stemming from the Music Library. The next subject of contemplation was a recording studio. Around that time, Seoul Recording in Dongbu Ichon-dong suspended business due to financial hardship. It was a shame that there was no longer a recording studio sizable enough to accommodate sessions. To this end, Hyundai Card teams visited the engineer who oversaw the sound consulting for the Beatles’ Abbey Road recording studio to assess the potential of the building. However, the idea had to be let go because the space had to be completely renovated to turn the building into a recording studio. Although both concepts—the recording studio and bar—failed to come to fruition, a swift decision could be made after attending to the question, ‘Is this an area in which Hyundai Card is accomplished?’ Because an exhibition space was an area in which Hyundai Card had already developed expertise, which could inform the planning, the project proceeded without difficulties.

In the Storage, Korean and international experimental projects are coordinated regardless of genre, including the Traces: The Origins of Hyundai Card Design exhibition as its opening show.

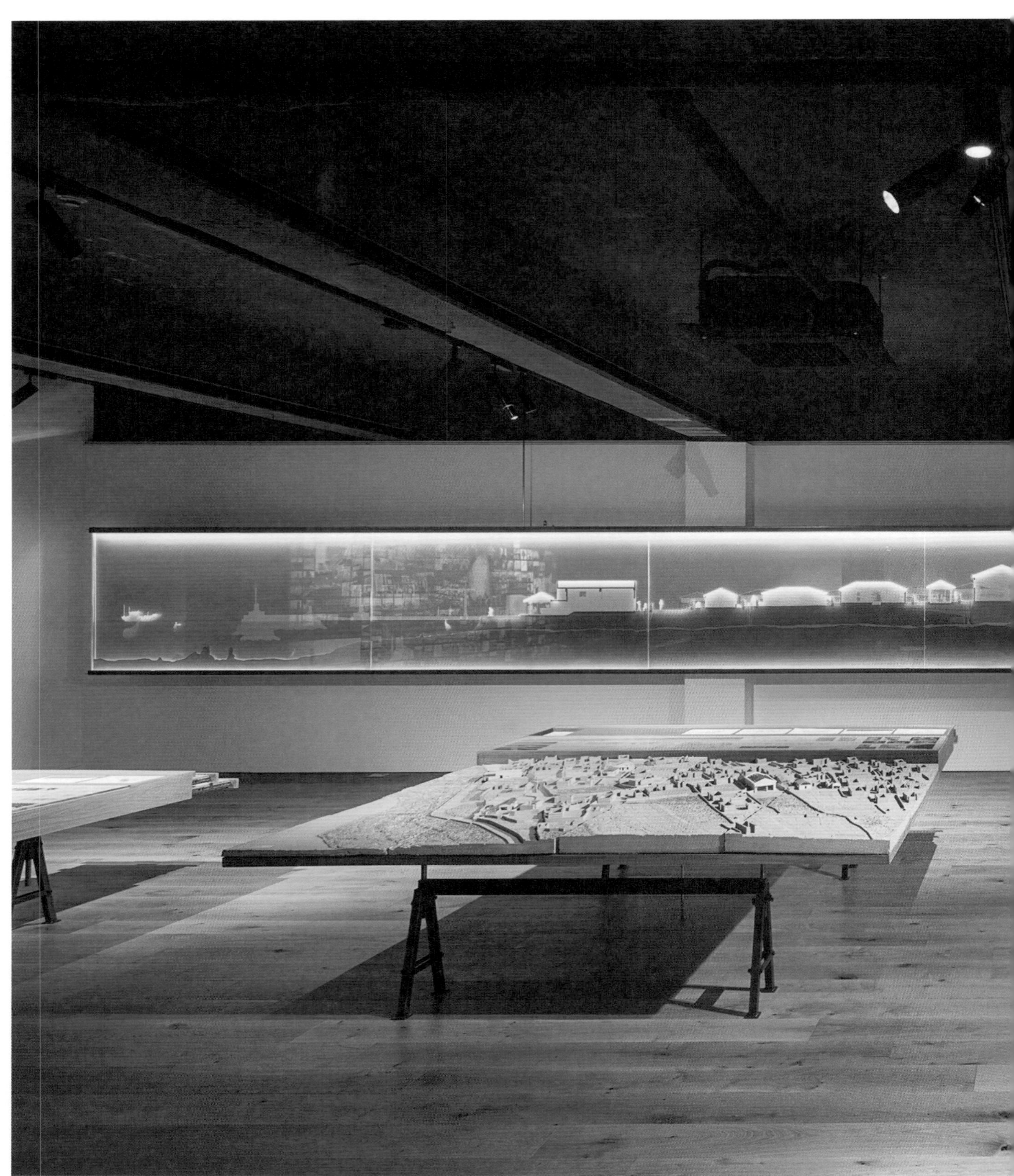

4.2.2
Point of Space

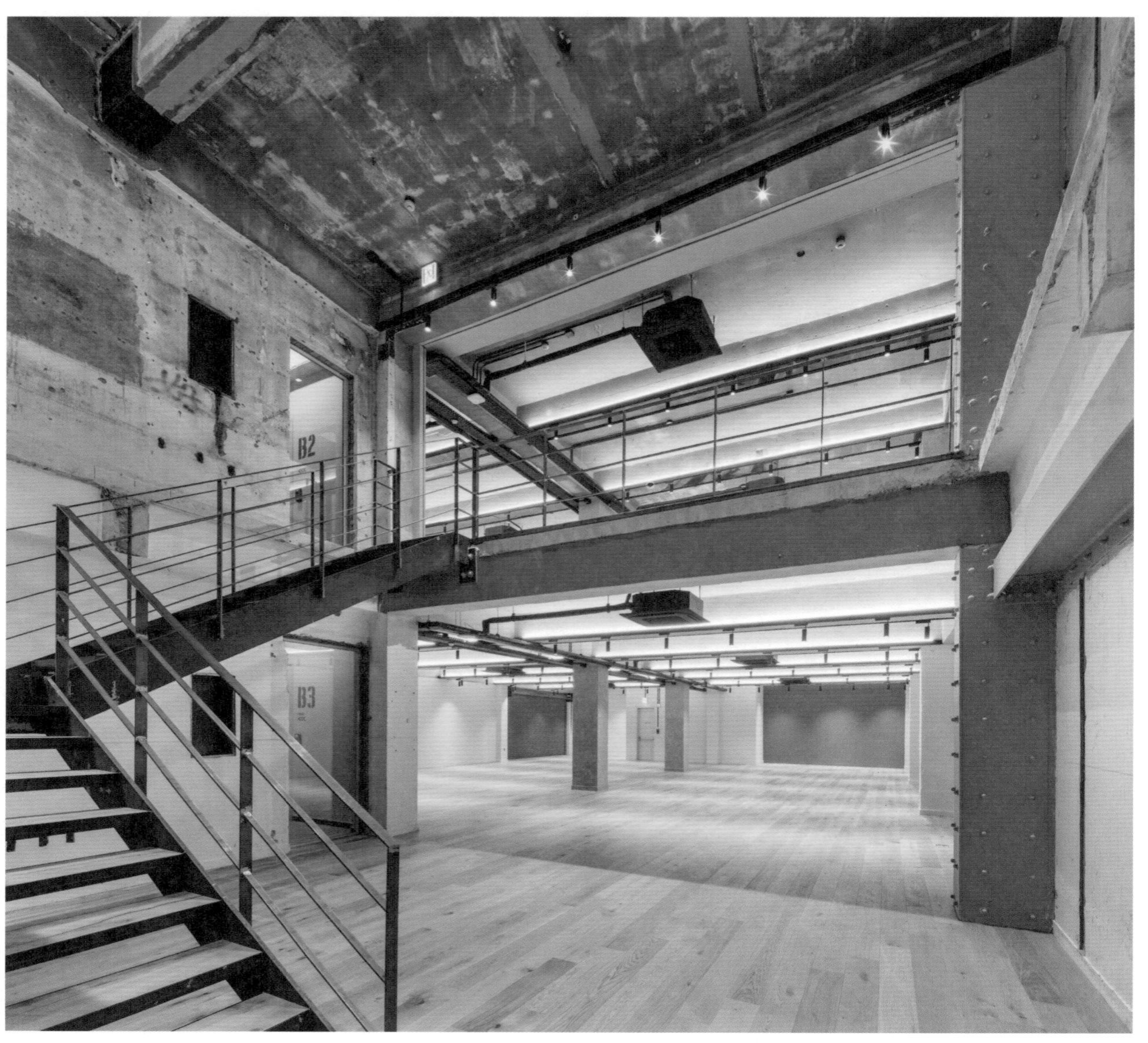

"In New York or Chelsea, one often encounters galleries that are rough but approachable. For the Storage, I wanted to create a strong bond with Vinyl & Plastic while making the space welcoming. The parts that had to be retained after demolition were extensively checked and agreed upon. The previous space, however, did not have internal stairs. The remaining narrow staircase was fit for use, but a new staircase was installed to connect the second and third basement floors without disrupting the flow between the exhibitions. The main protagonist of the Storage is the artwork. Thus, the staircase had to be a subtle one: an iron staircase."

Mary Spackman, Principal Designer of Spackman Associates – Storage Space Design

As visitors exit Vinyl & Plastic and descend the stairs located between it and the Music Library, the pink neon sign points them to the entrance to the Storage. Unlike general white cube-type galleries and exhibition spaces, any artificiality was minimized in the Storage's interiors. The unrefined impression was intentional, created by retaining the original wall texture. Even the graffiti discovered during construction was left intact. Throughout the construction period, significant prudence was exercised, as evidenced by the frequent changes to or preservation of the existing site, one aspect at a time. Interestingly, the Storage is spread over two to three floors underground when viewed from the main street of Itaewon but four stories above the ground when viewed from the back alleyways. As the gleaming street with its fashion stores and restaurants reaches the more uninviting spaces of the alleyways, the space becomes Janus-like with its unpredictable switches in aspect according to one's point of view. The parking lot facing the alley has been finished in hot pink, in line with the graffiti in Vinyl & Plastic and the signage in the Storage. The hot pink reinforces this connection while seizing one's attention. Moreover, there was an old staircase between the Music Library, Vinyl & Plastic, and the Storage. The staircase was narrow, had uneven stairs, and was difficult to climb. If the stairs were blocked, however, visitors had to make a detour of about 200 m. Noting this, Hyundai Card rearranged the stairs to improve their efficiency. The repair also connected the alleyways, adding more playfulness to the urban landscape of Itaewon.

Any artificiality was minimized in the Storage's interiors, and even the graffiti discovered during construction was left intact.

4.3 Studio Black

Studio Black
Location: 22, Seocho-daero 78-gil, Seocho-gu, Seoul
Area: 3,300m²
Design: 2016.03 ~ 2016.09
Construction: 2016.06 ~ 2017.02
Interior: Gensler
Contractor: Arco

The lounge floor of Studio Black operates as a hub like a hotel lobby. It is a space where members frequent to work, take a break, and communicate with each other.

In planning Studio Black, Hyundai Card wanted to experiment with a shared office format that would be equipped with various facilities such as a lecture room (the effectiveness of which had been proven in the Yeouido headquarters). Studio Black, armed with a nap room for brief breaks, a testing room for testing different OS devices, and a smoking room, functions as a pre-eminent shared office space favored by start-ups. The lounge floor of Studio Black operates as a hub, like that of a hotel lobby. It is a space its members frequent for work, to take a break, and to communicate with each other.

4.3.1
The Way We Build

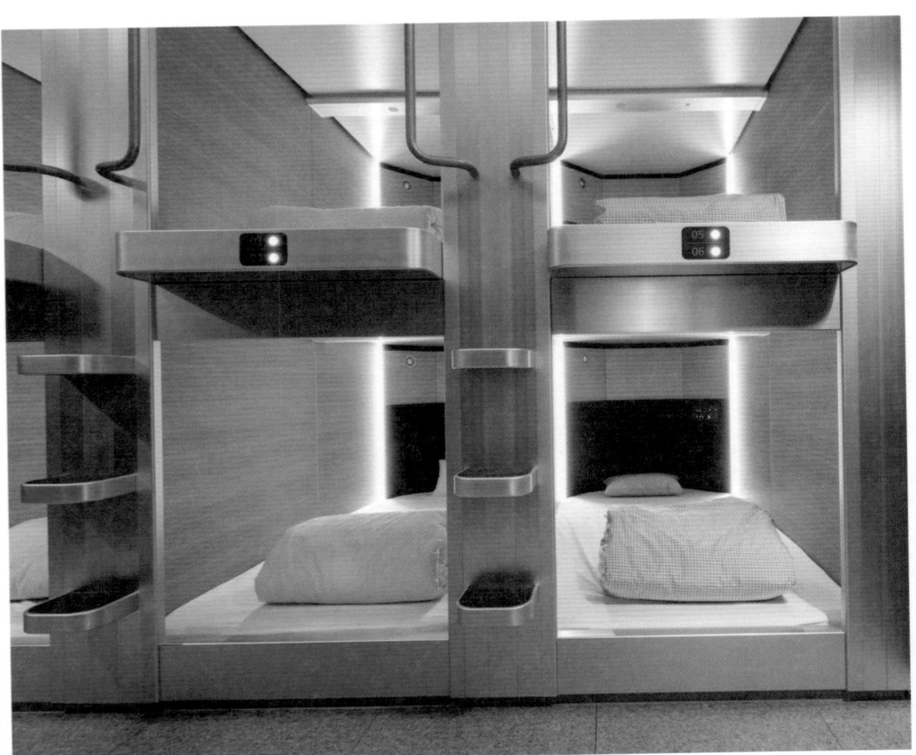

Studio Black was Hyundai Card's first shot at building expertise concerning branding, corporate culture, and spatial design in a new business area. In Korea, there was no attempt whatsoever to increase the value of a building, unlike that of other nations. As such, Hyundai Card increased the value of its space by applying the label, 'Operated by Hyundai Card', and experimented with the management of the building through its commanding grasp of the potential challenges. Relevant research was conducted under the leadership of the Corporate Culture Division by visiting major shared offices worldwide, including WeWork, Soho House, and Neue House. However, problems arose as the project was launched under an ambiguous concept, in contrast to the clearer visions that guided Hyundai Card's other spaces. It remained undecided whether to prioritize securing enclosed personal workspaces over enabling the potential of various business interactions and collaborative work via open spaces and vice versa. 'I hadn't had the chance to examine the project closely until it was almost complete. What I found back then was shocking. The Neue House, which I visited on a business trip to L.A. in 2017, is a perfectly open space. It is a space in which to communicate, influence each other, not a space in which to work. That is why there are no partitions and are many lounges. WeWork, on the other hand, encourages its occupants to work hard in an enclosed room. Their product is working spaces themselves. Unfortunately, Studio Black did not have a clear business model before beginning construction. It was to be an enclosed room with open vaulted spaces or one not quite adequate to be considered a lounge. This early indecision allowed me to insist to the staff that a clear and vivid concept and model has to exist from the outset.' After several revisions, the Studio Black was completed, including features such as a nap room for enjoying short naps, a testing room where one can try out various OS devices, and a smoking room, among others. To build a startup ecosystem centered on IT and creative groups, programs that facilitate tailored learning and exchange have had to be strengthened. Hyundai Card has also selected companies with whom to collaborate or support in each season. As such, Studio Black has functioned as a leading shared office favored by enterprising startups.

Studio Black has been equipped with a range of facilities, including a nap room for taking short naps, a testing room where one can try out various OS devices, and a smoking room.

"The rise of shared office spaces is a global trend which has significantly impacted the existing workspaces of companies. Studio Black promotes new ideas about the formation of communities within a workspace. Thus, the shared space was given greater priority over individual office spaces while continuing to revise the design. This was why the focus was on the lounge on the 9th floor, where encounters between individuals are made more readily and there are many for collaborative working. In addition to the Flex Room, purposed for regular sessions and lectures, the sofas and meeting tables can be moved as necessary, enabling diverse exchanges and the natural formation of communities of varying composition."
Sabu Song, Design Director of Gensler - Studio Black Space Design

4.3.2
Point of Space

Studio Black occupies five floors in total, from the 8th to the 12th, as well as the rooftop garden of the Hongwoo Building near Gangnam Station. These floors have been structured in a square with all sides of the same length, and the area of each floor is about 660m^2. The entire area is largely divided into a lounge floor plus the rooftop garden and studio floors occupying the remaining space. Studio floors feature a sophisticated modular design; all areas are composed of basically the same module, which is expandable or reduceable as required. The basic configuration is a room for up to 10 people, but this can be increased as needed. To remove vagueness or obscurity, enclosed spaces have been avoided as much as possible while shared spaces were added in greater numbers to modify the interior space. Each floor has a floor lounge, a common space, a meeting room, and a phone booth. There is one room left blank (deliberately not designated as a workspace) between the studio cells. These blank rooms tend to include simple exercise equipment such as yoga mats and exercise balls. The lounge floor functions as a hub like the reception in a hotel, endlessly frequented by the members who take breaks and conduct conversations with one another. The entrance features a concierge desk, while the sofas at the center and the meeting table on the right facilitate conversions between the members and visitors, and provide a space to take a break. In the adjacent flex room, diverse programs are offered to members. On the left side of the lounge floor, there is a hot desk and a meeting room. The hot desk is a membership workspace with one desk only for casual working in this open space. There is also a photo studio, a testing room, a nap room, and a shower room concentrated around the lounge floor. The testing room is equipped with a variety of OS devices and design tools. There, members can perform compatibility tests for applications they have created as digital devices from the old version to the latest are available, including smartphones, tablet PCs, and VR devices, as well as two 3D printers. In the Make Section on the opposite side, various tools for handiwork, such as monkey wrenches, hammers, paints, and adhesives are prepared, which also allows members to realize actual models and mockups.

The studio floor features a design that adopts a sophisticated modular structure. All spaces are composed of basically the same module, expandable or reduceable as required.

4.4 Castle of Skywalkers

Castle of Skywalkers
Location: 45-1, Seogyang-gil, Jiksan-eup, Seobuk-gu, Cheonan-si, Chungcheongnam-do
Site Area: 21,955m²
Building Area: 2,988m²
Gross Floor Area: 9,356m²
Coverage Ratio: 13.60%
Gross Floor Ratio: 34.11%
Building Scale: B1~4F
Design: 2011.05 ~ 2012.05
Construction: 2012.06 ~ 2013.06
Architecture: Doojin Hwang Architects
Contractor: Hyundai Engineering, Dawon ID&C, Kesson

The Castle of Skywalkers is a training facility for Hyundai Capital's male volleyball team. The spatial goal was to bolster the team's capabilities and strengths by providing optimal personalized spaces for the players. Since it has the finest facilities, internationally unrivaled, it is often used as a key reference point at home and abroad. It has also influenced other professional sports facilities in Korea and given cause for the required improvements. Recognized by numerous awarding bodies for sports marketing on its regionally sensitive marketing and team operations, the space showcases how professional sports can grow within their region.

Despite the difficulties of restarting the project using a completed design, it swiftly became a blessing in disguise as this world-class building could be executed to an unprecedented standard.

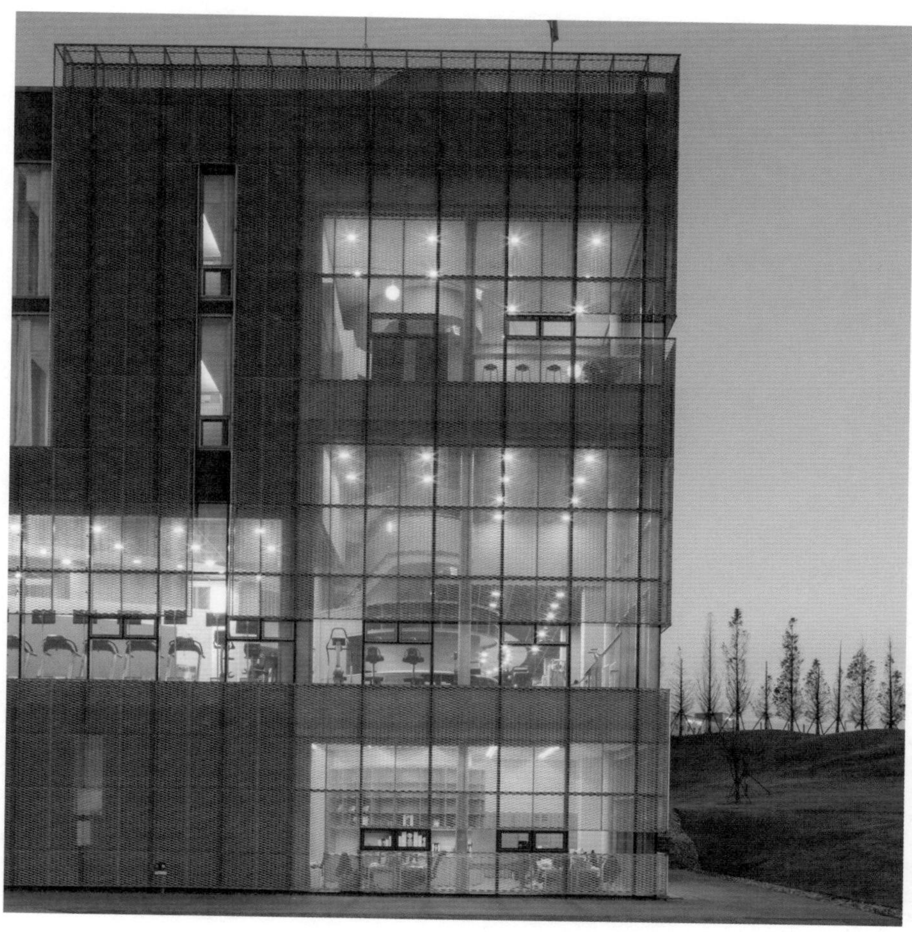

4.4.1
The Way We Build

The Skywalkers training facility project began in 2008 at the suggestion of the city of Cheonan. Cheonan, the hometown of the volleyball team, proposed the construction of a new training facility in the sports park near the city hall using the BTO (Build-Transfer-Operate) method. The architect Doojin Hwang took on the project as he was already involved in the spatial enhancement of the rooftop of the Yeouido headquarters. However, after the design was completed and right before on-site construction was to begin, the project came to an abrupt halt as Cheonan City Council objected to its commencement in December 2010. Nevertheless, Hyundai Capital was determined to establish this base for its professional team to develop a close bond with its home region. As few teams are based in their hometowns, athletes visit the region only to attend games. To overcome this limitation in Korean professional sport, Hyundai Capital decided to found this base in Cheonan for its professional team. Accordingly, in 2011, it resumed the project by purchasing a site on the outskirts of Cheonan. When the project site was at the center of Cheonan, the gymnasium and accommodation had to stand separate from the main building. The aim was to provide the athletes with an independent space for daily use in which they could find moments of repose. Subsequently, a number of concerns arose: the players' immunity is close to zero due to the physical fatigue incurred during training and game seasons and even the short commute to the adjacent buildingcould pose a threat; dividing the building into two would occupy too much land. After much discussion, it was decided to unite the gymnasium and dormitory into a single building. Volleyball games require high ceilings, and so the design shaped the accommodation to a surround to the 25 m-high central part.

This design was internationally unprecedented. The Castle of Skywalkers also features a courtyard. Modern sports players are like the knights of the Middle Ages. The players' physical conditions must stay exceptional, and a high moral code is required. The team of around 20 members is also appropriately referred to as the 'knights'. As a space accommodating these gallant athletes, 'castle' is not a misnomer. The exterior also reflects this character by using an expanded metal shell as the finishing material. In addition, the elements that do not cover the exterior skin, such as the balcony on the third and fourth floors, have been applied in between, resembling the 'battlements' of a medieval castle. The architect channeled the concept of the arena so that the court is viewable from all directions. The design also allowed the game to be visible from the hallway or balcony on the third and fourth floors. The training center, meanwhile, is variable between a games mode—using only one court—and a practice mode—consisting of two courts. By placing the grandstands on the court's both sides, a square has been formed employing a plane of approximately 50 m x 50 m. The key point was that players would be able to perform indoors in rain or in winter. The 150 m-long indoor circular track that loops the building creates a functional as well as a dynamic atmosphere. Elevators and stairs are located at each corner, with a central axis pointing towards the court's center. Defined figures fix key spaces in order, among which are auxiliary facilities. Along with the court in the middle, the physical therapy rooms, power analysis rooms, and saunas are present on the first and second floors. The stairs on the third and fourth floors feature other facilities such as kitchens and lounges as rest areas for the players. Most particularly, the weight training area, including the jogging track, was deliberately

The Castle of Skywalkers experiments with the multi-faceted analysis of the sport so much so that it could qualify as a lab solely dedicated to the study of volleyball. In particular, the facility takes a careful approach to the athletes' psychology. For instance, the weight training space is the best lit space and has the best views so that the trials of strength training can transcend bodily pain.

credited with the best view on the second floor to alleviate the hardships felt by the players during training sessions. In the case of the accommodation, it was first undecided whether to lay out the rooms as single or double rooms. Private rooms would ensure that each player had privacy even though the professional and the residential areas meet in this building. Ultimately, emphasizing the collaborative endeavors between a leading player and a supportive player, the rooms were designed to be two-person rooms. Still, the bedrooms were separate to ensure privacy. When a player walks out to the balcony of each bedroom, he encounters the balcony of the adjacent room. As such, players can go to the next room via the balcony, not the hallway, which led to the coined term, 'balcony-mate'. For the comfortable daily routine of athletes with an average height of 190 cm, the facilities underwent meticulous checks and analysis, right down to centimeter measurements. Increasing the ceiling height to 3 m high and the bed length to 240 cm laid the groundwork for these spaces. In addition, even the height of the door

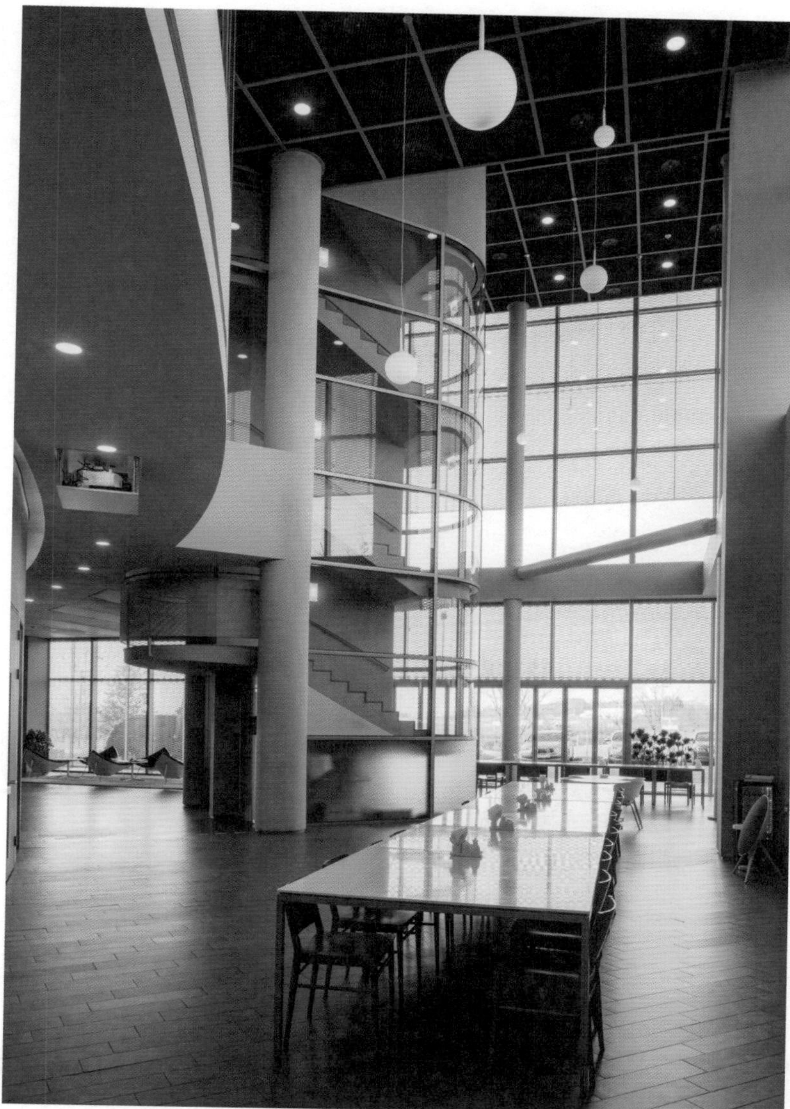

In addition to the 600-seat stand, people can watch games and practice sessions from the corridors on the third and fourth floors. Furthermore, when one side of the wall fully opens, the court is visible from the outside lawn, reminiscent of an opera house in which the interior and exterior have been united. On the second floor, a 150-m-long running track 'Blue Ring' facilitates all-weather training. The accommodation for the players is on the third and fourth floors where the vertically overlapping spaces are immediately discernible.

handle and toilet were adjusted at the players' suggestion by making space design mock-ups. To assure that the sounds from practice sessions do not interfere with the players' privacy, three doors were placed between the hallway and the bedroom. The third and fourth floors secure a load-bearing wall structure, unlike the lower parts of the building, to preserve this tactful division. Since the gymnasium and the dormitory have been combined, a clear cross-sectional view was a key consideration. Despite the high construction cost, the elevators were made of a transparent capsule to allow the players to see at a glance who is practicing and who is at the gym. Athletes running on a jogging track can see the worktable in the dining room's kitchen. When leaving their rooms, athletes can even see where the building cleaners are cleaning. The ability to see one another moving within this enormous structure strengthens a sense of belonging. The Castle of Skywalkers is much more than a space for living and training; rather, it spurs the volleyball team to grow and evolve. The honor of winning the V-League in the 2018-2019 season can also be attributed to this work of architecture and life running therein as planned.

4.4.2
Point of Space

Each of the square building's corners points exactly to the east, west, south, and north. No side faces the north in this arrangement, allowing sunlight to enter all year round. The expanded metal outer skin functions as an awning, blocking excessive light in summer and drawing light into its corners in winter. However, since its metal finish can bother neighboring buildings with reflected light, an optimal grade was applied by analyzing the angle of the sun and the inclined reflection angle of the mesh surface. Furthermore, the structural performance of the expanded metal was verified through numerous wind tunnel tests. The structural simplicity was assured by minimizing columns at the building's corners. Bedrooms, lounges, dining rooms, and offices are designed for the enjoyment of natural light while maintaining a close connection to the countryside. With its vertically overlapping diverse functions, the structure's configuration differs by floor. The basement floor is a reinforced concrete structure, and the first and second floors are supported by reinforced steel-framed concrete columns. Unrestricted use of the space was required on the lower floors where volleyball games and training take place. As such, all spaces have been connected by partitions made of a glass curtain wall. As the southeast side of the building sits on higher ground, the second floor is at the same level as the hill outside. When the walled boundaries disappear, the greenery outside and the indoor court takes on an easy association. In general, enjoying the scenery is possible only via a small window on the wall or the structure. On the contrary, in the Castle of Skywalkers, the walls and columns were removed in favor of a structure that suspends the upper floors, and electric folding doors have been installed. To open the 36 m-long sides fully, the electric folding doors used in airplane hangars have been used. When the wall disappears and the space fills with light and fresh air, the players who have long been inside can feel an unparalleled sense of freedom as the court space expands further and further away. For the same reason, skylights have

There are sufficient rest areas within the building. However, a streamlined annex was added to the building's outside for families or fans. Communion with nature outside can take place daily, but the artificial lighting perfectly simulates the environment of the volleyball arena once practice begins.

"The greatest desire for the building we felt internally was that it should stimulate and promote the growth of the volleyball team, existing as something beyond a mere space for living and training. In this building, one cannot hide. This is not about monitoring, but about the attention paid to communion in daily life—the sense that they all are part of the team, even including the kitchen team preparing meals, those cleaning the space, and those watering the lawns. To draw everyone into the team, we needed to build a three-dimensional space allowing cross-sectional eye contact instead of segregating each section of the building."

Doojin Hwang, Founder of Doojin Hwang Architects - Castle of Skywalkers Architecture and Space Design

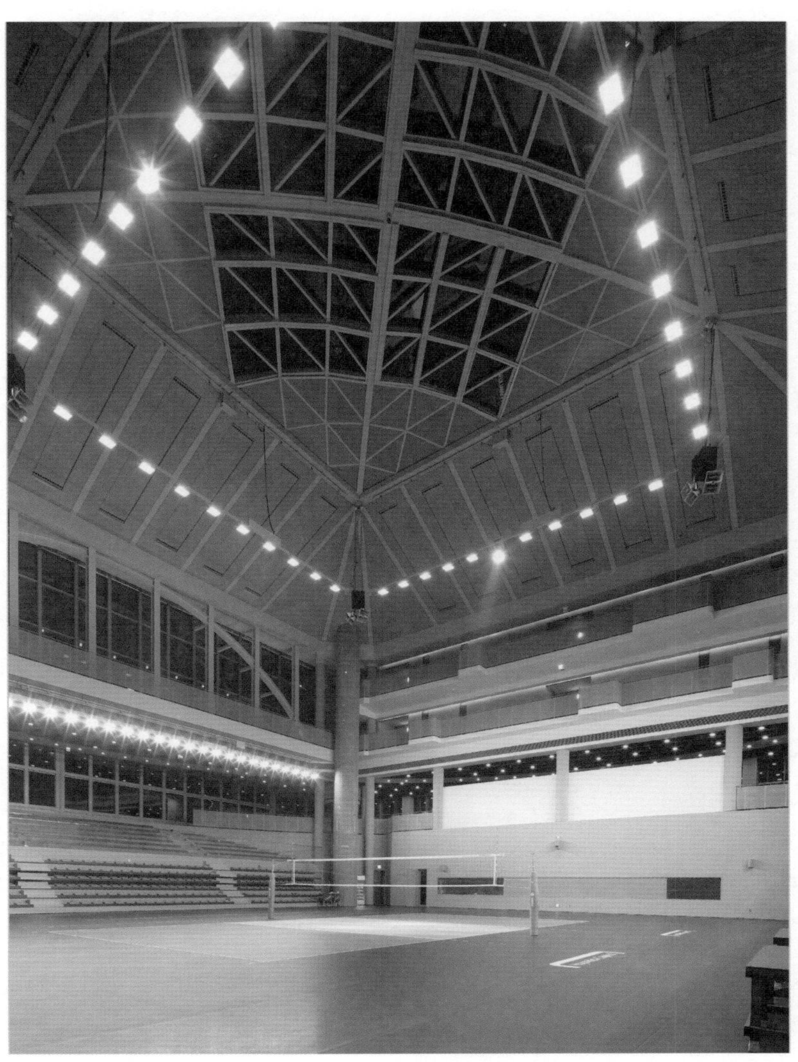

been installed to deliver natural light. To this end, a single-layer roof was employed, which is uncommon in stadiums. The truss of a typical material is double-layered and thick, not permitting light to enter easily. Using a truss was the original design plan; however, Professor Kyungju Hwang, who is also a structural engineer, proposed a single-layer structure. This enabled the light and functional ceiling in the present form. The exposed ceiling also provides a spectacle for volleyball players who frequently look up above due to the nature of this game. The court flooring material, Mondoflex, has been adapted to the Skywalkers team colors of grey and orange, instead of the more commonly used green and orange, to strengthen a sense of team spirit. The physical therapy room, which is accessible from the court, is equipped with soundproof material to ensure peace and privacy during treatment. Additionally, aqua-therapy and oxygen therapy devices help athletes to rehabilitate and recover. The hydrotherapy room is for rehabilitation programs using a central underwater treadmill that emits a current, equipped with state-of-art devices rarely found in Korea. This completes a one-stop facility for general training and rehabilitation training support to athletes so that they are in the best shape possible to adhere to their training schedule. What determines the outcome of the season is not a grueling training regime but the wisdom of strategizing. The key here is the strategy analysis room on the second floor. The cameras installed both around the court and on the ceilings record the players' movements 24/7. The recordings can be displayed at any moment on the court's electronic board to improve the players' movements. This is the motion analysis system (Dartfish). At higher ground sits the semi-buried bunker of the building, which was not in the original plan. However, mindful of the visits made by the families of the players during the season, a place to spend time together was conceived. This annex draws upon artificial landscaping with a natural curve and contrasts geometrically with the square-shaped central building. There is a small waterfall, the source of the sound of trickling water, which quiets the minds of the players. There is also a metal bridge crossing over the running water. Magnolias, which bloom in spring, have been planted around the perimeter to mark the consummation and sublime achievements at the end of a season. Office PARK-KIM, in charge of the landscaping, used expanded metal for the water landscaping, which also covers the building's exterior wall.

The system that opens one wall in its entirety compensates for the stifling atmosphere of indoor exercise. To this end, the walls and pillars were removed, and electric folding doors used in airplane hangars were installed. Outside the folding door, a small annex was added for family visits or fan meetings. The semi-buried bunker becomes a single area containing the court when the wall of the gymnasium is fully open. The bunker's streamlined exterior contrasts with the solid square building.

"Compared to other sports facilities, the Castle of Skywalkers is special in that it is a one-stop system that handles the diverse functions of every anticipated need, except perhaps that of a hospital ward. Immediately outside the dormitory is a practice field, allowing athletes to train individually at any time. All of the spaces have been tailored to the tall players and any shortcomings are hard to find as the team constantly checks and improves the space. The players say they are more comfortable here than at home."

Sungmin Moon, Player, Cheonan Hyundai Capital Skywalkers

4.5 House of the Purple

House of the Purple
Location: 23, Dosan-daero 49-gil, Gangnam-gu, Seoul
Area: 280.5m^2
Design: 2010.06 ~ 2010.10
Construction: 2010.10 ~ 2011.02
Interior: ...,staat
Contractor: Kesson

The entrance to House of the Purple is located deep inside without a signboard signaling its existence, underlining its exclusivity. A space that reflected the tastes of Hyundai Card, the Purple members closed its doors in 2018.

House of the Purple is a place in which a speakeasy bar—an unfamiliar concept at the time even in large cities such as New York and London—was contrived as a membership-only space for Hyundai Card the Purple members. The target audience was professionals in their late 30s, and the space aimed to offer a lifestyle space suited to them. It also functioned as a 'design salon' filled with a whopping 150 artworks by internationally renowned designers and artists.

4.5.1
The Way We Build

The House of the Purple project dates back to 2009. Vice-Chairman Ted Chung, who visited New York on an insight trip, took great interest in a speakeasy bar. The speakeasy bar may look like a simple hot dog joint but unveils a stylish bar when passing beyond the front door. At that time, while considering a number of exclusive benefits for premium card members, the idea of creating a membership bar where its members can enjoy the high life took flight. A creative agency based in Amsterdam (...,staat) was commissioned to plan and design the space. The firm, ...,staat, consists of 30 designers working across various fields such as graphic design, architecture, fashion, and film. First, Hyundai Card researched the defining character traits of it's the Purple members, based on their tastes and lifestyles. The conclusion was to look to the house as its basic concept. Hyundai Card envisioned a space in the form of a house that would celebrate the tastes and desires of the Purple members rather than devising a more informal hangout space. The first step was to hand-pick the objects that were to be made-to-order or were vintage or special editions, so that the space would harmonize the modern and the antique, through design products and rare works. Over 150 pieces were collected, including General Electric's speakers, one of the eight existing examples of the Silver Cloud Chair by Richard Whitten, and a Saporiti Cosmos set by Augusto Bozzi created over 70 years ago. House of the Purple closed its doors in 2018, and Vice-Chairman Ted Chung shared the following assessment: 'At the time, even the most cosmopolitan of metropolitan centers, such as New York and London, were quite oblivious when it came to the concept of a speakeasy bar. We embarked on the project with the idea that it would be fun to create a bar in the corner of Apgujeong under an inconspicuous sign. Then, as it progressed, the scale expanded beyond our expectations. The concept was planned to a bit of an excessive degree by the agency. This experience revealed to me the crucial importance of project management, as in the Library series. Still, House of the Purple was and continues to be an excellent and unrivaled shot at a bar project.'

House of the Purple was largely divided into a bar and restaurant. Hyundai Card wanted to realize a space with a domestic feel in which Hyundai Card the Purple members could demonstrate and share their tastes and privileged lifestyles with their close friends.

"The lack of an overt design concept is the idea behind House of the Purple. We thought that simply arranging a carefully selected artworks would create a superb spatial design. That is why the role of the curator was particularly important to this project. After visiting Korea for a presentation in 2010, the first thing we did was to ask and study the members, and to comb the world to find design works that would best reflect their tastes."

Jochem Leegstra, Creative Director, ...,staat - House of the Purple Space Design

4.5.2 Point of Space

From planning to completion, House of the Purple took about two years in total. The creative agency ...,staat was commissioned for the planning, Bourne and Korea Total Impact for the branding and CI design, and Fiction Factory in Amsterdam for the final interior design. Although it is located on the first floor of a building in downtown Apgujeong, House of the Purple covered the windows facing the road to signal its exclusivity and trigger the curiosity of passersby. There is no signboard, and the entrance was placed deep inside the alleyway so as not to be noticeable at all. Such decisions imply the secretive nature of the space. The closed glass doors could only be opened after checking in with a member's credit card. The building, with its exposed concrete exterior, has a flat L-shaped structure, with the living room and study divided around the entrance and the concierge. The whole space was decorated to convey warmth and a sense of relaxation, a place in which everyday accessories tell mesmerizing stories. As the members enter the building and turn left, they are met by the living room with its strikingly magnificent speaker hanging from the ceiling. It is General Electric's 18-inch horn speaker which sounds like those from the 1920s and 1940s. The speaker was airlifted from Japan and was so large that it could not be brought inside in its entirety, so it was cut in half and reassembled in situ, which required significant expertise. The living room exemplifies the high level of hospitality shown to its visitors, who are allowed to settle in the chairs and use the items worthy of museum collections without any restriction. On the north wall, there is an extended bar for serving drinks, providing comfort even to its solo visitors. The bar tabletop is made of marble with shades of purple befitting the name of the bar. Of note is that marble of that size and quality has never before been used in Korea until this project. The espresso machine, Elektra's Belle Epoch, boasts a distinguished elegance, only three of which were available in Korea at the time. The kitchen and the study were present to the right of the entrance, and the terrace outside was revealed through the windows. The study was designed as a more intimate and contemplative space, equipped with a variety of books and art books, enabling incidental encounters with books. Also placed here were small tables for face-to-face conversations, each of which features a chessboard. Beyond the windows in the north of the bar, there is a small outdoor space, known among the members as the most private and pleasant area within the whole setup. However, to realize something resembling the cozy feeling when visiting the home of a friend, creating a truly typical domestic space would be a mistake. Hyundai Card also believed that filling the space with numerous expensive items does not create grandeur. That was why the two-year time-frame was necessary to ensure magnificence and to refrain from crowing about it at the same time. Familiar design objects, like Enzo Mari's 'La Mela' and Dieter Rams' T1000 radio, have been casually juxtaposed with the sixth work in the 12 machine-light series that Frank Buchwald created by assembling 200 different parts, the Airborne Snotty vase made by Marcel Wanders who 3D scanned the saliva particles that spread upon sneezing, the splint made by the Eames for soldiers during World War II, Treescapes by Lee Myung-Ho, and Zaha Hadid's table. This space, in possession of over 150 design items, is truly a 21st-century version of 'The Room of Wonders, Wunderkammer' filled its rare collections.

The interior of House of the Purple was filled with the finest products, including General Electric's 18-inch horn speaker producing a sound quality evocative of the 1920s to 1940s and the Elektra Belle Epoque espresso machine, in addition to a range of artworks.

4.6 Air Lounge

Air Lounge 1
Location: Incheon Airport Terminal 1
Area: 278m²
Design: 2009.08 ~ 2009.11
Construction: 2009.12 ~ 2010.03
Interior: Gensler
Contractor: Kesson

Air Lounge 2
Location: Incheon Airport Terminal 1
Area: 284m²
Design: 2011.04 ~ 2011.06
Construction: 2011.07 ~ 2011.11
Interior: Spackman Associates
Contractor: Dawon ID&C

In place of the closed and disjointed layouts of traditional airport lounges, the space was designed as a circular structure in which users can selectively and sequentially use the services they need.

The Hyundai Card Air Lounge was planned as 'a preparatory space for travel'. While existing airport lounges are merely a space in which to pass the time, the Air Lounge differentiated itself as a space in which to tie up loose ends before traveling while browsing various facilities as if one was in an exhibition hall. The Air Lounge was the starting point for Hyundai Card's branded spaces, such as the Library series.

4.6.1 The Way We Build

The Hyundai Card Air Lounge is a space in which to prepare for travel. For those who do not yet know enough about their destinations or journeys, the space offers guidebooks such as Wallpaper's City Guidebook and Zagat. Moreover, visitors could check on and use things they may forget to bring, such as cellphone chargers, while waiting for their flight. General air lounges focus on providing rest before travel. Conversely, the Hyundai Card Air Lounge aims to provide a variety of tailor-made services according to the approach to travel of its users. Instead of the closed and rigid structure of traditional lounges, the Hyundai Card Air Lounge was designed as a circular structure where users can selectively and sequentially use the services they need. This was to maximize spatial efficiency while satisfying both leisure travelers and travelers on a tighter schedule. Air Lounge 1, opened in March 2010, was designed by Gensler. Air Lounge 2, opened in December 2011 as the number of visitors skyrocketed, was handled by Spackman Associates. The new lounge employed the more modern monochrome stylings in black and white. However, the functional aspect was not forgotten to complement Air Lounge 1 by maximizing the number of seats, while offering users unique and different experiences based on their disparate designs. Vice-Chairman Ted Chung remembers Air Lounge 2 as the most satisfactory of the projects he worked on with Spackman Associates. Air Lounge 2 maintained the existing design concept while amending it to be more expansive in scope: 'The space was truly magnificent that I was struck by the thought that the best lounge in Incheon International Airport is ours. It was a space that made the image

The Air Lounge distinguished itself as a space in which to tie up loose ends before traveling, while also offering the opportunity to browse various facilities as if you were in a gallery.

of Hyundai Card in 2011 come alive. Tom Dixon's spherical lights grab one's attention the moment one walks in, building up the excitement for the journey. I liked how it conveyed cleanliness with extravagance.' Another captivating feature of the Air Lounge was its vending machine. The idea of placing one of these machines stemmed from a newspaper ad encountered by Vice-Chairman Ted Chung on his business trip. The vending machines offered rental services or sales of various items necessary for travel, such as travel guidebooks, bags, umbrellas, mobile phone accessories, multi-adapters, or foldable travel bags. Users could draw upon these services with their credit card points. The machines also feature a three-dimensional arrangement of products by improving an old digital vending machine display's shortcomings. As such, Hyundai Card Air Lounge was at the center of public attention for its pioneering services. However, there were numerous variables to consider as the space was located inside a unique facility, an airport. The cost of maintenance was also excessive, including the airport lease fees, which forced the space to suspend operations in December 2013.

4.6.2 Point of Space

Gensler, in working on Air Lounge 1, aimed to distance the space from the monotonous spaces of municipal facilities as well as to limit the chaos of places of transit. Accordingly, it was composed with minimal design elements. The black furniture extending horizontally in the center played a key role, leading to other spaces including the information desk, general lounge, and VIP lounge. The Black Box, serving as both a piece of furniture and a wall, resembled an Innovation Trunk from the 19th century that contained travel essentials in a compact case. Walking around the Black Box, visitors used computers, ate snacks, read travel books, or shopped at the vending machine. All materials were selected to create a contemporary atmosphere while transcending a sense of the present time. With black and white as the main colors, the space is sophisticated, subtle yet bold. The space was finished with a restrained palette, meaning that travel products sold by Hyundai Card pop against their backdrop. In the relatively limited space of 250m^2, visual expansiveness was secured through a glossy wall reflecting the landscape like a mirror. The cove lighting, with its invisible light source, flows subtly along the wall to create a calm atmosphere in the lounge. The terrazzo finish and the ball chair by Eero Aarnio were elements that distinguish the Air Lounge from other typical airport lounges. The tailored FIDS (flight information display systems) extend over one side of the white wall, adding an artful character to real-time information. The VIP lounge also displays the video work 'Settle 2010, Within 2010' by the Japanese artist Hiraki Sawa. The surreal scenes, in which miniature planes fly around a generic family house, foretell the excitement at travel destinations. Spackman Associates approached Air Lounge 2 through the concept of a 'single, strong and clear volume'. This idea was visible from the bar over 18 m long in the center, overpowering the entire space. The pendant lighting by Tom Dickson, suspended from the ceiling,

captures the visitor's imagination while reflecting the surroundings with its silver exterior. The 85 lights embroidered to resemble the Milky Way in the night sky expresses the exhilaration in the heart of those about to embark upon a journey. The bar in the center had three functions: the side closest to the entrance was the reception area, and as one headed further inside, it turned into a 'Food & Beverage Service Bar', while the furthest inside was for accessing the Internet. Installed on the wall at the farthest end was the vending machine with travel essentials, which was developed for Air Lounge 1. The real-time flight information displays also contribute to the perfection of this branded space. The flight departure and arrival info was displayed in the 'Youandi' font, symbolizing Hyundai Card. The goal was to make the info look afloat in the air. Numerous experiments resulted in plastering special adhesive sheets to transparent glass onto which the video would be projected from behind. While Air Lounge 1 offers the opportunity to discover each of the divided spaces, Air Lounge 2 was intended as an open space in which the visitors could take any seat and use any utility at whim. The layout was devised to accommodate members visiting the lounge under the time pressure of travel after passing through the security checkpoints and duty-free shops. Air Lounges 1 and 2 transform the idea of the airport lounge. Evaluated as providing customer-friendly services with all spatial elements technically integrated, the space won the 2012 AIA Awards in the interiors category.

The 85 lights by Tom Dickson installed in Air Lounge 2 were rated as a spatial design capable of heightening the excitement of travel and embodying the image of Hyundai Card. Moreover, the vending machines were installed so that visitors could purchase travel essentials with ultimate convenience.

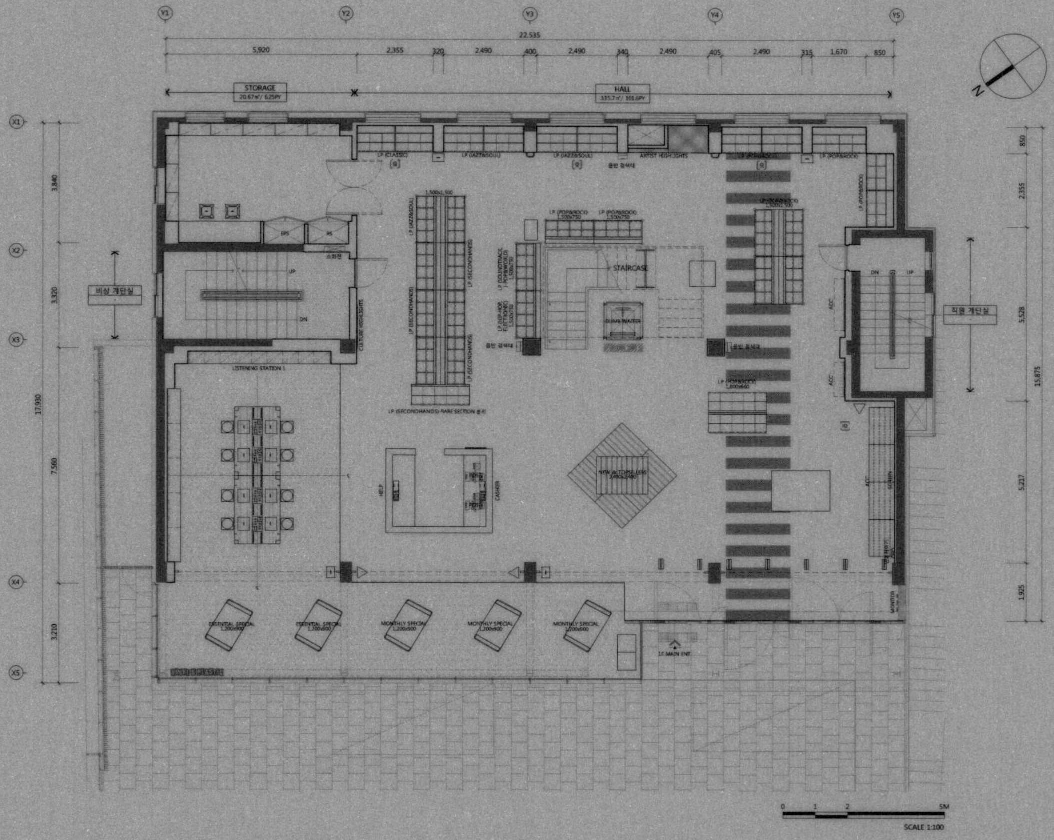

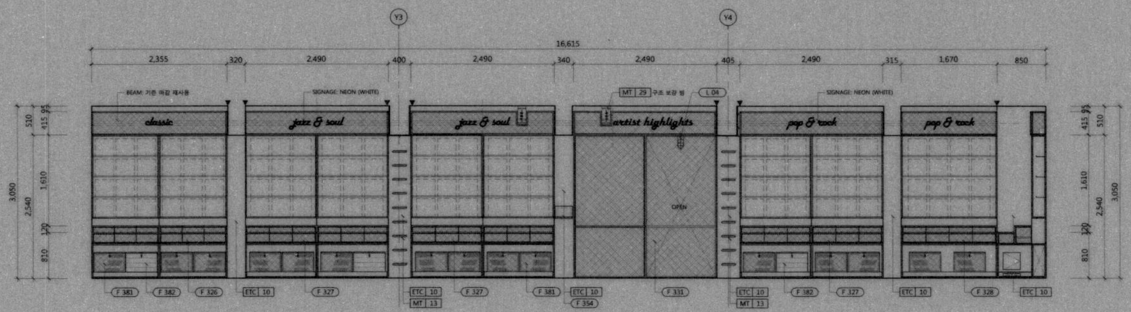

Vinyl & Plastic Floor Plan

Vinyl Display and Storage

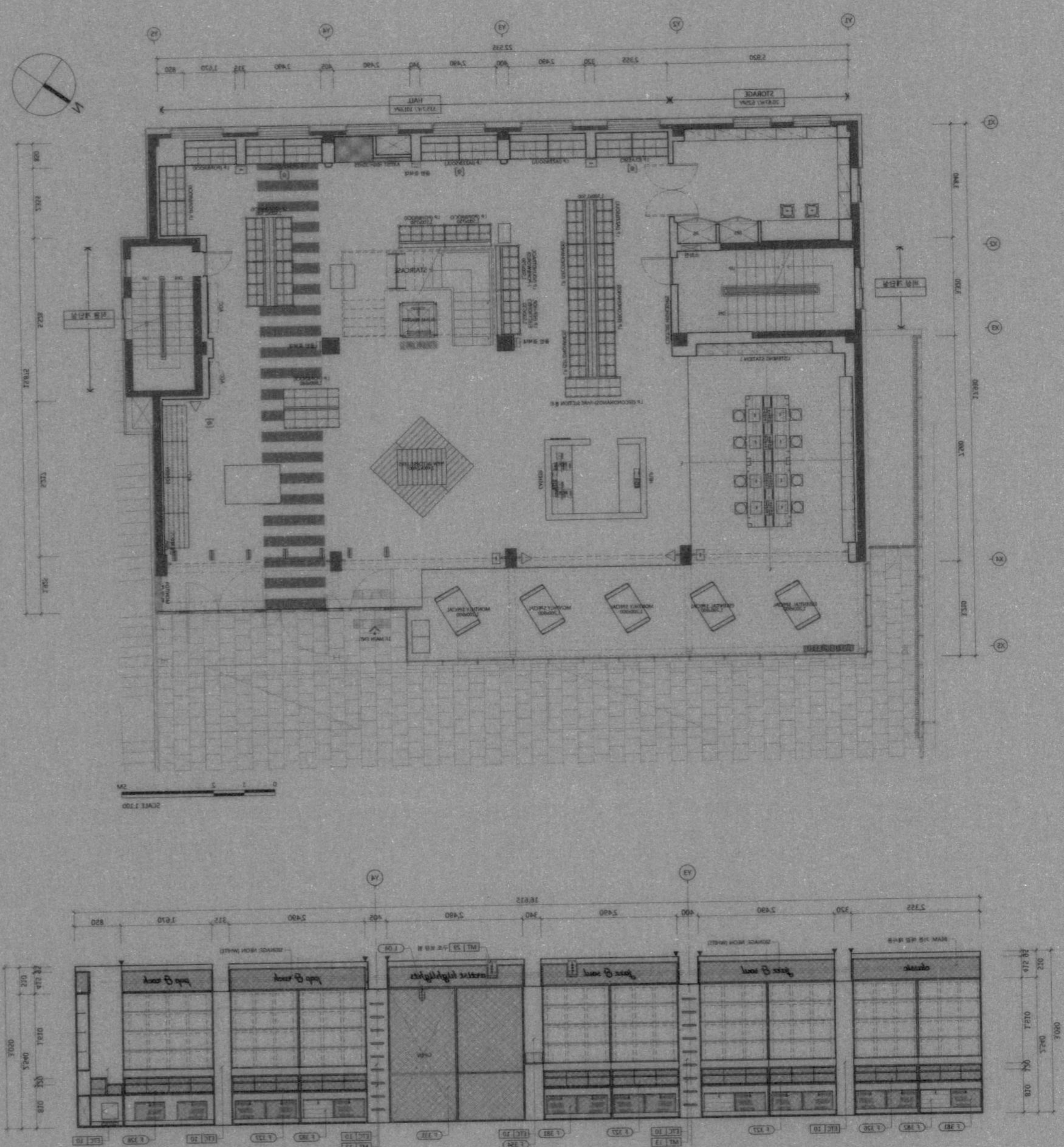

Vinyl & Plastic Floor Plan
Vinyl Display and Storage

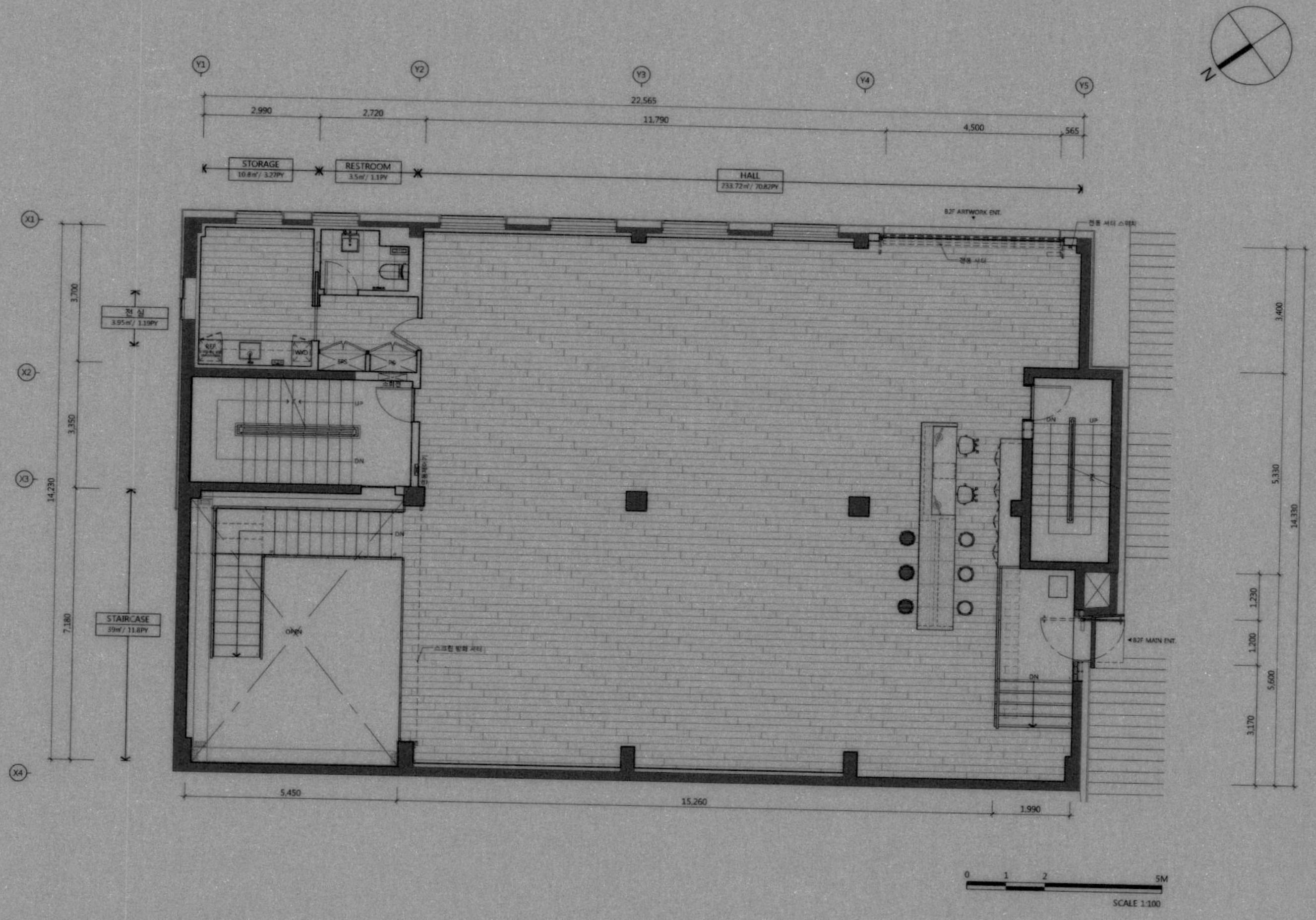

Storage B2 Floor Plan

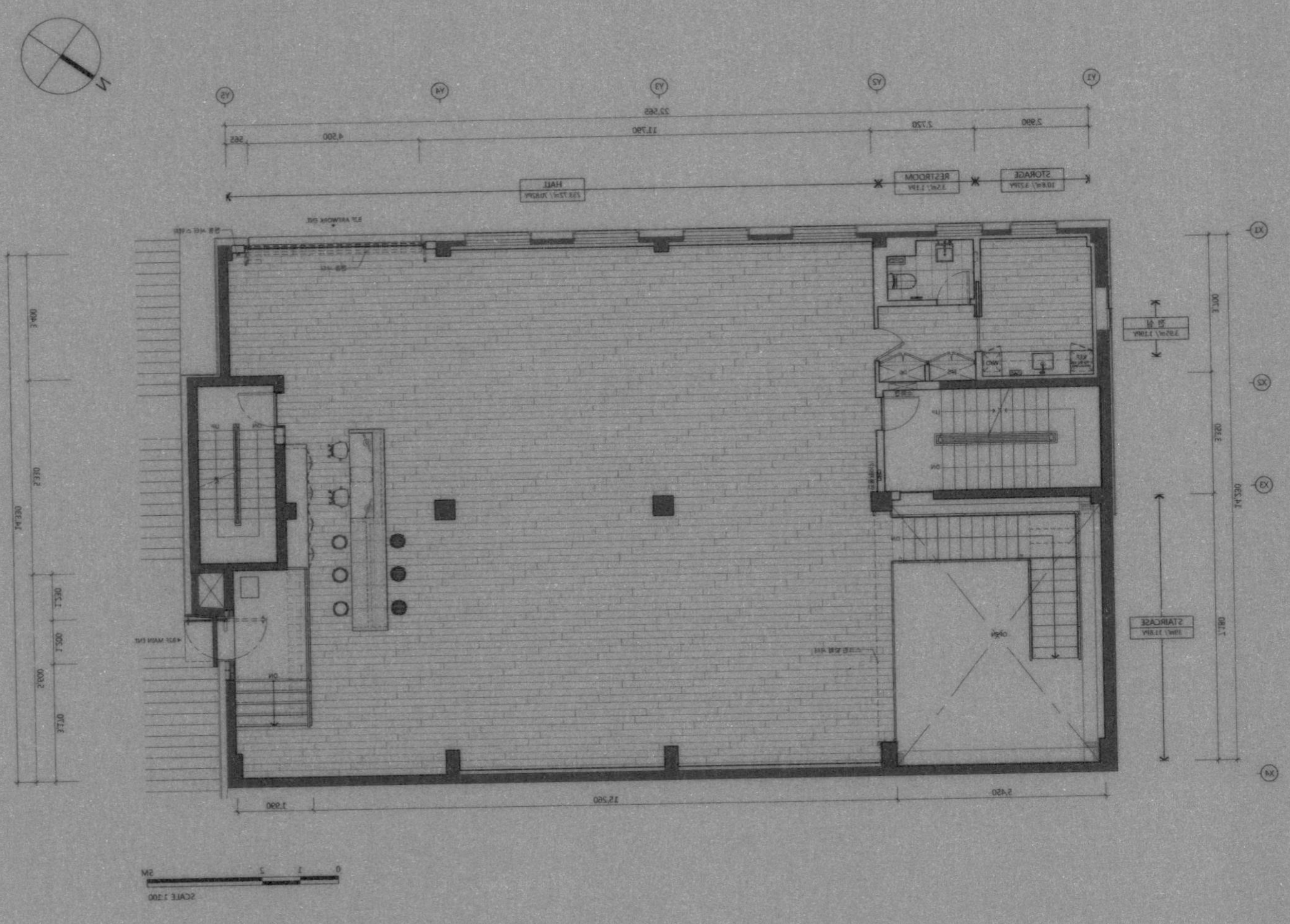

Plan/L8/L9/L11

0 1 5m

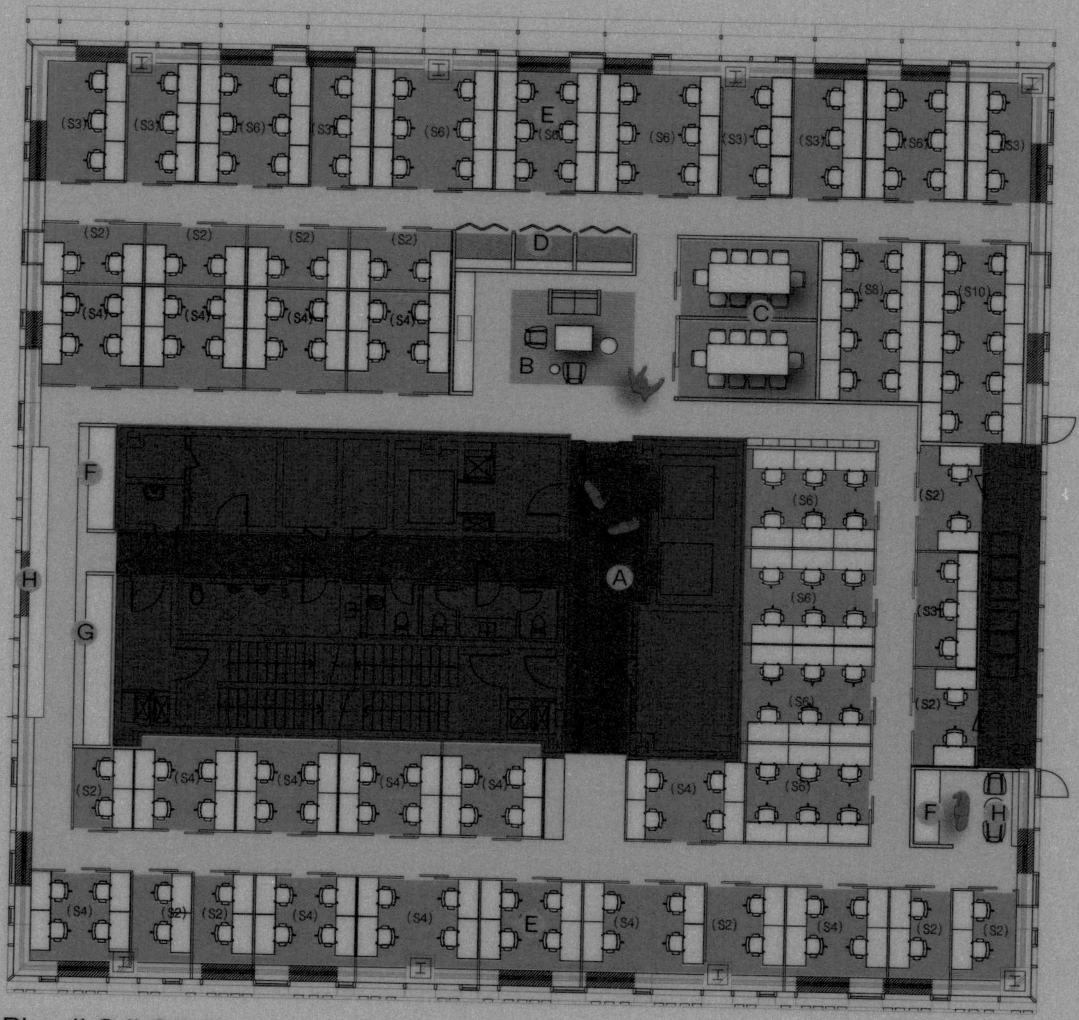

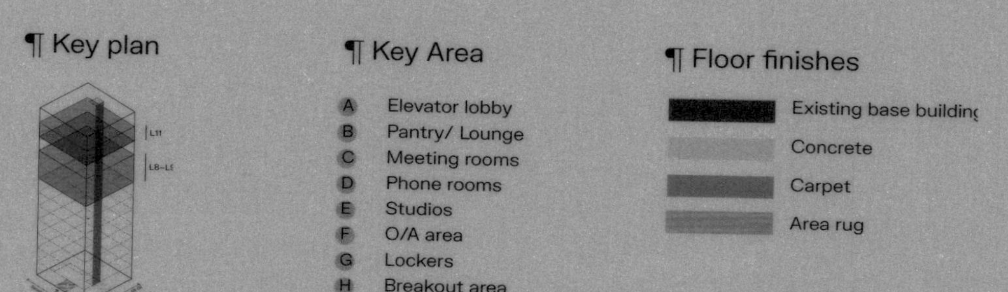
¶ Key plan
L11
L8~L11
¶ Key Area
A Elevator lobby
B Pantry/ Lounge
C Meeting rooms
D Phone rooms
E Studios
F O/A area
G Lockers
H Breakout area
¶ Floor finishes
Existing base building
Concrete
Carpet
Area rug

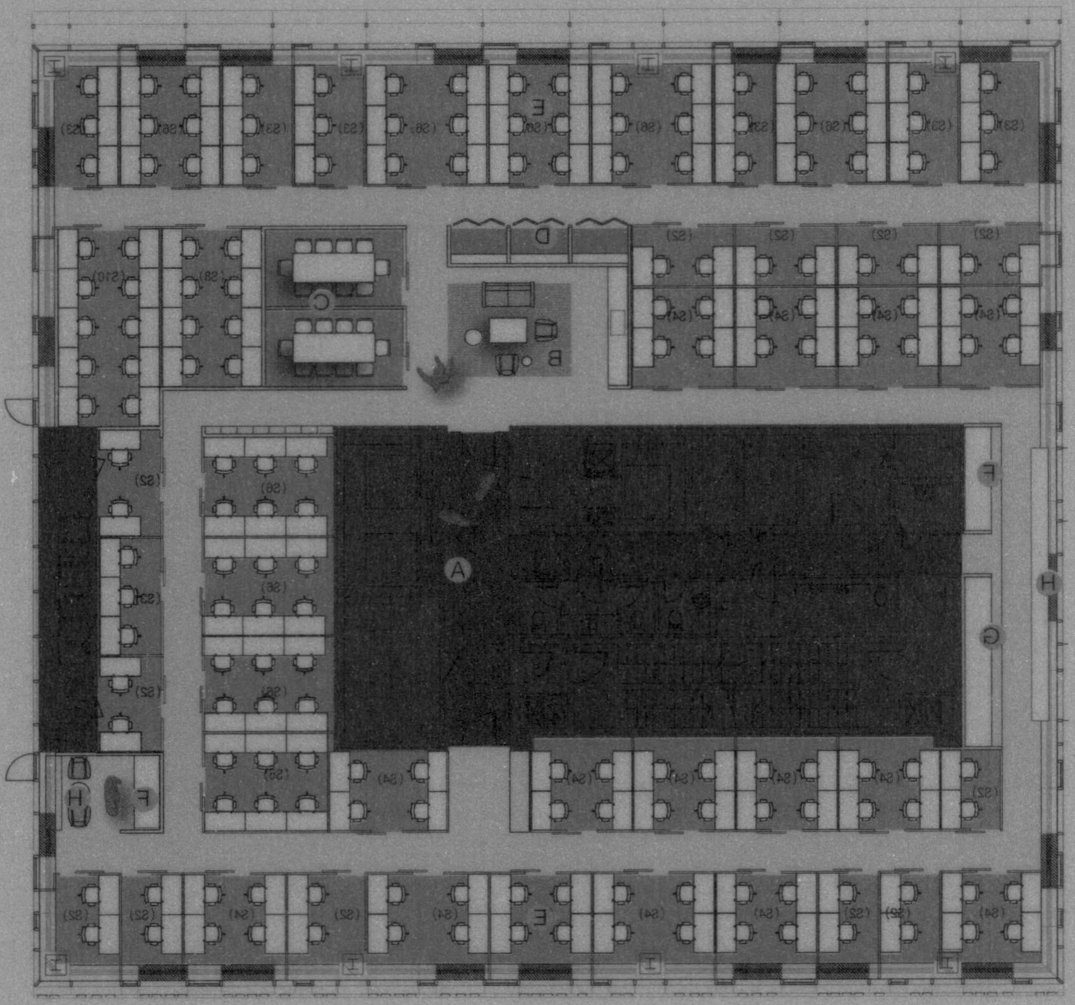

Plan/L8/L9/L11

¶ Key plan

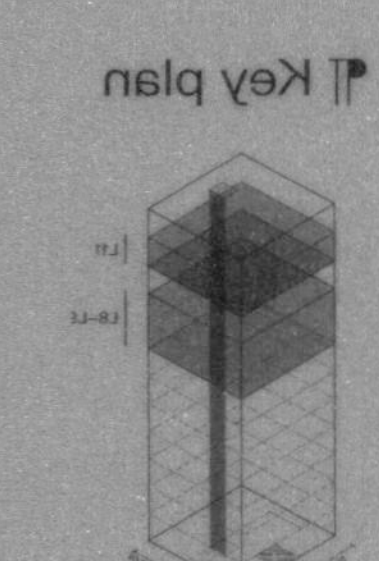

¶ Key Area

- A Elevator lobby
- B Pantry/ Lounge
- C Meeting rooms
- D Phone rooms
- E Studios
- F O/A area
- G Lockers
- H Breakout area

¶ Floor finishes

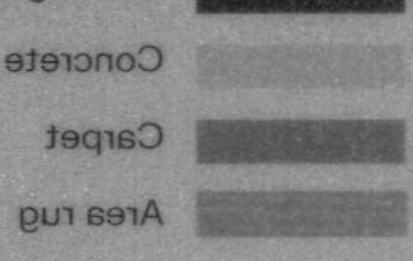

Castle of Skywalkers Sectional Model

Castle of Skywalkers Sectional Model

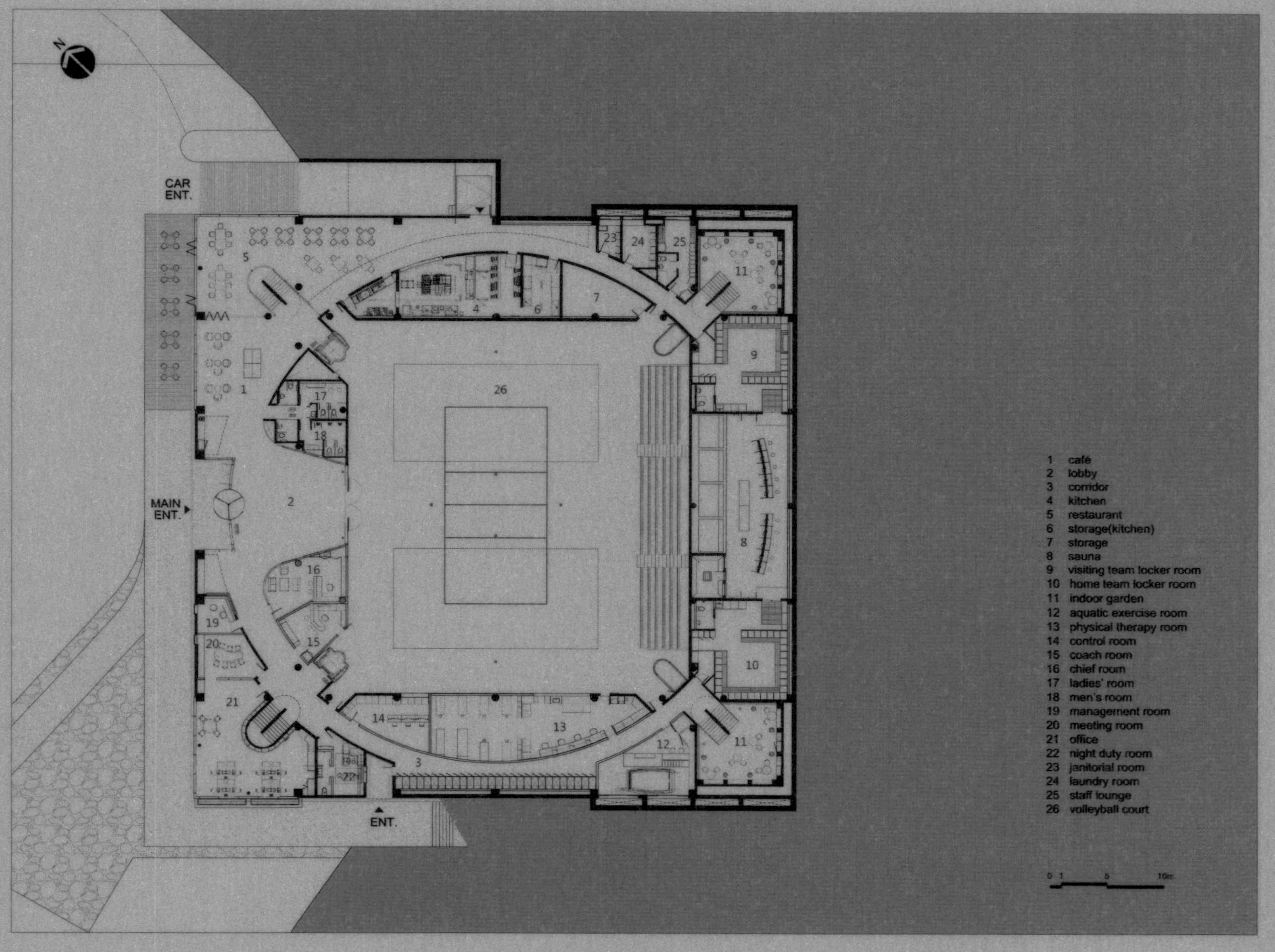
CAR ENT.
MAIN ENT.
ENT.
1 café
2 lobby
3 corridor
4 kitchen
5 restaurant
6 storage(kitchen)
7 storage
8 sauna
9 visiting team locker room
10 home team locker room
11 indoor garden
12 aquatic exercise room
13 physical therapy room
14 control room
15 coach room
16 chief room
17 ladies' room
18 men's room
19 management room
20 meeting room
21 office
22 night duty room
23 janitorial room
24 laundry room
25 staff lounge
26 volleyball court

Castle of Skywalkers 1F Floor Plan

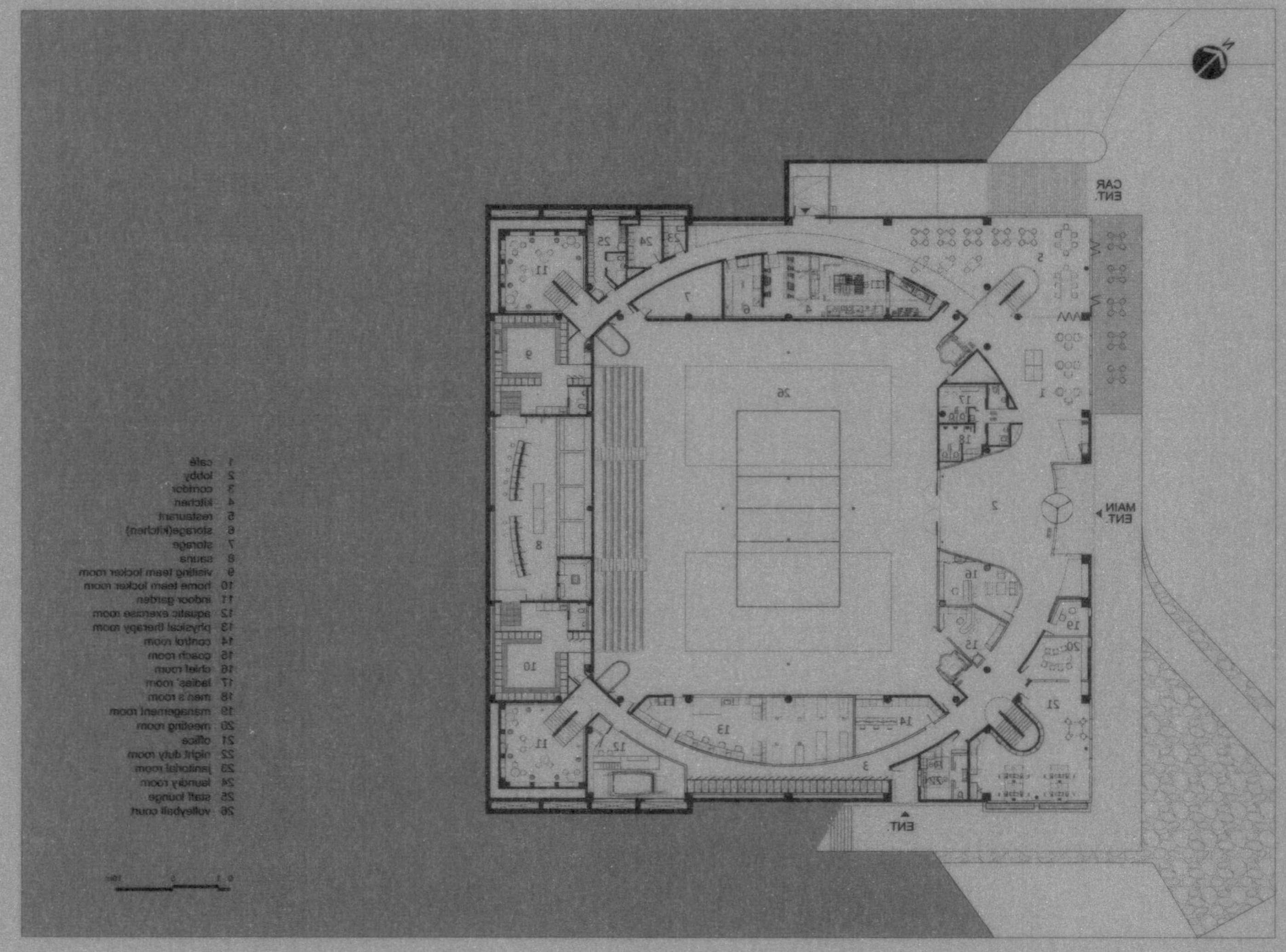

Castle of Skywalkers 1F Floor Plan

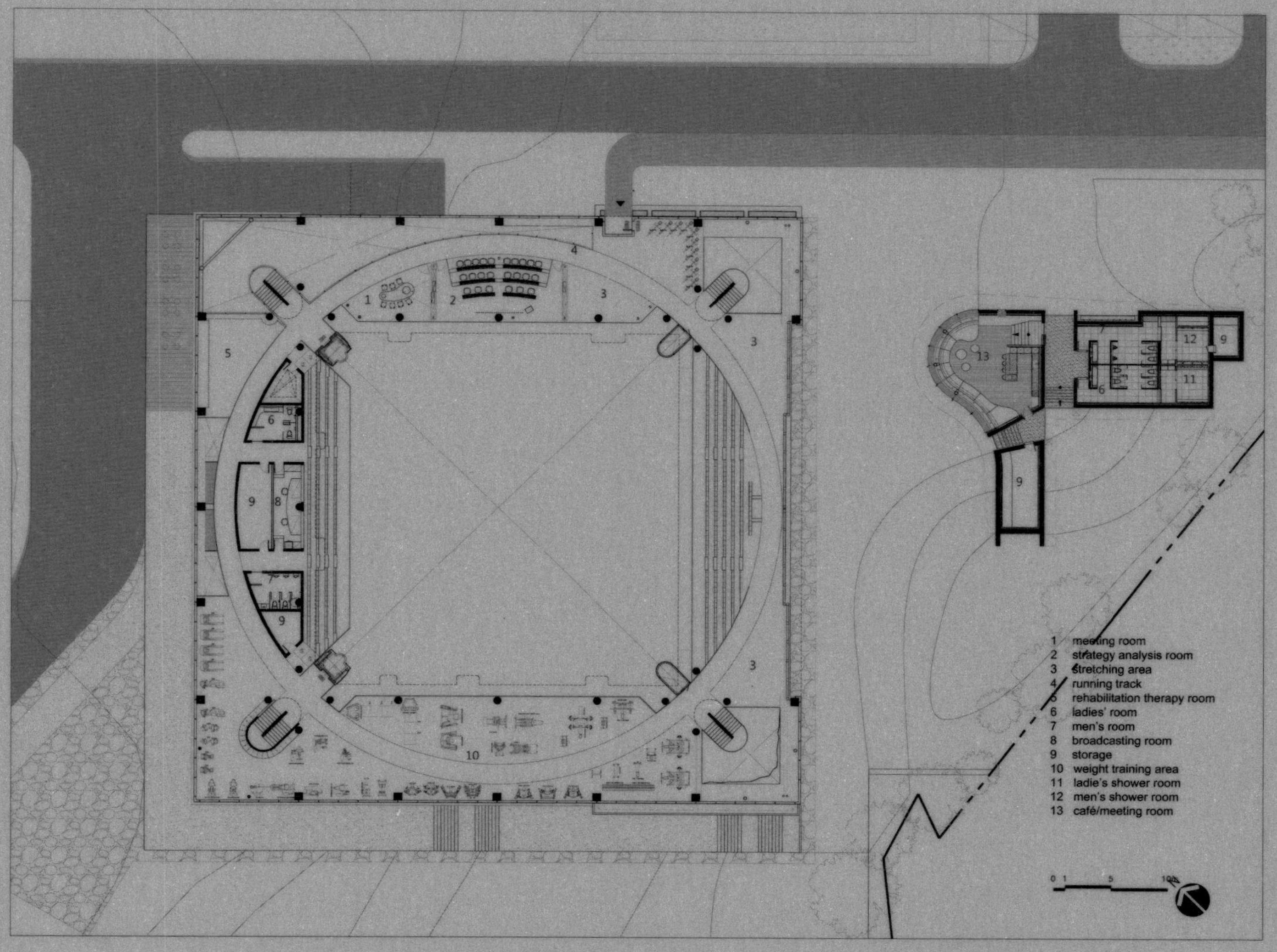

Castle of Skywalkers 2F Floor Plan

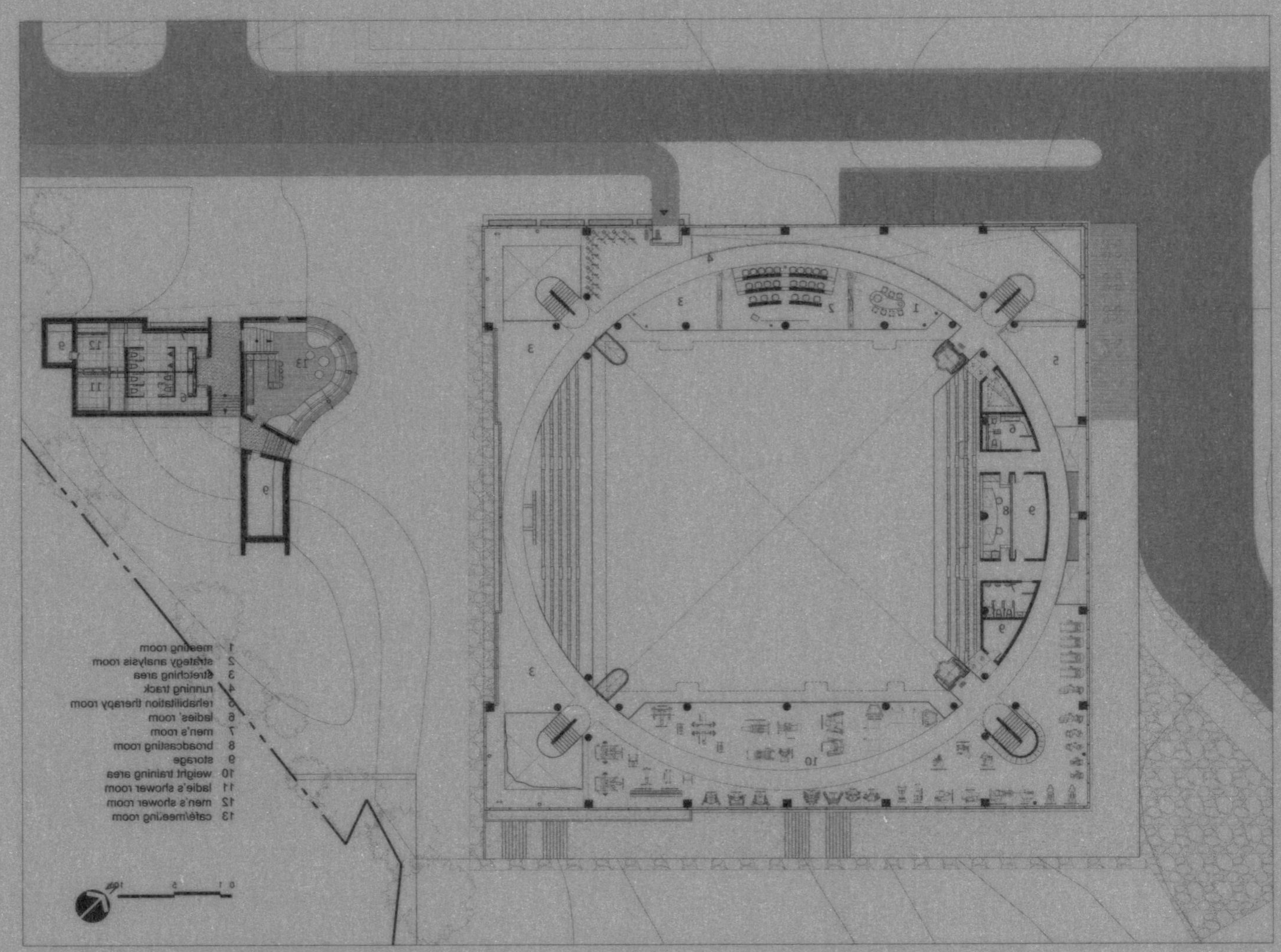

Castle of Skywalkers 2F Floor Plan

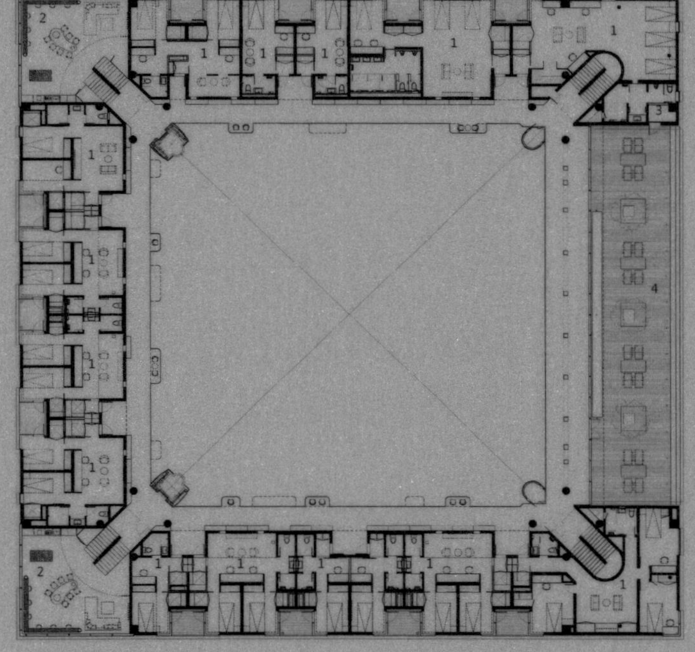

1 living quarters
2 lounge
3 storage
4 outdoor deck

0 1 5 10m

Castle of Skywalkers 3F Floor Plan

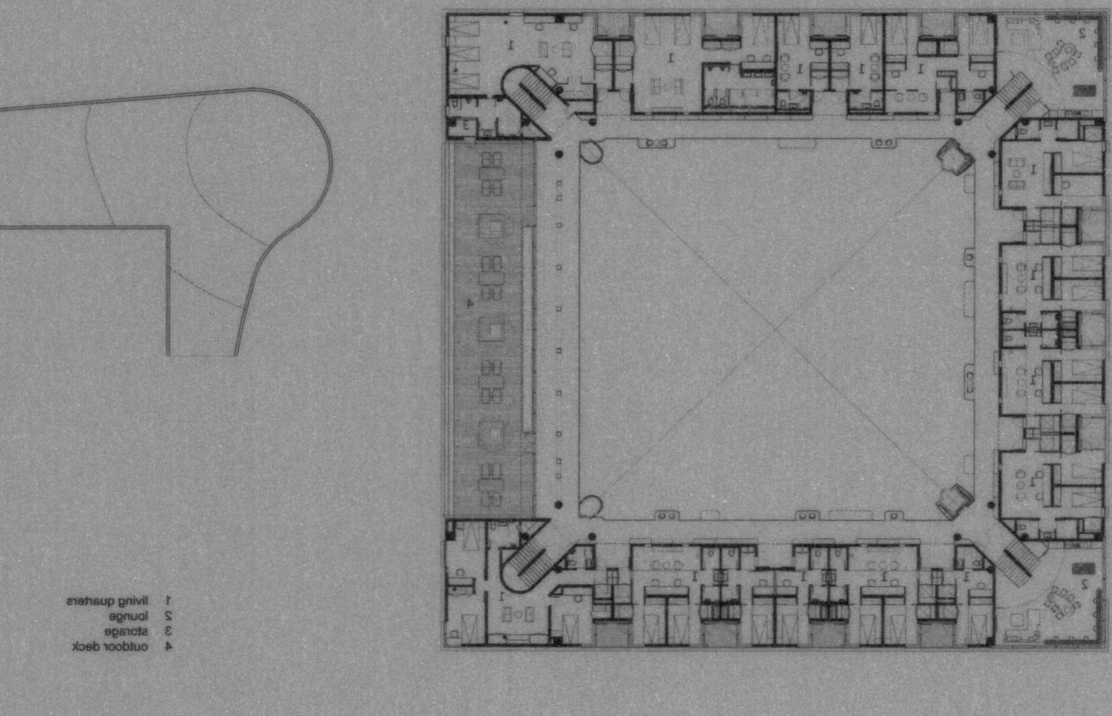

Castle of Skywalkers 3F Floor Plan

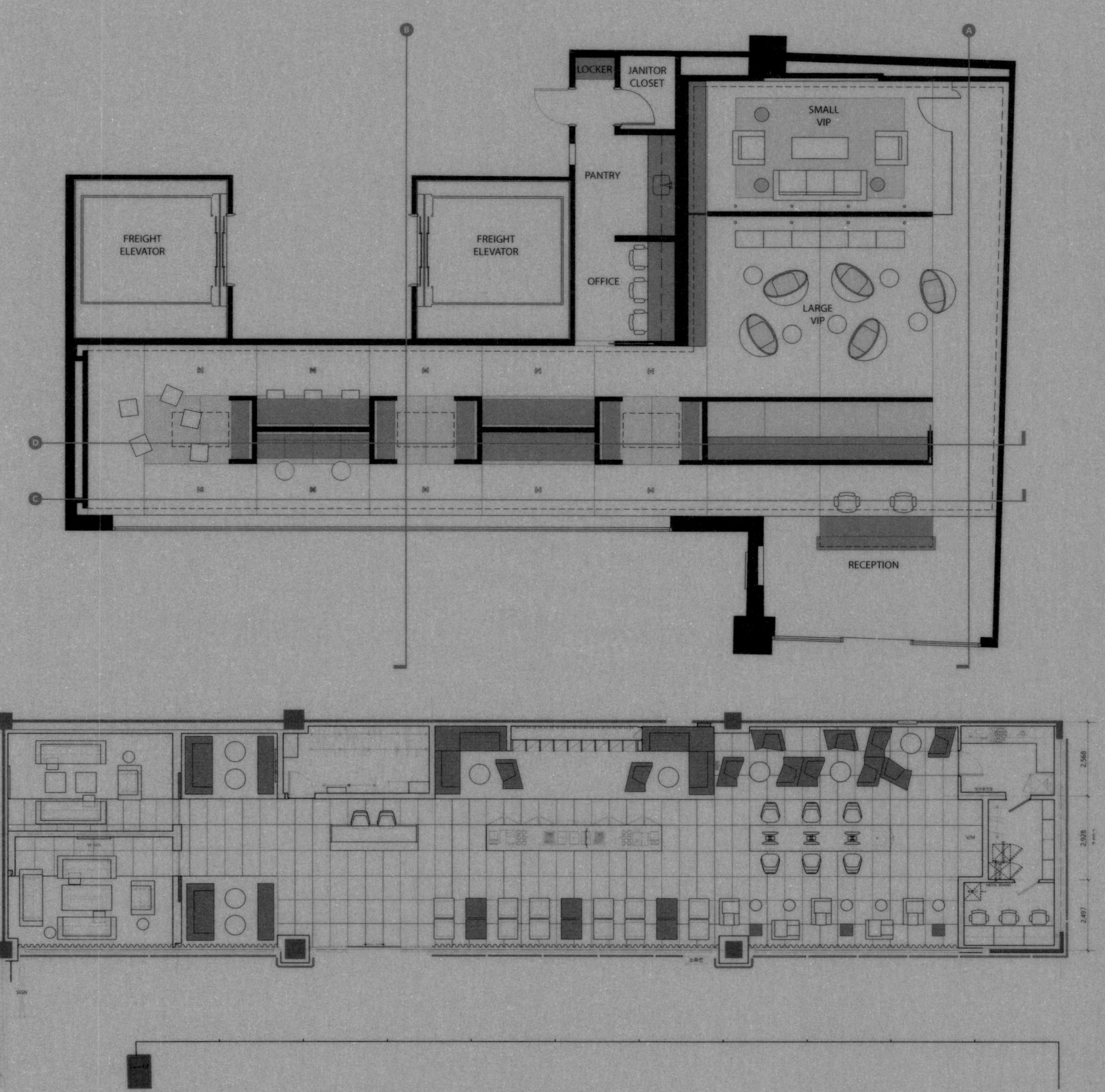

Air Lounge 1 Floor Plan
Air Lounge 2 Floor Plan

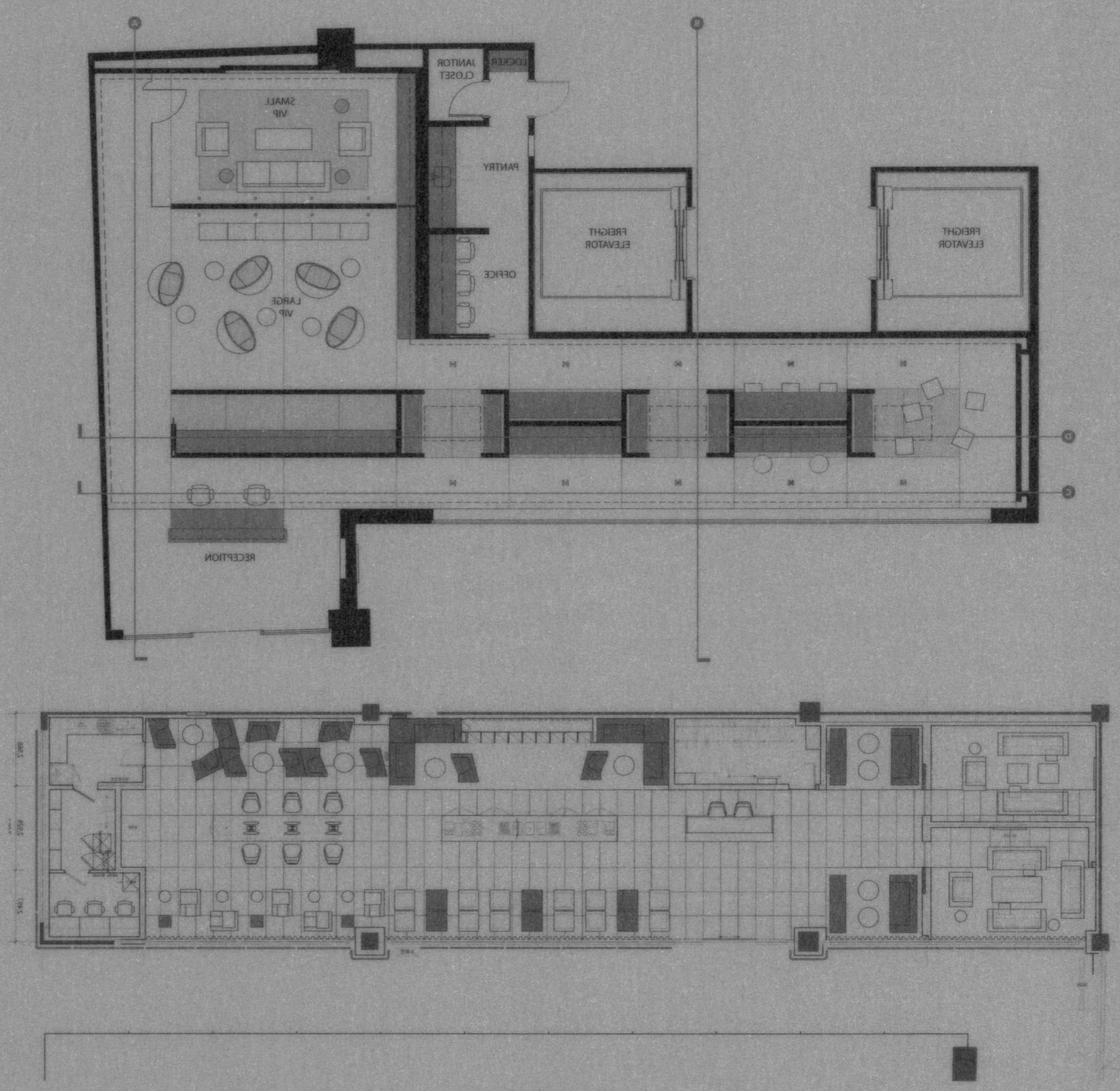

Air Lounge 1 Floor Plan
Air Lounge 2 Floor Plan

Regeneration
Projects

Hyundai
Card's re-
generation
initiatives in
line with

Corporate Social Respons-ibilities

Symbolic of success in local self-sufficiency, the Gapado Project involved planning an entire island. Bongpyeong Market in Gangwon-do and 1913 Songjeong Station Market in Gwangju were the projects overseen by Hyundai Card to protect the declining traditional marketplaces, making a notable contribution to the local community. The Gapado project was initiated by Hyundai Card, Jeju Self-Governing Province, and 101 Architects with the governing aim of repainting the reality of Gapado under a new paradigm. Under the theme 'Sustainable Island: Gapado', Hyundai Card made significant investments in the project over six years driven by the goals of recovery and conservation of natural ecosystems, the establishment of an autonomous economic cycle, and the coexistence of regions and cultures. The project has been praised as a self-sustaining regional development project worthy of iconic status in the history of Korean public projects. Meanwhile, the worldwide decline

of traditional regional markets is undeniable in competition with the accelerating growth and influence of department stores, large retail stores, and online marketplaces. However, there is a breed of traditional market that has evolved into a powerful branded space, one capable of preserving the tradition and culture of a region while maintaining its original role as a refrigerator and restaurant for the local community. The best examples are Bongpyeong Market, on which Hyundai Card and Hyundai Capital collaborated with Gangwon-do, and the Gwangju 1913 Songjeong Station Market, which was completed in collaboration with the Gwangju Center for Creative Economy & Innovation. Hyundai Card believes that the value of corporate social contribution depends on the issues attended to and ultimately solved by the company, not the scale of its investment or donation.

Gapado Island Project

p. 352

Instead of building anew from the ground up, every architectural structure in Gapado was renovated to its maximum possible extent guided by the fundamental design principle of using and regenerating existing buildings. To create an architecture that retains some sense of the rustic landscape so particular to Gapado, the architecture has excluded disruptive elements to the original terrain and landscape.

5.1

Bong-pyeong Market

p. 370

For the sustainable design of Bongpyeong Market, Hyundai Card contemplated devices that would produce sufficient effects without new constructions: VMD via tents with colorful identities, mini signs for each store, manuals on how to devise attractive stores.

5.2

1913 Songjeong Station Market

p. 378

As the result of considering how best to enhance the distinctive characters and competitiveness of individual stores while also highlighting the historic importance of the old Songjeong Station Maeil Market, a retro mood was proposed to communicate the vitality of the market in its prime era of the 1970s and 1980s. Another key selling point was the newly created signboards that embraced the history of these commercial sites.

5.3

5.1 Gapado Island Project

Gapado
Location: Gapa-ri, Daejeong-eup, Seogwipo-si, Jeju-do
Area: 339m²(Gapado Terminal),
563m²(Fishery Center),
965m²(Artist In Residence),
189m²(Town Hall),
84m²(Snack Bar, Diver Locker),
84m²(House A, B),
60m²(House C),
51m²(House D),
154m²(House E, F, Kitchen)
Masterplan: 2013.06 ~ 2015.09
Design: 2015.10 ~ 2016.11
Completion: 2018
Architecture: 101 Architects, Wook Choi
Identity: Hyundai Card Design Lab

Only the terminal and fishery center with a simple structure was newly built with concrete, and the rest are works of regeneration.

The Gapado Project is a socially engaged project that began with the will to break away from new-construction-centered regional development and to reorient efforts towards the determination to create a new paradigm for 'self-sufficiency' that draws on nature as its key asset. Hyundai Card has long contemplated the possibilities and shown commitment to establishing a sustainable ecosystem on the island by pouring all of its expertise into the regeneration, both structurally and socially, of Gapado Island.

5.1.1 The Way We Build

Nestled in the southwest of Jeju Island, Gapado means 'an island of many waves'. It has a low, flat topography with an elevation of only 20 m above sea level, as well as beautiful green barley fields that stretch to the horizon. Of around 170 inhabitants, over 90% work in the fishing industry. Despite its unspoiled and beautiful natural landscape, Gapado in April and May every year is a place to which over 60,000 tourists flock to attend the Green Barley Festival. This has given rise to several serious threats to the native ecosystem, as makeshift and temporary facilities have sprung up all over the area and the local commercial district has been unsettled by this influx. In light of this, and with Gapado's unique landscape in mind, Hyundai Card decided that the best approach to the redevelopment would be to shift the paradigm capable of promoting regional self-sufficiency based on the beauty of the island's nature. The project and its resulting interventions could not simply be stopgaps or a facility-oriented development. As such, new structures were built only when absolutely necessary, and the guiding principle was to regenerate and reuse abandoned structures and old houses. When the Gapado project was set in motion in 2012, a dedicated in-house team was formed which frequented sites in Seoul, Jeju, and Gapado, and discussed the project's direction with Jeju Self-Governing Province and 101 Architects. To conduct the necessary case studies and research at this early phase of the project, the team visited about 20 places worldwide including Berlin, Amsterdam, Stuttgart, Aomori, London, and Naoshima, to examine cases of regional regeneration. It took four years just to complete the lengthy administrative procedures, including numerous surveys, workshops for residents, public hearings, and meetings involving Jeju Island's Governor Heeryong Won. Throughout the process, Hyundai Card built trust by navigating the opinions of Gapado residents and cultivating a better understanding of their lives. 101 Architects also proposed a basic plan and invested a great deal of effort in these early stages, as evidenced by the book, Gapado, which collates research on Gapado before the realization of the project. As such, Hyundai Card and 101 Architects were meticulous in their analysis of Gapado from the perspective of ecology, the humanities, and the economy to finally settle on their most important aim: a sustainable island. The dream was to create a self-sustaining ecosystem based on more nuanced exchanges between nature and residents, and not an island that would merely be consumed by tourists. The most crucial tests in Gapado were the preservation and maintenance of the natural realm, the coexistence of the local community and its cultures, and the establishment of a sustainable economic ecosystem for the islanders. Considering that the island is only 20 m above sea level, new constructions were avoided and the recycling of

Hyundai Card refrained from building brand-new facilities to preserve the unique character of Gapado. To this end, new constructions were minimized, and the regeneration of abandoned structures become the top priority when creating facilities such as snack bars and accommodation.

"The world-renowned fame and scale of 101 Architects did not prevent them from investing passion and appreciable efforts into a small and local project such as Gapado. I truly appreciate their level of enthusiasm for the enterprise. I am also grateful to the employees of Hyundai Card who clearly fostered a deep love for Gapado. However, the project raised the most questions and doubts of any conducted by Hyundai Card. For whom and for what have we dedicated the last six years of precious time and effort? The good intentions of the company were not enough. Of course, evaluating the success of any project in an arbitrary way is problematic, especially when numerous interest groups are involved, from a corporation, to local government, to the local neighborhood. The Gapado project, however, was the largest of Hyundai Card's architectural and space projects, and it resulted in a great deal of disappointment notwithstanding the scale of the efforts invested."

Ted Chung, Vice-Chairman of Hyundai Card · Hyundai Capital · Hyundai Commercial

abandoned houses was favored. Another aim was to make Gapado a place to visit year-round, outside of the popular Green Barley Festival season. These goals became the background to which accommodations, a passenger ticket office, a snack bar, and restaurants were built, primarily to lengthen the period of a visitor's stay. Moreover, an Artist in Residence program was introduced to enrich the cultural value of Gapado. The Artist in Residence program encourages Korean and international artists, writers, and humanities scholars to immerse themselves in all sorts of creative activities. The program laid the groundwork for developing the local community via an influx of members of the younger generation and a millennial culture. Restaurants and snack bars that sell locally produced agricultural and fishery products are also operated by the residents of Gapado, contemplating a more virtuous economic cycle whereby the profits would be returned to the community. In addition to the architecture, a range of other aspects such as the natural cycles beyond human control, the needs of the residents, and tourist attractions were to be considered. With its opening ceremony on 12 April 2018, the Gapado project leaves behind an archive of a wide range of documentation based on four books created for preliminary research by 101 Architects and an exhibition held at the Storage. Moreover, the project was introduced in a special edition of Domus (vol. 1302, Feb. 2019), and became widely recognized by the international architecture community. As the project is a rare and distinctive public project, numerous local government officials in Asia, as well as academics from the University of Lausanne in Switzerland and Columbia University in the U.S., have visited the island to conduct on-site surveys. Gapado was planned as a self-sustaining regional development project and has taken on momentous significance for the history of public projects in Korea. Nevertheless, Gapado posed significant roadblocks in determining the scope and coordinates for a corporation committed to contributing to the local community. These roadblocks required critical examination of the efforts made by participants, points that succeeded or were overlooked, and problems that prompted failures in ambition. The sustainability of the Gapado project now depends on the extent to which the local residents and administers are cognizant of the contradictions between their reality and their willingness to solve the ecological and economic issues the island faces.

The Artists in Residence building, the most prominent structure in Gapado, was designed to support a program that would enrich the cultural value of Gapado and attract members of the younger generations.

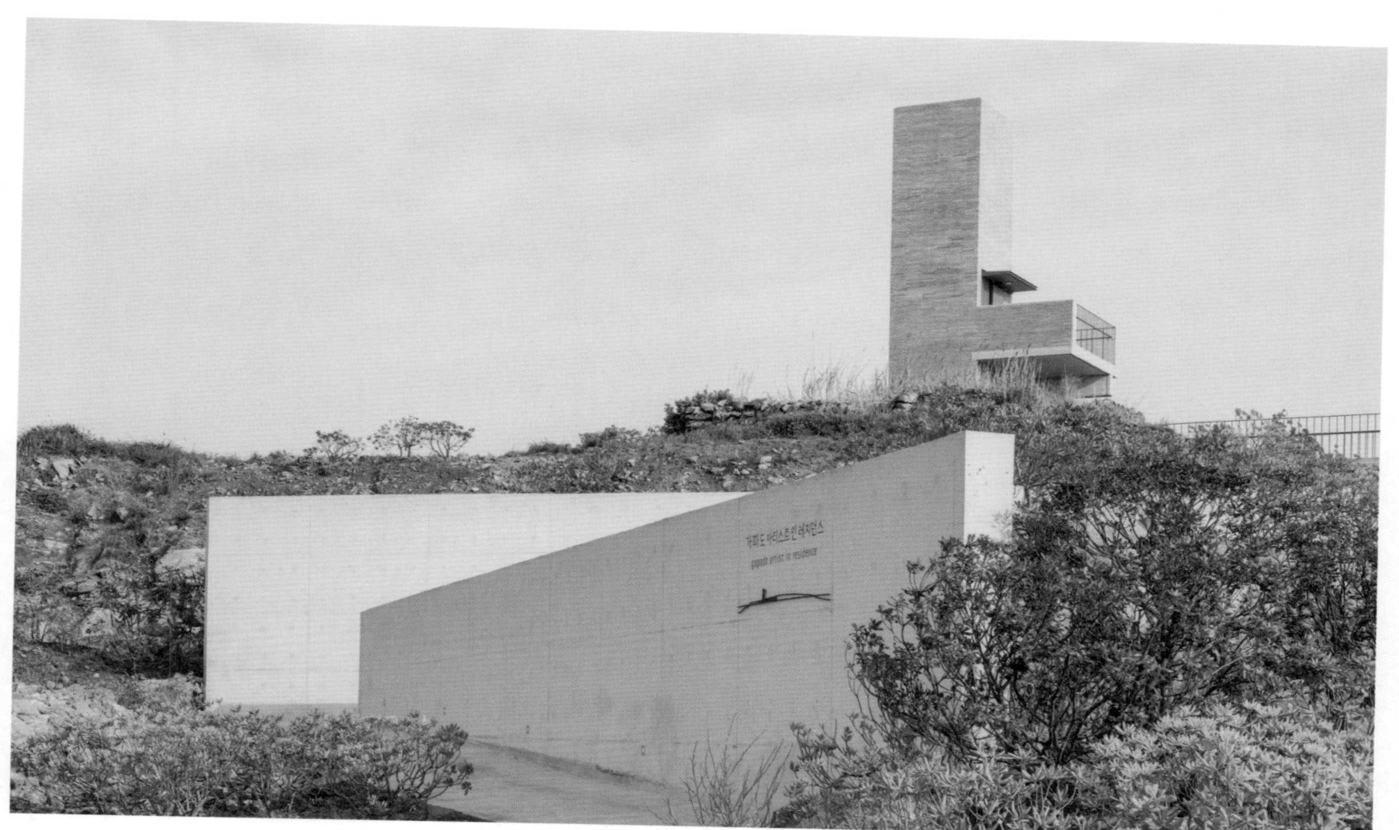
가파도 아티스트 인 레지던스
gapado artist in residence

5.1.2 Point of Space

Every architectural structure in Gapado was recycled to the highest standard under the guiding principle of the regeneration of existing structures. The building of brand-new structures was shunned. By adopting this approach, unsightly abandoned houses disappeared. When there was no other option than to erect a new structure, the rule was that the low, flat terrain unique to Gapado must not be disturbed. As a result, only the terminal and fishery center with its simple new structure was built using concrete, while the rest of those touched by the project were works of regeneration. Artists in Residence, the most prominent building in Gapado, withstood the IMF economic crisis 20 years ago. Most of the building is underground, with a little sitting above ground, and is a renovation of this existing structure so as to avoid any undue interventions within the landscape of the island. For new constructions and renovations, simple processes and materials were used, bearing both the harsh natural environment and the residents' ability to maintain the structures in the coming years in mind. As Wook Choi, Head Architect of 101 Architects, explained, 'Following contemplation of what suits Gapado, we concluded to make the building invisible to preserve the simplicity of the landscape. That is, nature would be incorporated into the architecture and the architecture would eliminate any elements that were at odds with the original terrain and landscape.' Accordingly, the process of dismantling and

"After the Gapado project began, the residents held individual expectations which varied greatly. As such, we had to communicate with each of the 170 residents to navigate their opinions and draw them into alignment. To set the Gapado project in motion, we worked with the Jeju Self-Governing Province to establish a task force team (TFT) and made ordinances, trying to establish the systems. Neither side can unilaterally make decisions on a local project that touches the lives of islanders like in the Gapado project. What we wanted to convey to the residents was that our approach offered an alternative way of life, not that it would improve their lives. What Hyundai Card did in the Gapado project was to help islanders act of their own volition."

Yong Namgoong, Hyundai Card Brand Planning Team - Gapado Project Management

rearranging makeshift facilities and neglected structures was more important than the construction of new ones. Gapado House, which recruited an empty house, was carefully adjusted – even its entrance – not to interfere with the lives of the residents. It is also not in conflict with the existing housing style of the village. When creating and arranging new structures and artifacts, the horizontal landscape unique to Gapado was protected. Part of the coastal road, which was covered in cement, was removed to restore native vegetation. Moreover, while considering the direction and color of the stone wall, 101 Architects wanted to maintain the landscape, not its buildings. In addition to architecture, Hyundai Card Design Lab was responsible for signage and graphic design, including the Gapado logo. The Gapado logo, consisting of two fine lines drawn simply, reflects the topography of Gapado as is. It could easily be drawn by anyone. Hyundai Card won the Main Prize in the regional branding category at 2019 iF Design Award for the Gapado project.

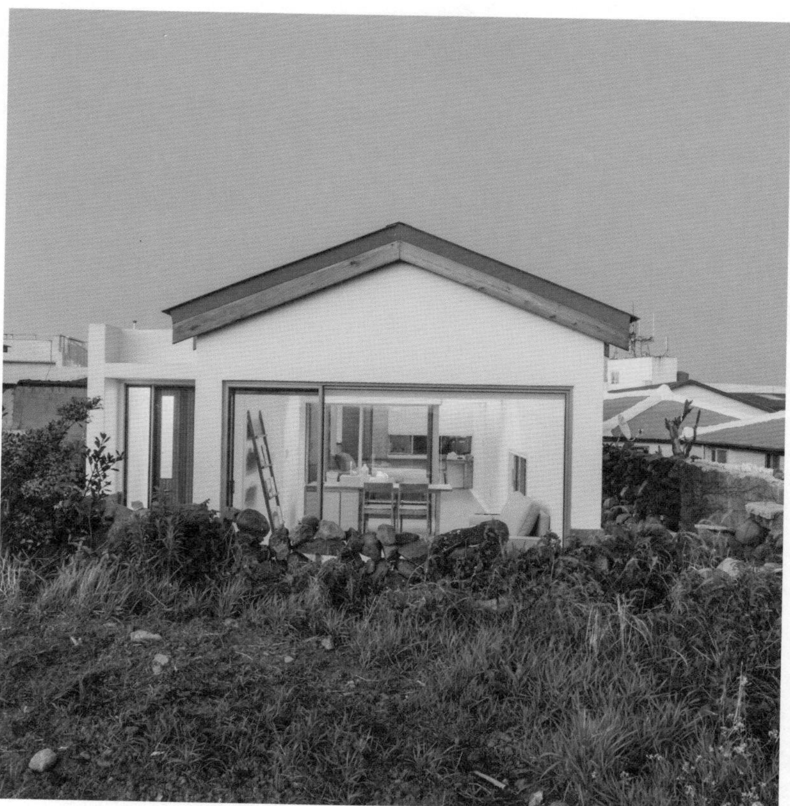

Artists in Residence is a building remodeled from a structure that managed to withstand the IMF economic crisis. Its form buries deep into the ground to live in harmony with the island landscape while optimizing the use of the existing structure. Gapado House is also a renovated old building, reconstructed without disturbing the prevailing character of the village.

가 ga 파 pa 도 do

"The Gapado residents once came to our office. At that time, the chairperson of the female divers (haenyeo) association asked if I knew whether there is a spring, summer, autumn, or winter out at sea. When I heard that, I felt the need to understand them better. To build architecture is to make interventions in the lives of others. To refrain from interfering too deeply in their lives, we ensured the number and scale of architectural projects and the nature of the materials were subdued. Adding a new element was less important than demolishing and reorganizing existing structures. In doing so, we made every effort to leave the landscape intact. Of course, many things had to be put off and endured, such as various restrictions and administrative procedures unique to public projects. Yet, Gapado was something that had to be done for the next generation."

Wook Choi, Head Architect of 101 Architecture – Gapado Project Architecture

The Fishery Center & Restaurant is the backbone of Gapado's economy. It has been newly renovated as a space for fishermen and female divers (haenyeo) to perform necessary chores such as cleaning nets. There is also a restaurant where tourists can taste the seafood dishes typical of Gapado, as well as the Haenyeo Grill where female divers cook seafood.

"Transition was part of the Gapado design from the beginning. For example, the logo can be drawn by anyone thanks to its two simple lines, allowing the residents to make it their own. That is why the efficiency of communication and the seeking of understanding from the islanders was more important than the eventual design outcome. For instance, we explained how new packaging would contribute to sales by making the seaweed portable and presentable instead of simply beautifying the package design. The packaging was described as 'great for wrapping the seaweed for sale, for transportation, and for giving as a gift'. For the logo, we also clarified that 'there is a line, but as there are no strict guidelines, you can do whatever you want. You can color it or write your store's name under it'. It was not about whether one followed the guidelines or not—it was about having the greater fixity of purpose for the option we suggested. Then, in no time, the islanders understood the value of the logo and how best to use it."

Noah Shin, Hyundai Card Design Team 2 – Gapado Project Designer

5.2 Bongpyeong Market

Bongpyeong Market
Location: 14-1, Dongijangteo-gil, Bongpyeong-myeon, Pyeongchang-gun, Gangwon-do
Site Area: 11,697m^2
Store Area: 3,000m^2
Number of Stores: 100
Design & Construction: 2013.03 ~ 2014.04
Completion: 2014
Identity: Hyundai Card Design Lab

The most visually striking thing in Bongpyeong Market is a colorful temperament that guides visitors through a wide range of products by appointing different tent colors for each product group. In addition, various signboards were introduced and upgraded to facilitate customer communications.

Bongpyeong Market is a project on which Hyundai Card and Hyundai Capital collaborated with Gangwon-do as part of the 'traditional market revitalization project'. Of the many markets, Bongpyeong Market was selected because it was the most appropriate local market to serve as a prototype for use thereafter throughout Gangwon-do. The goal was not to create one thriving market but to create a model that would inspire other traditional markets.

5.2.1 The Way We Build

Bongpyeong Market in Gangwon-do, with its 400-year history, was once well-known as one of the largest traditional markets in Korea. Even today, there is a daily market opening at Bongpyeong Market. On the second and seventh of the month, when the five-day market is on, about 100 sellers flock into the marketplace. Bongpyeong Market's fame can be partly attributed to the symbolic Korean short story, When Buckwheat Flowers Bloom, by Hyo-Seok Lee. When the buckwheat flower season arrives, the Hyo-Seok Lee Literary Museum, which holds a literary festival, is packed with tourists. On the contrary, located only 100 m away from the Museum, Bongpyeong Market was almost empty. It exemplifies how a traditional market can be forsaken for failing to keep up with the rapidly changing times. This was an issue commonly found in many traditional markets across the nation, not only for Bongpyeong Market. In response, the Gangwon Provincial Government, which was contemplating ways of revitalizing the traditional market, approached Hyundai Card for help. After a year of preparations, Bongpyeong Market opened its new doors in April 2014 as a result of the collaboration between Hyundai Card and the Gangwon Provincial Government. Of over 57 markets in the region, Bongpyeong Market was selected because it was the best case to serve as a prototype for use thereafter throughout Gangwon-do. Another charming point of interest was that the market possessed a few modernized arcades and convenience facilities. That is, the form of a traditional market was preserved in relatively full fashion. As such, the value that Hyundai Card emphasized was the 'change to protect'. Instead of modernization, of turning the space into cutting-edge facilities or building anew, Hyundai Card has improved dilapidated facilities while highlighting the old market's unique qualities. Hyundai Card considered it important to establish the market as a place of greater joy despite the inconvenience of certain facilities, elevating the market beyond a place to procure a meal or purchase food. Surely, convincing merchants who had long-established ways of doing business was not an easy task. There was, at first, a backlash. However, Hyundai Card took a step-by-step approach to reveal how popular opinion of the market could change through conversations. To this end, communication channels were opened with the Market Prosperity Association, the Bongpyeong Market Merchants Association, and the five-day Market Merchants Association. As for the merchants, 22 roundtable meetings were held, in addition to conducting VMD and customer service training. The process was prioritized over the design. In short, Bongpyeong Market was a project where elements auxiliary to architecture and design, such as communications and content design, were far more important. Since the new opening, Bongpyeong Market witnessed a 2.5 times increase in the number of visitors and a 30% increase in sales.

The color classification of tents by product group was a crucial branding strategy, considering that most of the visitors are tourists. As such, visitors can find and select items sold by the stores through the colors of the tents: green for agricultural produces, orange for snacks and cooked food, blue for fishery produce, purple for clothing, and yellow for miscellaneous goods.

5.2.2 Point of Space

The Hyundai Card Design Lab had to come up with a sustainable design approach without, if at all possible, creating new structures. The solution was to enhance VMD. Specifically, VMD was improved through tents of varying colors, mini signs for each store, and manuals on 'how to create an attractive store'. These strategies were effective without the need to establish new structures. Hyundai Card summarizes the governing principles behind Bongpyeong Market as follows: first, respect the traditions of Bongpyeong Market and local merchants but reinterpret them in a new way. Second, create a space desirable to today's consumers by improving convenience, the lack of which caused the traditional marketplace to momentarily stagnate. To this end, a rest area and information desk were introduced, as well as a car-free street and improvements to signage, all of which allow consumers to navigate Bongpyeong Market with greater ease. Third, secure the trust of customers towards the market merchants and products. To achieve this, photographs of merchants holding their products were pasted onto mini signboards and business cards. The photographs meant that the merchants would put their face to their products and conduct their business with honesty. Last but not least, encourage communication with customers. A word-of-mouth strategy was set in motion by uncovering stories to be put on the mini signboards, such as the buckwheat pancakes made by a grandmother who opened the very first store in the market, a grain shop run by a mother-son duo for a decade, and a

60-year-old bedding store. A Bongpyeong Market Newsletter for promotional activities was also founded. In addition, buckwheat pizza, buckwheat hotteok, and buckwheat canapes have become the must-try delicacies of Bongpyeong Market, with recipes developed by the chef in Hyundai Card's kitchen. The Hyundai Card social engagement team members personally met with 114 merchants for over a year and recorded their stories to reach these innovative ideas. The most visually striking thing in Bongpyeong Market is its vibrant and colorful identity. This was a key branding strategy, particularly considering that most of the visitors are tourists. As such, visitors can find and select items sold by the stores through the colors of the tents: green for agricultural produces, orange for snacks and cooked food, blue for fishery produces, purple for clothing, and yellow for miscellaneous goods. Their identity is further strengthened with the aprons and money belts of the respective colors worn by each merchant. This also added an element of visual spectacle to Bongpyeong Market. Aerial photographs of Bongpyeong Market venerate the beautifully colored vibrant tents flowing along the alleyways of the mar-

The VMD of Bongpyeong Market has been enhanced with mini signboards for each store and 'how to create an attractive store' manuals.

“At first, they thought Hyundai Card would establish a new building in the market. We designed tents and stalls and made props for the Bongpyeong Market, but they were not enough to persuade the merchants. Accordingly, we had a lot of conversations with them to explore and explain the proposed design changes, so that they could personally inspect and discern the impact of our design. We selected two stores for use as examples to offer a preview to merchants of how the stores in Bongpyeong Market would be transformed. We did so out of the idea that making sweeping changes to the traditional market would cause chaos. As such, the process of discussing possible developments with the merchants was crucial, earning their trust and making our integrity known.”

Youngkwan Kim, Hyundai Card Entrepreneurship Support Team – Bongpyeong Market Project Management

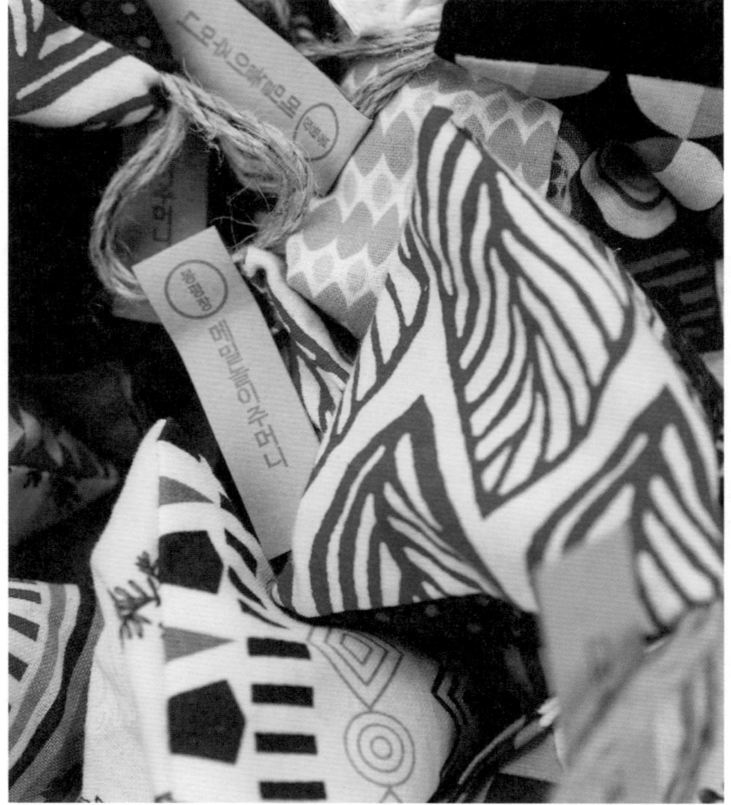

For better communication with those visiting Bongpyeong Market, souvenirs such as a buckwheat bag of luck and stamps have been organized, as well as developing an exciting range of packaging methods.

5.3 1913 Songjeong Station Market

1913 Songjeong Station Market
Location: 970-10, Songjeong-dong, Gwangsan-gu, Gwangju
Site Area: 3,775m²
Store Area: 2,530m²
Number of Stores: 55
Design & Construction: 2015.03 ~ 2015.09
Completion: 2016
Identity: Hyundai Card Design Lab

1913 Songjeong Station Market is located across from KTX Gwangju Songjeong Station and boasts a history of over 100 years.

Established as a landmark in Gwangju, 1913 Songjeong Station Market, was a project carried out by Hyundai Motor Group and Hyundai Card in collaboration with the Gwangju Center for Creative Economy and Innovation. Unlike other traditional market revitalization projects, focusing on modernizing market facilities, the 1913 Songjeong Station Market project attended to the influx of young merchants based on the unique history and traditions behind the old market.

5.3.1
The Way We Build

Arriving at KTX Gwangju Songjeong Station, one is greeted by the 1913 Songjeong Station Market right across the street. It takes less than 10 minutes to walk from the entrance to the end of the market street of about 170 m. Within two weeks of its opening in April 2016, the market set a new record of 50,000 visitors. Notably, several traditional markets have since been modernized. Still, the 1913 Songjeong Station Market has gained palpable popularity on social media, becoming a must-visit place in Gwangju. Even only a few years ago, it was relatively unknown that the market stood across from KTX Gwangju Songjeong Station for over a century. Despite this long history, Songjeong Station Maeil Market (the predecessor of 1913 Songjeong Station Market) was in decline like any other traditional market. Of the 55 stores, 17 were vacant, and the market had practically lost its purpose, with only a few stores open daily. Hyundai Card contemplated more far-reaching solutions to increase the market's competitive edge. One-off support or the extension of a modern facility was not in the plan. The key to revitalizing a market was soon understood as the preservation of 'history' and 'tradition'. Hyundai Card believed that the unique atmosphere of the bustling traditional marketplace would draw people in, and so placing the traces of this history on full display became the central goal. Hyundai Card was determined to avoid building new structures and establishing monumental new hardware such as installing an arcade. As such, the points of differentiation that would inform the transformation of the market into one that would be self-sustaining were found elsewhere. Since opening under its new name, the participation of young merchants contributed to the colorful and lively feel of the 1913 Songjeong Station Market. Developing this concept and the key marketing points was also crucial. However, ultimately, the essence of market revitalization lay with the individual stores and the merchants in the market. One of the more common issues facing traditional markets is the aging of the merchants. From the beginning of the project, Hyundai Card concentrated on attracting young merchants as a driving force behind building the new market. Additionally, discerning that the market would not be able to compete head-to-head with new large malls, Hyundai Card positioned the market as one focused on culinary experiences. This would be a breath of fresh air; the market's old stores have received an upgrade through hand-crafted beer, croquettes, bakeries, and other shops run by young merchants. This strategy introduced a dynamic synergy between the old and the new. As such, the 1913 Songjeong Station Market did not aim to establish itself as a tourist destination or as a bureaucratic triumph. The crux point was to prove that traditional markets could also be competitive and motivate other traditional markets facing similar challenges. It is not difficult to produce one or two success stories, as long as there are ample finances and a talented workforce. However, such success cannot be the only alternative as it can create a sense of unfairness between the new market and other traditional markets nearby. Accordingly, the following principles were set: first, that the total budget would be less than 1 billion won; second, that the project period should be shorter than 12 months; third, that the total input should be by fewer than five employees; and fourth that there should be an emphasis on existing traditions. This was to demonstrate that the end result could be achieved with a reasonable level of investment and ingenuity. However, contrary to the initial intentions to support the community, it was difficult to avoid gentrification as an inevitable part of the process of the revitalization of the dilapidated market. Furthermore, the project left a question and a point to contemplate: whether a corporation is truly the most appropriate agent to conduct a local regeneration project.

Hyundai Card wanted to emphasize a sense of historic heritage while also enhancing the competitive edge of each store. After much deliberation, they decided to stage a more retro ambiance so that the vitality of the 1970s and 1980s, a prime time in the life of the market, could be revived.

1913
송정역시장
문화와 추억의 창조전통시장, 꿈의 나눔터

5.3.2
Point of Space

Each of the stores inside the 1913 Songjeong Station Market has been reborn by celebrating its eccentricities and unique histories.

The project began by changing the market's name to 1913 Songjeong Station Market. The origins of the old name, Songjeong Station Maeil Market, dating back to 1913 when Songjeong Station was built, and a street market formed around the station. By putting the market's birth year in the new name, Hyundai Card intended to remind visitors of the market's 100-year history and instill pride in its merchants. After contemplating how best to address the site's history along with enhancing the competitive edge of each store, Hyundai Card decided to stage a more retro feel so that the vitality of the 1970s and 1980s, a prime time in the life of the market, could be revived. Signboards with the stores' history were newly designed and installed. The stories behind the stores with significant histories were unearthed to attract visitors. As such, the stores in 1913 Songjeong Station Market were reborn as inimitable places, each with a 'one-and-only' story. The location of the KTX Gwangju Songjeong Station could provide an important competitive edge as an average of 12,000 passengers pass through the station every day, which is only a three-minute walk from the market. This location led to the concept of 'a Second Waiting Room with a Nostalgic Look'. To attract KTX users to the market, an abandoned space was converted into a vaulted shelter and public toilet. Also installed was a real-time train information board resembling the display at an airport. In addition, unmanned storage facilities were also installed for users to browse the market without the need to carry their things around. Sales methods have also been upgraded. Small-packaged items are labeled with the 1913 Songjeong Station Market logo so that KTX users can purchase products as gifts. The night market held every Saturday evening features street performances based on the local culture. This was introduced with the significance and competitiveness of today's traditional markets in view, driven by the suggestion that what they lack is the energy and entertainment provided within. 1913 Songjeong Station Market is not just a place in which to buy and consume food but is also fast becoming a base for local culture and a playground.

쑥's 초코파이
국수집

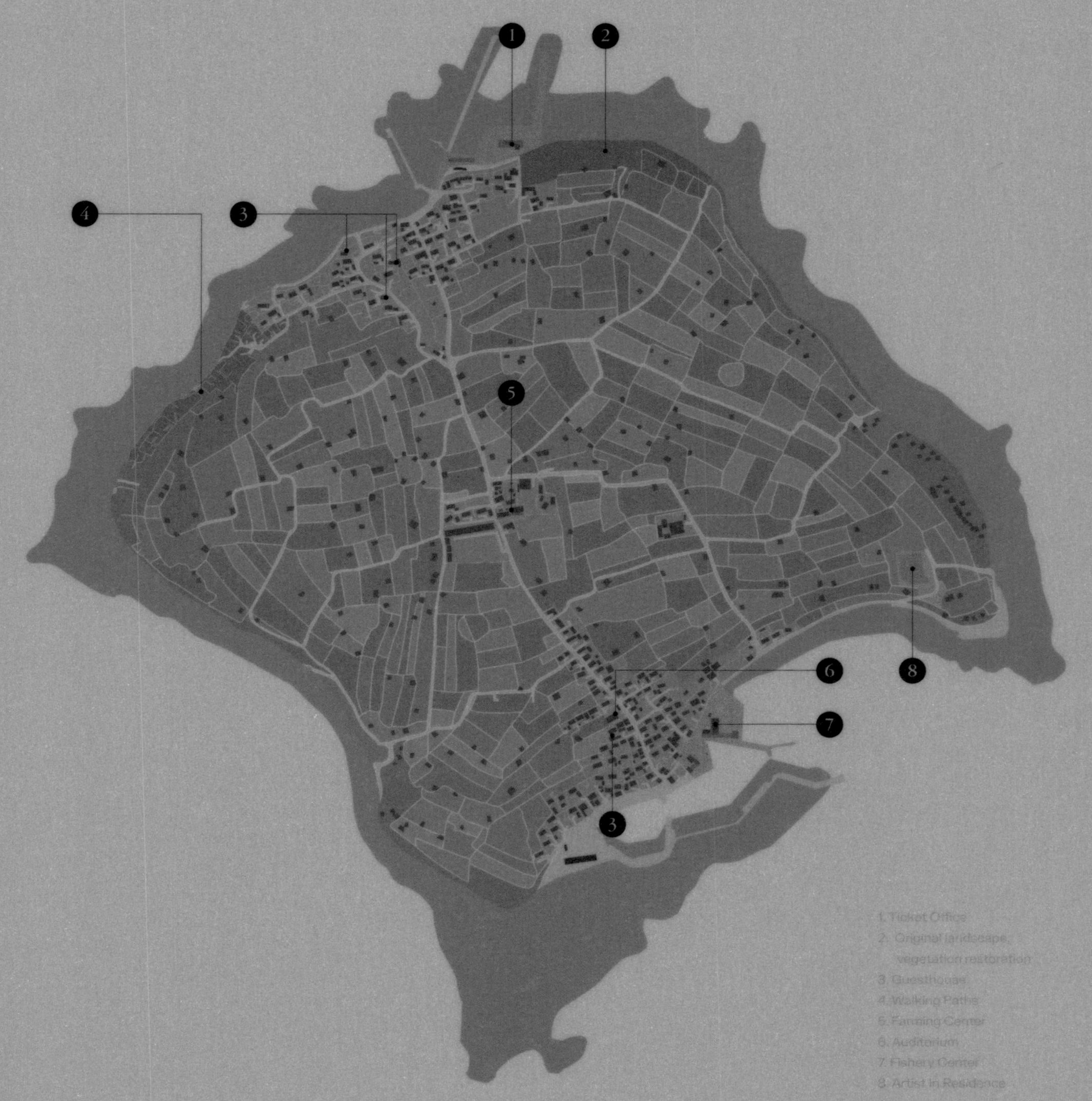

1. Ticket Office
2. Original landscape, vegetation restoration
3. Guesthouse
4. Walking Paths
5. Farming Center
6. Auditorium
7. Fishery Center
8. Artist in Residence

Gapado Map

1. Ticket Office
2. Original landscape,
   vegetation restoration
3. Guesthouse
4. Walking Paths
5. Farming Center
6. Auditorium
7. Fishery Center
8. Artist in Residence

Gapado Map

Gapado Sangdong Site Plan

N
0 10 20m
Ticket Office
Ticket Office
Pergola
Original landscape,
vegetation restoration
Guesthouse
Bathhouse
for haenyeos
(female divers)
Guesthouse

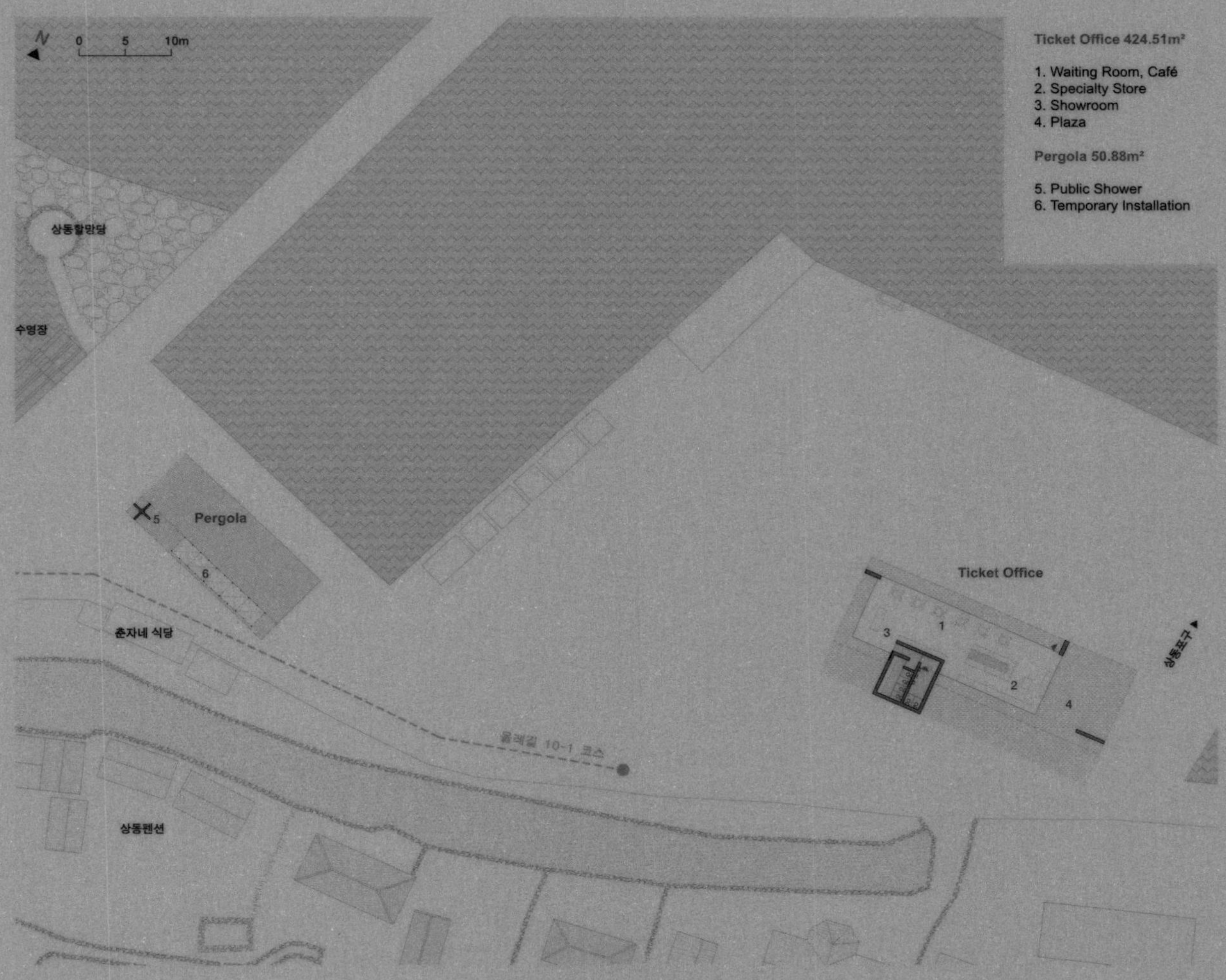

Gapado Ticket Office Plan

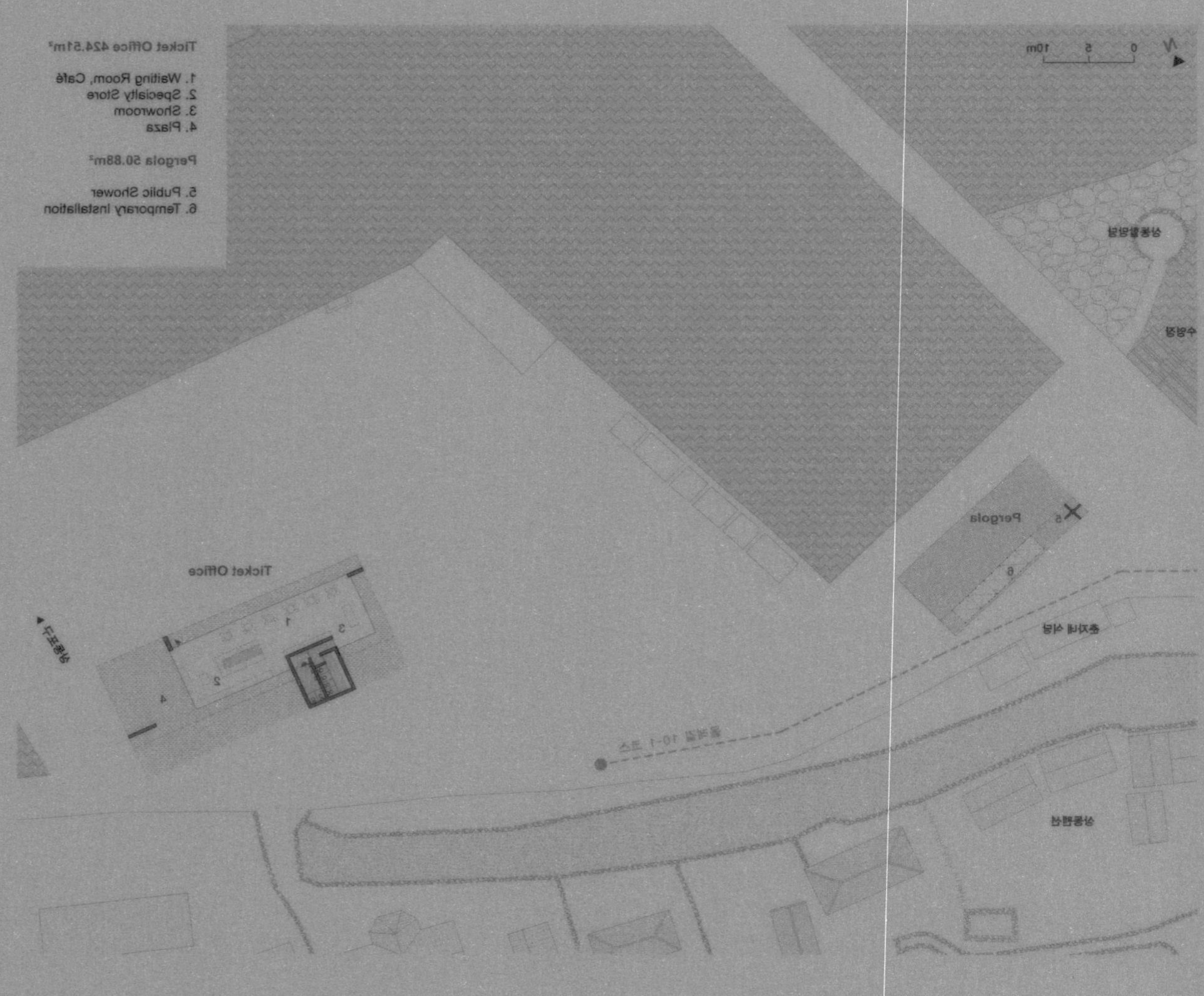

Gapado Ticket Office Plan

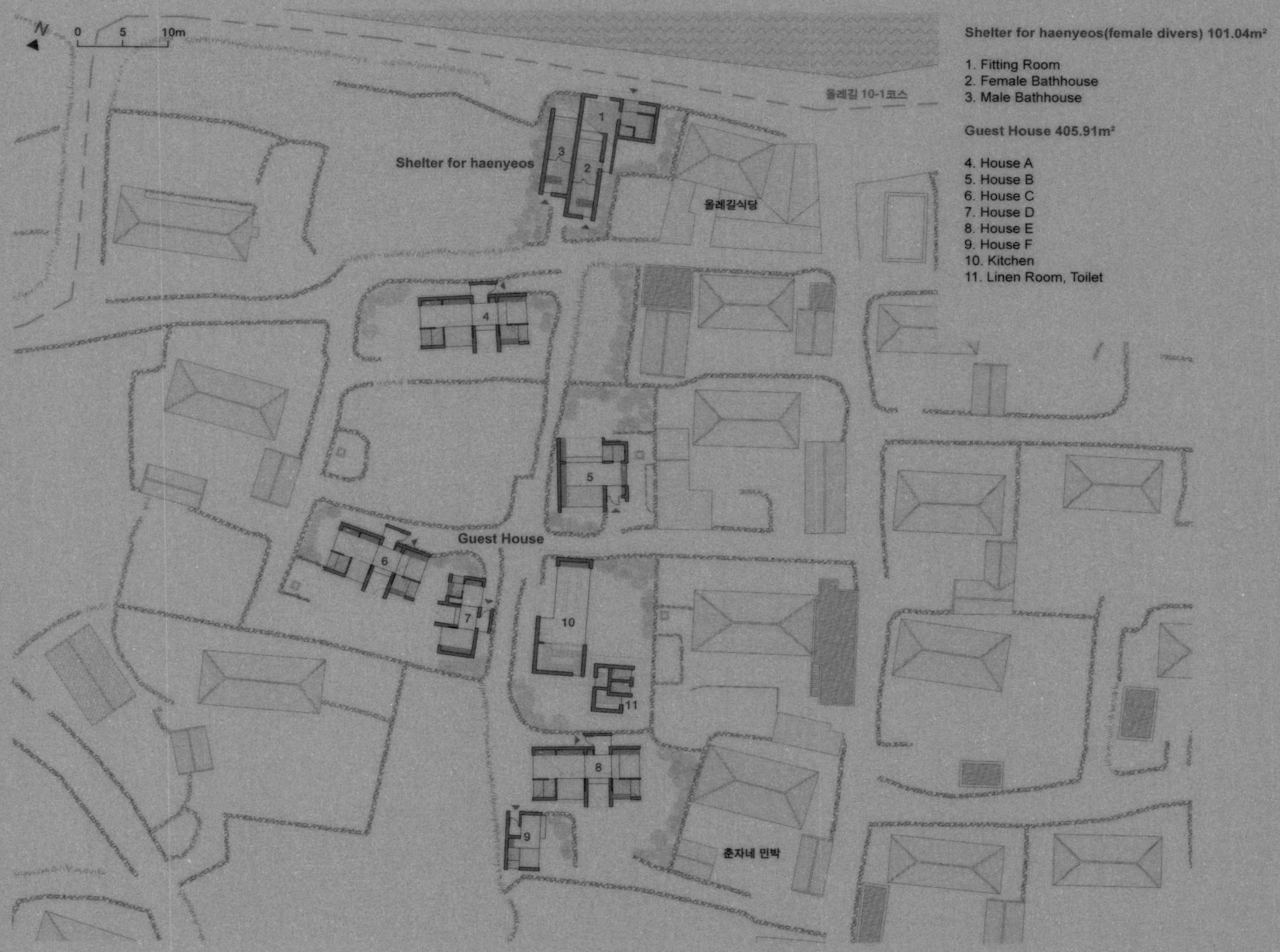
N
0
5
10m
올레길 10-1코스
Shelter for haenyeos
1
2
3
올레길식당
4
5
Guest House
6
7
10
11
8
9
춘자네 민박
Shelter for haenyeos(female divers) 101.04m²
1. Fitting Room
2. Female Bathhouse
3. Male Bathhouse
Guest House 405.91m²
4. House A
5. House B
6. House C
7. House D
8. House E
9. House F
10. Kitchen
11. Linen Room, Toilet

Gapado Guesthouse Site Plan

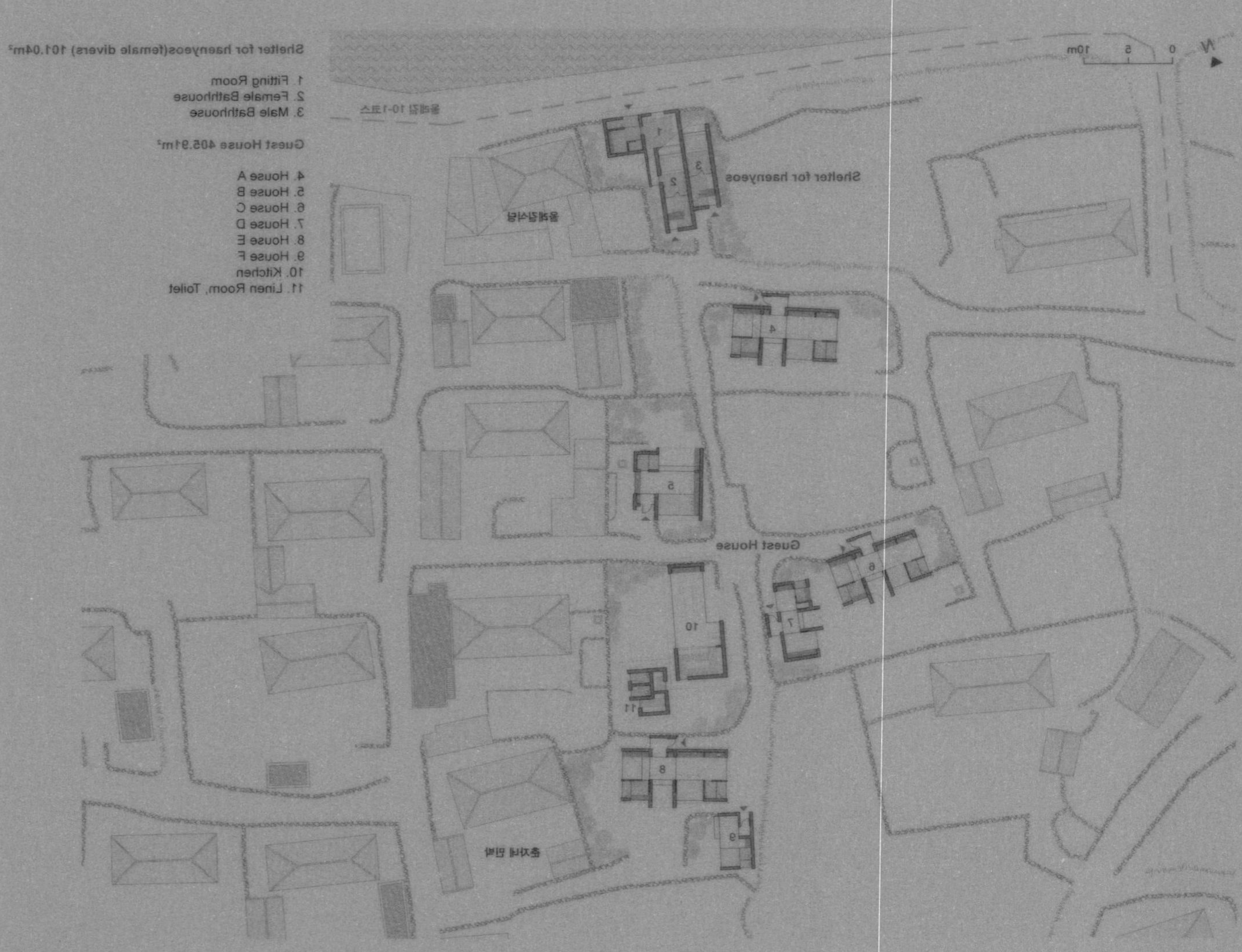

Shelter for haenyeos(female divers) 101.04m²

1. Fitting Room
2. Female Bathhouse
3. Male Bathhouse

Guest House 405.91m²

4. House A
5. House B
6. House C
7. House D
8. House E
9. House F
10. Kitchen
11. Linen Room, Toilet

Gapado Guesthouse Site Plan

Words on Hyundai Card Spaces

Wook Choi, Head Architect of 101 Architects / Yeongdeungpo and Busan Offices, Design and Cooking Libraries, Gapado Project

How did you first come into contact with Hyundai Card?
Our relationship began when our firm received a request to participate in the competitive process for the Yeongdeungpo Office in 2012. Afterward, we were tasked with designing a number of projects, such as the Design Library, Cooking Library, and the Gapado Project, in addition to the Card Factory in the Yeouido headquarters.

What was your most important idea when working on the Design Library?

The concept of Gahoe-dong became of top priority. In the rather relaxed atmosphere of the area, I hoped the space would transmit a certain nostalgia in which one would be prompted to focus on their own histories and inner self. This place had traditionally been home to residences of the nobility, and the concept of the study—the most important place for scholars—naturally presented itself. One may recall images from the 'House within a House' on the second floor of the bookshelves, or from the Gioheon, the small study room built where a water tank once stood. It was important to provide a moment's pause within the space to view the sky from the square-shaped courtyard, a form unique to the traditional hanok. So, we removed all obstacles and left the space as clear as possible. Hanok architecture implies warmth and care; as we intended, the librarians and visitors of the Design Library feel at ease here.

101 Architects has a robust sense of interior and exterior architecture, including that for furniture and interiors as a comprehensive system.

It would be better to say that we do not consider the interior and exterior of architecture to be separate. In much the same way as a lining, outer material, labels, and buttons are all considered constituents of clothing, proper details are required to form an architectural concept in a single mass. I do not pursue details through specific works of furniture or interiors. The detail itself is not crucial—I simply unite the appropriate details necessary to draw a space into the fullest realization of a whole.

What kind of client is Vice-Chairman Ted Chung?

He is a client with an exceptionally intuitive understanding of architecture. I am reminded of the time we visited Casa del Fascio, a hexahedral box-style modernist building in Como, Italy. Built in the early 1930s, it is the work of architect Giuseppe Terragni. When I briefly explained that the concept behind the building was limited to and controlled by mathematical modules, he immediately understood this design principle. He is also involved and active in expressing his opinions and supplying prompt feedback. In the case of the Cooking Library, he wanted the space to be more than a space in which to read, envisaging one that would also encompass the act of cooking, where the scent of bread would rise through the building. Furnished with this idea, I immediately explored his concept with an illustration. He was always delighted to see the Library, and I think his love and interest in architecture has grown over time through his personal involvement in the design and building of the structure.

You have worked on various projects with Hyundai Card, including the Yeongdeungpo and Busan offices, the Design and Cooking Libraries, and the Gapado project. What is your secret to collaborating with a company on such a diverse range of projects, if any? If there is any one architectural principle that you have developed during your collaboration with Hyundai Card, please share it.

The reason 101 Architects was able to work on numerous projects with Hyundai Card appears to be because our architectural language aligns so closely with that of Hyundai Card. Basically, we refrain from building unnecessary structures. This is why our architecture has a very refined geometric form. Also, if possible, we use unprocessed, raw materials. For example, we do not cover the raw structural materials by applying wallpapers or veneers. We try to use a low-iron glass of high transparency. Attention must be paid to the surrounding environment and to accessibility, in consideration of the public passing through our spaces.

Although Hyundai Card is a financial company, it oversees numerous projects based on its interest in design and architecture. As an architect, what do you think of Hyundai Card's architectural projects?

Hyundai Card was already tens of years ahead of its peers. They recognize the importance of architectural and spatial branding and are pioneers in the field. Hyundai Card's projects serve as architectural references for other companies and contribute to raising the international reputation of Korean architecture.

'Eschewing unnecessary structures, Hyundai Card's architectural interventions have significantly refined geometric contours.'

Hyundai Card has also contributed to the public domain by donating its architectural and design talents, not merely making monetary donations. The case in point is the Gapado project.

Hyundai Card is a unique company. Few businesses worldwide carry out architectural and spatial projects in the way it does. Regional projects, such as the Gapado project, were originally a task for the government, and so businesses would simply make financial donations to show their support. On the contrary, Hyundai Card takes charge of projects, which requires far more effort and investment than simple one-off donations. The Gapado project was built around the respectful enthusiasm of Vice-Chairman Ted Chung.

You mentioned that the Gapado project is an ongoing project. Please tell us more about any upcoming alterations or additions.

Local governments take Gapado as the gold standard in regional redevelopment. Professors from the University of Lausanne in Switzerland and Columbia University have also visited Gapado to personally witness the implementation of our plans and were surprised that a project of this size and scale has managed to check off all of the required regulatory steps with great efficiency. Although the standards vary from country to country, public projects tend to be heavily regulated. There were two goals when embarking on the Gapado project: the first was to make a case for regional development in collaboration with Hyundai Card, and the second was to obviate the bureaucratic contradictions so often faced by public projects in Korea. In terms of providing a starting point for the broader discussion of public projects and their societal role, I believe that the Gapado project has achieved its intended purpose.

You are also the publisher of the Korean edition of Domus, a world-renowned architecture magazine. Domus, founded by the architect Gio Ponti, is also famous for employing top architects as its editors-in-chief, including Michele de Lucchi and Alessandro Mendini.

At the time, Domus was looking for a Korean partner. I have memories of reading Domus while studying when I was in school, and I harbored a great fondness for a magazine. Still, at first, I was scared. As much as I could not bring myself to imagine how complex and difficult the role would be, I knew I was not ready for a position of this magnitude. At the same time, however, I had a desire to contribute what little I could to the Korean architecture scene. Thankfully, Hyundai Card provided some support, and I decided to make an investment that I could happily afford. At first, I was all over the place, even doing the Italian translation myself! But now, things are finding their order, one issue at a time. It is wonderful to have the opportunity to meet a wide variety of new people through Domus.

Wook Choi
Head Architect of 101 Architects. After graduating from the Department of Architecture at Hongik University and the Università Iuav di Venezia, Italy, he established 101 Architects in 2000. His representative works include the Hyundai Card Yeongdeungpo and Busan offices, Design and Cooking Libraries, Gapado project, Hakgojae Gallery, Dugahun, and (former) Seoul Mayor's Office. He was invited to the 2006 Venice Biennale and the 2007 Shenzhen-Hong Kong Biennale and oversees the publication of Domus Korea. He won the 2013 DFAA (Design for Asia Awards) grand prize for the Design Library and the 2014 Kimm Jong Soung Architecture Award for the Yeongdeungpo office building.

Philippe Paré, Director of Gensler Paris/ Convention Hall, UX Lab, Workspace, HCA, BHAF, HCUK, Air Lounge

How did you first come into contact with Hyundai Card?

As the head of design, I worked at Gensler's offices in Los Angeles and London for 16 years, and now I am working in their newly opened office in Paris. We have collaborated with Hyundai Card and the Hyundai Capital teams for nearly 13 years. The first project I oversaw was Hyundai Capital's U.S. headquarters in Southern California, for which we won the design competition. However, after seeing the project proposal Vice-Chairman Ted Chung asked us to rethink some of our decisions, as these had been made when we were not as familiar with the Hyundai Capital brand and ethos.

How did you decide to integrate Hyundai Card's corporate identity into the space?

I think Ted Chung understands the power of a space to express a company's brand in physical terms. If you think about the most powerful ways of creating an in-house corporate culture, the workspace is probably the best place to start. A culture equips employees with the knowledge to make the right decisions for themselves because everyone in the company shares the same established values. What is of particular interest is that culture and the space framing it are not fixed but constantly evolving.

Could you speak about some of the specific changes made?

Looking back at the first project in Irvine, we pursued a serious global image so that the brand would build trust. We took a very clean, simple, and minimalistic—almost museum-like—intellectual approach. Hyundai Card's brand has evolved to occupy a position of high esteem, capable of setting the terms for the entire industry. We have transitioned to the attitude that the space should make admirable, but employees shouldn't sacrifice themselves for that cause. We focus on providing a great experience for employees.

Do you think the way a space is defined also affects the way people work and communicate? Is this what you did in the Digital Workspace project?

The office space we designed can inspire innovation. Greater communication takes place between employees in spaces that invite interaction. They also perform at their highest when they find an environment that matches their psychological state. That is why we have combined a lot of different working environments, which in turn have a huge impact on employees' performance and their relationships with one another. Offices now need a social space in which teams can connect outside of work. A decade ago, this would have been seen as an unproductive initiative, but in reality, when not working, team members get to know each other better and build greater trust. I think that this is most particularly the case with Digital Workspace, as it still has the atmosphere of a startup company. This is a comfortable place even for people who work late, so much so that it almost feels like working from home.

Why do you think Hyundai Card is so focused on branding not only in office spaces but also in cultural spaces, such as the Music Library?

When we experience something, we use all of our senses in the creation of that memory. I think that memory is the most powerful and inspiring form of communication and can be used to communicate any message. Companies can also communicate with their customers intuitively and emotionally; instead of simply associating a company with a service, customers recall their positive experiences.

'Hyundai Card works like they are a member of our team, unlike other clients.'

What is Ted Chung like as a client?

He is truly great. He is very strict and demanding, which is in line with the level of authority and responsibilities he carries in his position. He has an extremely clear architectural vision, and sometimes we had to make sure we understood it correctly. At the same time, he has sufficient wisdom to understand that we should be given the space and license to perform our best work as designers. He drew us a clear roadmap but gives us the creative freedom to find and implement the best possible solution. In addition, he knows which projects are the best suited to the company, depending on the strengths of each designer and architect. Therefore, he is very selective about who he appoints to which project. It was indeed fun to have candid and enriching conversations with him as our relationship developed.

undai Card is now known as a digital npany. Many financial companies want provide digital-related services. Do you nk these changes will affect the space d its corporate culture?

Yes, I think so. As a digital comny, they are required to hire employees h new breeds of expertise and to emasize with them properly. You need to ow where they developed their skills d what type of working environment y prefer. To be attractive to them as an ployer, you need to create environments t they would consider to be familiar d intelligible. I often think they have a ferent sensitivity to lighting; employs in digital teams tend to prefer darker aces because they look at the screen a long time. As such, it is important create an environment that embraces ch conditions. Furthermore, as working urs are long, it is essential to create a t area for games and social activities employees to relax and de-stress.

Hyundai Card employees are involved in every aspect of the project: they write requirements, place their orders, select a company. What do you think of this kind of participation?

We are like a team working together. Hyundai Card is more like a design company than other companies in its way of approaching details. I have never seen a company that even makes its own corporate badges and water bottles. Furthermore, it is committed to doing the right thing. Thus, we employ smart ways of completing projects without compromising on quality. Each project is also a chance to accumulate greater knowledge. As such, each project is likely to improve on the last as one trusts each other and learns from mistakes. What Hyundai Card has achieved so successfully is that thousands of decisions continue to reinforce an original idea. To do that, a team that understands the initial concept is essential. We are very grateful to have such a team.

Philippe Paré
Paré received his bachelor's degree from the University of Montreal and continued his studies in architecture at the University of Guanajuato, Mexico, and California Polytechnic State University. Working at Gensler for 16 years, he has built a global portfolio of workplace spatial design that has driven interaction and inspiration. As Director of the Gensler Paris office, Paré oversees the growth of the Paris office and business throughout France. His approach to design is idea-centered, drawing attention to crafted details. He has won over 100 design awards, including the honorary award of the AIA Institute for Interior Architecture, and his work has been featured in publications such as The New York Times, Wallpaper, Financial Times, Time, and Frame.

Mary Spackman, Principal Designer of Spackman Associates / Auditorium, Lecture Room, the Box, Café & Pub, Air Lounge 2, Vinyl & Plastic, the Storage

How did you come to work on projects with Hyundai Card?

Vice-Chairman Ted Chung, who took over at Hyundai Card in 2003, observed the project on which we had been working on for a year and commissioned us to work on the employees' cafeteria in Building 1 of the Yeouido Headquarters. At that time, the building did not require full renovation, but at the very least he wanted to create a new cafeteria for employees. Starting with this project, we were commissioned for the spaces in Building 1, and in 2009, we carried out the overall design of Building 2.

You have worked on various architectural and spatial projects for Hyundai Card. To which project do you feel the most attached?

I have worked on many projects with Hyundai Card over the past 18 years, but I feel the greatest attachment to Yeouido Office Building 2 completed in 2010. Then in 2020, we renovated a few more floors of Building 2. As we have designed a range of office space with Hyundai Card since the beginning of 2003, large and small, we naturally understand the corporate character pursued by Hyundai Card. Building 2 was a project intended to encapsulate Hyundai Card's corporate identity and culture informed by our impressions up to that point. We were not held back in our desire to fill the space with such conceptual ideas. The standout symbolic design element in Building 2 is the 'glass box'. The Auditorium, the Box, and various functional spaces such as the kitchen on the first basement floor, have all found expression as a glass box. The glass box conveys transparency—the lifeblood of a financial company—and creates a vibrant atmosphere by naturally exposing the daily movements of the employees, the driving force in the company. The overall tone and manner are 'sophisticated but restrained'. We sought a clean and elegant style that was neither exaggerated nor flashy. We considered this to be the style that would best suit the confidence of Hyundai Card.

Were any of your design ambitions made possible thanks to a Hyundai Card project?

Hyundai Card is a client that provides full support to the realization of a design once the precise intention, purpose, and reason have been established. That is why we have been able to implement numerous design initiatives that other companies cannot afford. As such, Spackman Associates considers many of its Hyundai Card projects as its best work. For instance, one of the design concepts at the time of designing Building 2 of the Yeouido Complex was 'Linking the Two Headquarters'. To make this concept a reality, not only did Buildings 1 and 2 have to be physically linked underground but Building 2 also had to face Building 1. In light of this, I suggested adding the Box, a glass structure, which would extend towards Building 1, and an open-style lobby by eliminating half of the slab on the second floor. Due to the magnitude of the construction, following my proposal, it must have been a difficult decision. Still, because Vice-Chairman Ted Chung understood that the proposal was absolutely necessary, the plan could be implemented.

u have also worked on numerous work-ce projects. Do you see any governing tures or trends in Korean office design?

I have noticed that most Kore- companies are pivoting away from a id corporate culture to a more informal e. Therefore, in many cases, instead of lecting the present corporate culture the space, companies imbue their of-es with their future visions. As the em-yees frequent and use such spaces, ew corporate culture begins to form. st corporations in Korea believe that creation of a free and fun space will turally facilitate creativity and promote re effective communication between ployees. Of course, the spatial trans-mation has a considerable impact on behavioral and psychological profiles employees. However, without a change he corporate culture and its connected tems, there can be no success. Hyundai rd did not stop at making the space agile d free—it also experimented with flexible rking hours and policies that would allow ployees to move between departments, rating the imaginations and the design A of the employees. In the early days, undai Card also focused on defining corporate identity and the direction its corporate culture, creating a more pressive branded space. Conversely, ay, Hyundai Card's design expression s become more relaxed, and it is now perimenting in various ways with spatial sign to redefine its corporate identity d culture.

at kind of client is Vice-Chairman Ted ung? What requests does he tend to ke when embarking upon a project?

Based on my years of experience rking in collaboration with Hyundai Card, ve come to understand that the compa-is 'always evolving'. It is always evolving cause it is led by a CEO who is not scared change but quite the opposite, proac-ely embracing it and is highly curious out new methods and experimentation hout apprehension. At the beginning of a ject, there is little that is off-limits, and signers are given a significant leeway. However, excessive or frivolous designs without a given rationale are not tolerated. Practicality and stability are prioritized. Hyundai Card is a client that understands the importance of expertise across each field. One of the best things about working with Hyundai Card is that you can draw in experts for more nuanced collaborations. I believe Vice-Chairman Ted Chung possesses a strong sense of which designers and architects are suited to which project. He is a person who discovers joy in the process overall. As the saying goes, people who enjoy their work, work better.

Despite its status as a financial company, Hyundai Card has overseen a wide range of projects based on its interest in design and architecture.

Hyundai Card is sensitive to the trends that drive the market and wider society. Hyundai Card is also attentive to the needs and desires of customers, clearly recognizing what is important to maintaining their lifestyles. Based on this understanding, Hyundai Card is committed to creating an emotional connection with its customers, even if this has no obvious connection to the finances or increasing the number of customers. Sometimes these attempts are cultural spaces, while other times they are social contribution projects. I believe that such wide-ranging ventures are made in an attempt to communicate more organically with customers, approaching them in a uniquely Hyundai Card way. Perhaps it is because Hyundai Card has no boundaries in its ideas and thoughts. Looking at Hyundai Card, I see one thing that has remained consistent; in everything it does, it strives to become a corporation that is attuned to its customers' lives, a business that carries an eternal significance in their lives.

'Hyundai Card provides its full support to a design once the precise aim, purpose, and reasoning have been clarified.'

Mary Spackman
Principal Designer of Spackman Associates. She graduated from the Smith College Department of Political Science and the New York School of Interior Design. She subsequently worked at Rockwell Group and Space, overseeing a wide range of projects such as New York's first W hotel, numerous office buildings and restaurants. After establishing Spackman Associates in 2002, she developed a strong reputation for creating unique corporate identities through cogent communication. Her representative works include office buildings for Amorepacific, Samyang, Shinyoung, Dongwha, SK, Kim & Chang, and Yulchon, among others, as well as spatial design projects for Naver and Seoul Somerset Palace Hotel.

Sabu Song, Design Director of Gensler/ Music Library, Studio Black

What do you think of Hyundai Card as a client?

It has been a remarkable opportunity and a wonderful experience to work with a client like Hyundai Card that values and supports a holistic approach to design. Just like any other relationship, good chemistry, trust, and communication between a client and architect are essential in maintaining a successful and dynamic relationship. As an innovative company, it has always been important that the Hyundai Card brand value and its authentic character are conveyed and integrated into the design. Hyundai Card is a company that excels at brand marketing and understands that it can be a powerful tool for communicating the possibilities they can offer their customers, employees, partners, or any other group they want to reach. Over time, our work together has presented an opportunity for us to show how architecture can become a brand signature. Having worked on numerous Hyundai Card spaces, we take an in-depth look at how end users engage with the built environment or space through emotional, cognitive, multi-sensory, and values-based experiences.

How do you think Hyundai Card's space projects have affected the Korean architecture culture?

It has been over a decade since Gensler began collaborating with Hyundai Card. Whether it's designing a workplace or a cultural venue, the idea starts with a thorough understanding of the local context and culture. Hyundai Card is one step ahead in designing for future experiences and anticipating the needs of its end user. This has propelled the brand to be a leader in Korean architecture and around the globe.

What is your favorite Hyundai Card space?

The Music Library and Understage: This project offers the most visibility and accessibility to a public audience while capturing the spirit of the local neighborhood and its historical context. It blurs the boundaries between public and private spaces. I remember when the team first surveyed the neighborhood surrounding the project site. We recognized how densely the buildings were built—side by side—with each building maximizing its footprint within its zoning lines. This was uninviting and unapproachable for people. We designed the Music Library and Understage to be the complete opposite. While the building itself is semi-public, the spaces visually speak to each other, whether you are inside or outside. This made for an approachable and engaging experience that appeals to a broad cross-section of people.

What is the design concept of the Music Library and Understage?

The Gensler team was inspired by the music counterculture of the 1960s and 1970s, which forms the core of the project concept, with inspiration drawn from street music, art, and culture. The goal was to create a sense of time and place, a comprehensive, visceral experience for users through architecture, environmental graphics, and an art installation program that is fully integrated while tying different spaces together into one consistent, layered experience.

What were some specific requests from Hyundai Card?

The Gensler team was tasked with creating an immersive and engaging experience for musicians and Hyundai Card members that would elevate the brand as a purveyor of unique, cultural experiences. The branding for the space needed to flow seamlessly, offering context and organization in some areas and providing streamlined directional information in others.

Are there any design elements that Gensler had to push to convince the client?

For the artwork content on the underside of the exterior canopy, the JR and Gensler team carefully selected a photograph, originally taken by Bill Owens. The image is from the infamous "1969 Altamont," a fascinating collection of photographs of the historic Altamont festival. The team saw the photograph as a powerful representation of a turbulent chapter in countercultural history. The image was somewhat provocative due to nudity, and there was some hesitancy about whether the photograph would be culturally acceptable in the publicly exposed, proposed location. However, the billboard-sized mural now serves as the backdrop to an outdoor performance space, providing street recognition and immediately engendering a strong sense of place within the emerging neighborhood. The mural can be viewed from many different perspectives: from the upper street, down the street, inside the library, and the café. Each view frames it differently and creates perceptual surprises.

'The goal of the Music Library is to create an approachable experience that would appeal to a broad cross-section of people to build an emotional connection.'

scribe the reasoning behind the materials and details selection.

The design solution relied on raw d industrial materials to capture the tone the street music culture and encompass e "beautiful imperfections" of vinyl within e Music Library and Understage spac- The desire was to create an honest, eless environment that is authentic but o superlative. Beauty is very perceptive. e approach was to capture the natural gression of a specific type of beauty d align it with the stories of musical olution, thereby capturing the spirit of e underground music scene and analog dia, which have a special richness in ms of audio quality and physical tactil- The team's desire was to recreate this ture and imperfection that contains its n beauty, emotion, and authenticity. Seting materials that would age naturally s just as important as finding street ists to create authentic yet beautiful works in this site-specific, public space.

scribe the reasoning behind the Corten el (red weathering steel) on the façade.

The design team was interested in ating a hub of activity and a catalyst inrating the rich fabric of the surrounding ghborhood. The design team's approach the entry façade was to deflect attention m the surrounding cold, modern building perties and create an unexpected invern of weight and void, with ambiguities scale, using naturally aged Corten steel sitioned as a marquee.

Describe any difficulties experienced during the project.

The biggest challenge for the team was finding the most effective solutions for the existing framework, as the team was brought on mid-way through the construction process. Gensler was tasked with re-designing the interiors and exterior positioning within the framework and on-going construction.

Hyundai Card as a possible client

For the first time in many years, with the world moving faster than ever, there has been an increasing appreciation of nostalgic analog contents, such as vinyl albums. This suggests a slow but steady shift from digital downloading to analog listening and from large-scale arenas to intimate concert spaces, where the audience has a closer connection with the performing artists. Hyundai Card's vision has taken the notion of vinyl revival, quite literally, to a whole new level. "A great project starts with a great vision." We think the success of the project was only possible because Hyundai Card's vision supported and was aligned with the architect's aspirations and vision. The project aimed to give back to the community by providing an approachable experience that would appeal to a broad cross-section of people. It was designed in such a way as to get people interested in art and music— people who, perhaps, would have paid little attention before stepping foot in the place as well as connoisseurs, who may make very satisfying discoveries.

Sabu Song
Design Director at Gensler L.A. She is in charge of creating concepts and strategies for design projects through to completion, spanning architecture, interiors, product development, and branding. She has built an award-winning global portfolio with extensive design experience in office space, entertainment spaces, commercial establishments, airports and transportation, health and wellness, technology, and the creative industries. Her solutions, based on meticulous research, are functionally and aesthetically effective imbuing customers and her team with confidence.

Masamichi Katayama, Founder of Wonderwall / Travel Library

I would like to hear the philosophy that permeates Wonderwall's designs.

We aim for designs that surpass the simple expression of 'newness'. It is crucial to present a reawakening in the way the space exists and how it is presented, one which overturns common sense. In addition, we emphasize the 'landscape' created within the completed space. Another important matter when designing a space is effective communication. Design is a tool that facilitates communication between brands and their consumers. Therefore, we carry out projects after securing an accurate picture of the desired message of each brand.

How would you describe your experience working with Hyundai Card?

Even internationally, few corporations are as conscious of design as Hyundai Card. Hyundai Card is an ideal client for designers because it places no restrictions on design and willingly accepts complex proposals. Working with Hyundai Card, I came to see how a company's understanding of design can positively influence the progression of a project. A library completely themed around travel had never existed before, and so it was a superbly interesting proposition throughout the project. Japan also has no such space. Hyundai Card's brave experimentation, informed by its belief in the power of design, was truly inspiring and stimulating.

Has the Travel Library had an impact on its surroundings?

When we began the project, we first identified the nature of the area, Cheongdam-dong. It was a very extravagant center tailored to young consumers. We wanted to create a space in which people who come to shop or for gourmet experiences can experience the invigoration of 'escape in the middle of the neighborhood'. The Travel Library followed an unprecedented design that surprises visitors from the outside, and, once you enter, it keeps drawing one further inside. Creating a space of this kind was our primary goal. Triggering the will to travel is one of the opening gambits of the Travel Library. I hope in taking the time to visit the space this will become the spirit of travel itself.'

'Triggering the will to travel is one of the opening gambits of the Travel Library. I hope in taking the time to visit the space it will become the spirit of travel itself.'

ould like to hear your impressions of the y features of the Travel Library.

We wanted to create a place that ould kindle the desire to leave on a jouray, enabling the sharing of memories and gregating every type of curiosity. The cus was on creating a space in which one-time visit would be insufficient to lly comprehend the character and imications of the space. Therefore, to give e feeling of entering a mysterious cave afted out of books, we created a rather mplex space. Inside, there are no straight es. Every visit leads to wildly different ndscapes as all of the lines are wavering. ne hidden rooms open up the possibility discovering new things on every visit. aybe one will think of it as an amusement rk packed with travel knowledge and as space that will provide a new perspective ery single time.

Various props and objects related to travel also generate an exotic appeal.

From the chairs in the café to structural elements, you can appreciate the differences between various cultures around the world. The period of production and cultural backgrounds all differ between each piece of furniture — the chair imported directly from Africa, the chair demonstrating a Victorian inheritance, and the chair with a mischievous backrest.

Masamichi Katayama
Born in Okayama, Japan, Katayama graduated from Osaka Designer College. In 1992, he co-founded H. Design Associates, and in 2000 he established Wonderwall. He respects Japanese traditions and design philosophies and maximizes the brand vision of clients through clarifying vague and conceptual ideas. He has designed spaces for various global brands such as Nike, Thom Browne, Uniqlo, and Dean & Deluca. Completed in 2014, the Emporium was selected as one of the nine most beautiful shopping malls in the world by Architecture Digest. Also, in 2020, the Architecture and Interior magazine Frame awarded him with a Lifetime Achievement Award. Since 2011, he has been a Professor in the Department of Space Directing and Design at Musashino Art University.

Jean Nouvel, Founder of Atelier Jean Nouvel / Design Lab

The Design Lab of Hyundai Card is a relatively small project for a world-renowned architect to take on.

This project was one of my favorite projects. It truly excited me because it allowed me to actualize a concept, I had presented at the 2013 Milan Design Week. When I encountered Vice-Chairman Ted Chung, I was surprised at his depth of knowledge and intuitive understanding of what I wanted to achieve when sketching out ideas during our initial conversation. What I also perceived was a strong will to act following our discussion on future designs. Creating this space taught me or confirmed to me that the principles we set forth could be realized. I believe that the implementation of them was perfect and significant.

Please share your sense of an office building as a 'space'?

Today's offices are too rigid. The obligation of functionality imposed on those elements that constitute the office interior tends to lead to an excessively repetitive and not particularly interesting or stimulating atmosphere. Employees spend a significant portion of their day in an office building. More often than not, the workspace functions as a space of encounter as well as for conversation. Therefore, it needs to be more free-spirited than ever. Through the project with Hyundai Card, I wanted to convey the message that contemporary office spaces are capable of conveying greater freedom than those of the past.

How does the Design Lab differ from other office spaces?

The design approach we took was to coordinate in precise terms with what the members of the space would do. To put it differently, the Design Lab has been segmented to cope with the programs tailored to the particular needs and requirements of individuals. The professional tasks were considered individually for their specific contexts and conditions, such as clearing the inventory, using large design equipment, and securing highly classified information. Furthermore, the space allows flexible composition—enabling the swift formation of workgroups of two, three, four, or five. The pleasure of collaboration exists in the process, and employees feel far more comfortable in a space that fosters connection. Employees have the room to select their own working location, form relationships with their peers, and gather in specific environments. Ultimately, a decision to perform certain tasks prompts dynamic movement within the space. When those movements align with the architectural elements of the space, productivity will only increase.

‘The actual implementation through the Design Lab is superior to the simulation in the Design Week exhibition hall.’

› you have a message for the potential ers of this space?

Architecture is a window into the ltural expression of a particular moment. at is, ‘architecture’ is the expression of a ltural moment and the concretization of ltural space. I hope all designers working re ‘enjoy, create, share, exchange, and perience the utmost freedom’.

Jean Nouvel
Born in 1945 in Pümel, France, Jean Nouvel graduated from the École de Beaux-Arts in Paris in 1972 and received an honorary doctorate from the University of Buenos Aires in 1983. He founded Nouvel & Associates in 1984, worked for JNEC in 1989, and then established Atelier Jean Nouvel in 1994. The originality of all his projects makes all his work representative, and he presents state-of-the-art technologies and materials based upon the context of the urban humanities. His major works include the Arab World Institute of Paris, the Cartier Foundation, the Doha Tower, the National Museum of Qatar, and the Louvre Abu Dhabi. After receiving the Aga Khan Architecture Award from the Arab World Institute in 1989, he received the Golden Lion Award at the Venice Architecture Biennale in 2000, the RIBA Gold Medal in 2001, and the Pritzker Prize in 2008.

Doojin Hwang, Founder of Doojin Hwang Architects / Castle of Skywalkers

How did you come to design the Castle of Skywalkers?

Hyundai Capital took interest in our work following the opening of the Choonwongdang Hospital of Korean Medicine & Museum. Hyundai Capital wanted a precise and machine-like building from the outset, but they also wanted a space that would encourage interaction instead of prioritizing a structure solely built from a management perspective.

What is the advantage of the building's arrangement on this new site?

The floor plan of the building is an exact square. Our firm called it 'the square box' during the design process due to its simple proportionality. Each corner of the square was laid out to point to the north-west-east-south. This orientation was possible only because the building was to be built within the expanse of prairie, not within a city center. The light passing through the expanded metal fills the space. The design resembles the inside of a hanok structure finished with traditional Korean paper. We have also heard that the structure limits the amount of sunlight in the summer while absorbing ample light during the winter, demonstrating its energy efficiency.

Had you ever designed sports facilities before?

Never. Thanks to the cooperation of the Skywalkers team and other professional teams, we were able to inspect similar facilities. This led to one conclusion: such a space is a miscellany of functions, nothing else. Sports facilities once failed to reach generally agreed first-rate standards of design. As long as it was functional, the project was complete. For example, most weight training facilities are located in basements, which must have contributed to the negative impressions athletes have of weight training as taking place in damp, sweaty basements. Conversely, we decided to position the weight training facility in the spot with the best views, while the residential areas were directly located above. Case studies on other existing facilities have therefore become far less important.

What are some of the facility's most interesting features?

Its spaces are accessible from millions of points. Since the volleyball court is in the center of the building, many pathways are leading to this focal point. Natural light is not employed to light the games but is useful for training purposes. The space is very flexible in terms of light: hundreds of blinds throughout the building can compel the space to be extremely bright or dark as needed. Moreover, as we combined the square, rhombus, diagonals, and circles, all of the rooms and the walls belong to a certain geometric order. However, organizing a space according to geometric principles does not guarantee how efficiently it will serve its occupants or users. At some point in the process of such deliberation, we must reach a compromise, and in this building, we chose to match the players' movements by drafting an endless number of dynamic drawings.

The athletes basically live here. How did you incorporate the idea of 'a living space' within the building?

Our team spent a night with the players. They told us that they needed a washing machine and a refrigerator, and it was also interesting to hear that they wanted to be able to keep a puppy. The main concern was the provision of comfortable relaxation spaces and to maintain privacy when a training facility and a dormitory occupy the same building. Since volleyball is a very loud sport, you cannot relax in your room if you can hear others practicing. As such, although the initial proposal was to build two-person rooms, I strongly insisted on separating the bedrooms. You must pass through three doors from the court to the bedrooms. When they are all closed, the doors provide sufficient soundproofing so that outside noise cannot penetrate.

'For the Castle of Skywalkers, the greatest challenge was to make an unemotional and geometric realm satisfy both functionalities as well as emotionality.'

Although the resulting structure is the product of cutting-edge technology and rational development, were there any examples from the classical architecture that you used as a reference point?

The challenge of organizing and orienting the corner spaces is crucial in a courtyard-type building; the spiral corner staircase of the medieval castle offered one model. This idea helped us to position the elevator in the corner. Other references were the centripetal Hagia Sophia Cathedral, the Phillips Exeter Library by Louis Kahn, and the Stockholm Public Library by Gunnar Asplund. In terms of the overall feel and composition of the building, the structures I kept in mind were Professor Jongsoung Kimm's weightlifting arena. His work is imprinted on my memory as the most profoundly sublime works of architecture that transcend the mundane. In felicitous fashion, the Castle of Skywalkers won the Kimm Jong Soung Architecture Award in 2018.

hat were the architectural challenges nd achievements? Numerous issues must ave arisen when designing the space?

Usually, when one goes to a gym, ree-dimensional structures such as spaal frames and trusses give shape to the ace. This was not what I wanted. Since e nature of the game forces players to ok up, I pictured a ceiling resonant with eaning. As such, I demanded a smooth iling with no exposed structures. In reonse, Professor Kyungju Hwang sugsted attaching four convex domes, a yle which slightly resembles the dome ucture of Hagia Sophia in Istanbul. The of of the Castle of Skywalkers, after all, nctions as an outer and inner shell at e same time.

Could you elaborate on the inspiration you took from the hanok?

When one enters the courtyard of a hanok, it feels empty and unsettled, with no one clear point of contact or line of sight. No single element dominates. Only after that, one begins to notice the surrounding pillars, daecheong maru (the main raised wooden floor) lying beyond the pillars, and the space under the eaves. In the end, one's gaze is drawn to the mountains and the sky arcing above the hanok. In Castle of Skywalkers, I aimed to weave an impression of the whole without over-emphasizing individual elements. If there is a difference, this space forges a robust vertical relationship which is not typical of the hanok.

Doojin Hwang
Hwang graduated from the Department of Architecture of Seoul National University, and the Graduate School of Architecture of Yale University. As an architect fluent in both modern and traditional elements, he has taken on numerous hanok projects. His architectural interests can be condensed into three major themes: layered geometry, porosity, and high-density complex architecture, to which he has given the name 'Rainbow Cake Architecture'. His goal as an architect is to complete a work of architecture that will integrate those themes. His major works include Castle of Skywalkers, Choonwongdang Hospital of Korean Medicine & Museum, Won & Won 63.5, Korean Gallery, Museum of Far Eastern Antiquities, Stockholm, Sweden, and Tongin Market Art Gate.

Moongyu Choi, Professor at Yonsei University / Music Library

How did you become involved in the Music Library project?

Originally, the architect commissioned was Kazuyo Sejima. However, as the use of the building changed in the middle of the project, I was engaged to take over. I have a personal connection with Kazuyo Sejima, as my senior colleague at Toyo Ito Architects.

The decision to empty over half of the space that sits above-ground is truly daring.

When I saw the site, I was so struck by the magnificent views that I thought it would be senseless to build a structure here. However, the more pressing issue was to consider who the view was for and who was able to take ownership of it. As such, with all due respect to Hyundai Card's ambitions for the library, I embarked on a construction project that would only minimally impose upon the landscape, which was considered to be a bold approach.

Do you value public architecture?

Today, the contributions to be made to the lives of city dwellers are considered to be of greater value than the completion of a remarkable work of architecture. I believe that my role as an architect is to persuade clients to explore and enjoy their public role. The building of such structures is surely profitable for companies and individuals—however, I hope that the talk-of-the-town will increasingly be how the building enriched the neighborhood, how it made the surroundings a good place to live.

Even with corporate office buildings, the issue of public good arises the moment the building enters the context of the city. Interestingly, in this case, pursuing the public good did not intervene with the interests of the client.

Yes, the Music Library is a case in point. Hyundai Card was able to achieve its intended goal of brand expression. Back then, Hyundai Card had established a sophisticated corporate image in terms of its design and lifestyle. However, following the opening of the Music Library, it became known for its care and attention for the public good. Indeed, it is very difficult for a financial company to be a good company. Hyundai Card, however, made it possible, and people seem to recognize its value.

The fact that the city landscape can be viewed at a glance through this frame, the Music Library, is fascinating.

In the Music Library, you eventually encounter the sky. People naturally take a moment's pause during a long day to look up at the sky. The most significant question posed by the Music Library is, what attitude must architects adopt when building structures in cities? Some architects want to leave their imprint on a city and its buildings—the desire for which is revealed in the choice of materials, forms, or details. In my case, my fingerprints tend to be colorless and invisible.

Hyundai Card and Vice-Chairman Ted Chung have a profound and informed understanding of architecture. Is there any aspect that has evolved or changed when sharing opinions?

In the first place, Hyundai Card did not make any specific requests. Upon completion, Hyundai Card added the red corten steel and JR mural on the façade. The building may undergo modifications but that does not mean its original outline and purpose have changed. Therefore, the additions were appreciated. The structure feels like a real building that is in use, and I believe that is what architecture should be.

'The Music Library demonstrates the best in the public nature of architecture.'

nust have been difficult to design the pe from the entrance to the café on the st floor. What was the most important int to consider?

Safety was the most crucial point. is was because the design followed the turally sloping terrain. One is rarely able create a safe and sound structure on ch a slope. The advantage of slopes is at they instinctively excite those tripping wn their decline. Maybe this is due to a btle physical trigger when passing over ound that is not flat. In fact, sitting in e café on the first floor also prompts a ique feeling.

ficient operation of the programs is as portant as filling the architecture with rtinent content. Do you think the ele- nts you had in mind during the planning ase are now fully operational?

The concept of an LP library was deed smart. If it had been a library filled th books, the building might not have d the impact it has had. Planning content d operating the space is one of Hyundai rd's greatest strengths.

Were any new methods or new materials applied to this project?

When metal panels are used, in general, the modules are divided and attached to the size of the prefabricated elements. However, we developed a 12 m large glass plate for this project. It was of a size that was not used in Korea at the time. We initially proposed glass for the ground floor as a single large plate of glass. However, due to budget constraints, we decided to use two pieces. Yet, after attaching them, it was suggested that one piece would be better, so the two were removed again. It took four months for the glass plate to be delivered from China, extending the construction period by one year. The exterior may also look like concrete from the outside. However, when Japanese architects saw it, they were surprised at how we had managed to make the part so flat (because, in general, exposed concrete, the trace of a form tie is left behind as round patterns). As such, state-of-the-art technology and significantly complex stories are concealed within the Music Library. The case in point was the largest metal panel of any panel that could be produced in Korea at the time and a unique structural system to be employed in the building.

What is your definition of good architecture?

As an architect, you encounter places that stimulate the imagination, but I prefer comfortable spaces. I judge buildings not by the quality of their design but through the benefits they provide to their occupants Ultimately, good architecture, in my opinion, is where people feel comfortable. When you see people smiling brightly, being happy, and feeling joy within a space—that is good architecture.

Moongyu Choi
Professor, Department of Architectural Engineering, Yonsei University. He graduated from the Department of Architectural Engineering and the graduate school of Yonsei University, and Columbia University, New York. After working at Toyo Ito Architects, Hanul Architects, and City Architects, he established Ga.A Architects in 1999. He designed the Jeonghansuk Memorial Hall, Ssamji-gil, Soongsil University Student Center, and Hyundai Card Music Library. He also won the Korean Architecture Awards, Um Duk-Moon Architecture Award, Korea Architects Association Award, and Seoul Architecture Award. He has published many publications, including Doubt is Power, which he co-authored with Professor Hyung-Min Bae, and Questionology in Architecture to commemorate the 20th anniversary of Ga.A Architects.

Heeyeol Yoo, Musician & CEO of Antenna Music / Understage Curator

Of the many spaces created by Hyundai Card, which one have you frequented the most? And what unique impressions have you discovered while there?

I have been to the Design Library, and I often visited the Travel Library, but, of all of them, it's the Understage I've visited most often. Small theatres are popular venues, but the threshold is higher than one might think. The threshold is not for the musicians but for the public, who are somewhat distant. In contrast, the Understage is open as a space can be for communication. The space is tailored to artists' performances as a variety of events can take place. For experimental musicians who are just embarking on their careers, Understage is an oasis. Now, the playing of vinyl has become a popular part of contemporary cultural life that does not require explanation. When the Understage opened, however, Hyundai Card was truly a pioneer. Furthermore, Itaewon was set somewhat apart from the cultural experience. However, it has now transformed into a place where cultural experiences are available 24/7. It is common for a corporation to take on spaces under a lease with a high-profit margin. The fact that Hyundai Card emptied the center of the building to strengthen its cultural life was truly impressive. Since Hyundai Card was welcomed into the area, the landscape of Itaewon has been transformed.

How would you describe working with Vice-Chairman Ted Chung?

He is truly a diligent person, but with an eccentric side. He is even more interested in music than I am; many people enjoy music, but few expand their passion through a specific theme or genre. Vice-Chairman Ted Chung constantly comes up with ideas to provide practical help. Even though he is a businessperson, he invests in culture and is never consumed by the business outcome. I do not have many acquaintances in the business world, but as a person who belongs to the music industry, I think the way he acts and works is eminently desirable, and for his I am grateful. He is a person who has feelings for the music itself and for its creators.

What impact has Hyundai Card's had on the city and upon the public realm through its creation of public spaces?

Space, as a concept, essentially implies a congregation of people. The first step in realizing any design interest or embarking upon a project is securing a space. Physical space is essential for those who attempt to form a community by connecting people through their common interests. Design, travel, music, cooking, whatever it may be, cultural experiences require a physical space to facilitate the unfolding of these stories. For instance, since the digital space is also a 'space', an online platform like Sound Cloud can also be a reference point. The platform itself serves as a space to exhibit the full spectrum of music, from the music of amateurs to that of professionals. However, exporting this to an offline space is hardly possible. The performing spaces like the Understage are almost the only places where people can be face-to-face and share music. The intention behind the creation of public spaces originated from the desire to create such nuanced spaces

'Physical space is essential for those who attempt to form a community by connecting people through their common interests. The intention of creating various public spaces stems from the desire to create such spaces.'

u also work as a curator at the Underage. Do you follow any curatorial princies or have any goals?

I have three principles: first, ether the space is truly necessary for e musician; second, what does it mean r the musician to perform here to his or r future endeavors; and lastly, whether e musician's music is suited to this space.

terms of popular music, is there a parular project on which you would like to llaborate with Hyundai Card?

I hope to create an alternative osystem. Some people begin their mu careers as professional musicians, but any amateurs do not even investigate ents spaces in which to perform. With ose who do not even feel capable of tak the first step, it would be great to plan annual project. Some people dream of coming musicians and creating popular usic. Young creators make experimental usic among themselves. The experience performing at the Understage, even only ce, can have a greater value than one ar of practice at home. An audition pro- t for selecting and discovering creators uld also be possible.

Do you have any further comments?

Hyundai Card has always distinguished itself in terms of scale, as evidenced by its Super Concert. I often felt that the way they spend money is based on a different aim. The Understage considers the distribution of revenue from the musician's perspective, unlike other music venues. In addition to the cost of renting the space, various expenses such as sound, lighting, and guiding staff are calculated to stage a performance. These can be burdensome charges for new musicians. Providing practical help concerning such details is a very thoughtful gesture. Needless to say, the symbolic impact of Hyundai Card on popular music is widely known. However, the details thereof are not well known. If you look at what Hyundai Card is doing as an industry insider, they provide considerable support regarding every detail. This is not easy. The support is always greatly appreciated.

Heeyeol Yoo
Singer-songwriter and music producer. In 1992, he won the grand prize at the Yoo Jae-Ha Music Contest. Since 1994, he was a member of the project band, Toy. He has released a total of 7 albums and pursues music that is not world-weary and contains warm emotions albeit lacking a definitive climax. In 2011, he established an entertainment agency, Antenna, and has been running You Hee-Yeol's Sketchbook, a music program, for 12 years.

DJ Soulscape, Musician / Music Library Curator

You have been active as a curator since the early days of Hyundai Card Music Library...
Yes, before the opening of the Music Library, I had discussed several music-related projects with Hyundai Card. I was preparing a project that would showcase musicians worthy of legendary status in the history of Korean music, shedding new light on them, and connecting them to performance spaces. Meanwhile, the Music Library project began. Naturally, I became involved at the brainstorming stage, which led me to my role as a curator.

What is your main responsibility as a curator?

From the beginning, I thought that the simple recommendation of a few of my favorite albums would not be the essence of my role. Since I have experienced a range of music-related exchanges with Hyundai Card, we were able to set the direction and principle of music curation based on this trust. The overriding principle determined at that time was to use big data to exclude personal preferences, building an archive that would inspire and stimulate today's Korean music scene. I also thought it could be an ongoing project conducted in association with the Understage. We hoped that the foreign artists who visit Korea would naturally use this space for interaction and that Korean artists would be inspired to attain new levels of creativity in their songwriting.

When you first began curating the Music Library, as a consequence of the number and variety of musical genres, it must have been difficult to decide what to focus on.

The Music Library's location, Itaewon, was important. Itaewon is the first area to which Western pop music was introduced. Mr. Joong-Hyeon Shin, one of the oldest and most accomplished guitarists in Korea, also began his career in a US military base in Itaewon. After the Korean War, Itaewon became a cultural melting pot. The curation began with the idea of reflecting the unique qualities of Itaewon, where every genre of Western music was first heard and was acclimatized to the Understage and the Music Library.
Using big data, we extracted all of the landmark musical moments from the early 1900s. In doing so, however, the crucial point was 'counterculture'. I believe the Music Library's current collection reveals how genres of subcultures like funk, hip-hop, and electronic music infiltrated mainstream culture. The Korean popular music scene needs to understand how Western pop culture was inspired by countercultural genres which gave rise to new and radical creative powers.

It is impressive that the curation is ongoing. How is the curatorial work updated?

First, LPs are very easily worn out. The idea is to replace albums every five years. Of course, simply replacing worn-out albums does not mean that the curatorial work is ongoing. Even now, I am still working on discovering newly released albums that would not have been found otherwise or replacing albums with more original discoveries every year. In addition to introducing rare albums from new perspectives each month, I also work on adding objective commentary to all of the collected albums. Note that this commentary is not a mere note of recommendation but the outcome of extensive research.

'I am confident that the Music Library is a one-and-only institution in terms of its breadth, depth, and width to inspire musicians.'

at do you think is the most valuable ect of the Music Library? What kind xperience do you want people to have en they visit the space?

Of course, I hoped that the space uld inspire the Korean music scene. vever, that doesn't mean that the space designed exclusively for enthusiasts musicians. I want it to be a space where tors would be able to get closer to the sic, to get as in-depth an impression as y wanted. It is a place that can satisfy ryone's tastes with their respective ationships to music, from those who nt to listen to trendy music to those study music professionally.

Viewed from a musician's or a curator's perspective, what is your favorite thing about the Understage?

Above all, when viewed from a musician's point of view, it has to be the perfect performance facilities. All processes, including sound systems, are close to the most ideal composition possible. Not long ago, I collaborated with female artists, who told me that they had never experienced the same scale of musical imagination, which varies due to the height of the ceiling, until the moment they set foot in that space. Almost every artist who has worked or performed here will experience some kind of leveling up. Therefore, I have no choice but to invest in my role 100% as a curator with more affection for this space.

What does the Music Library mean to a professional musician?

As there are few physical archival spaces for musicians, even at a global level, I am convinced that the Music Library is the only institution that has the breadth, depth, and scope to inspire musicians. More importantly, the inspiration the Music Library provides can be amplified to a specific outcome thanks to the Understage. The Music Library influences the work of artists. Above all, as musicians gather for comfortable conversations or play sessions at the Understage, the interaction can lead to a more advanced collaboration. The Music Library, Understage, and Vinyl & Plastic form a perfect triad in the newly formed and constantly shifting music content like a single ecosystem.

DJ Soulscape
Based on his vast record collection, he introduces various music genres to the public, crossing the past and present. He founded the Studio 360 label to produce albums and music content and is active across a range of fields spanning film, broadcasting, advertising, performance, and media art. In particular, he has been reminding the public of forgotten Korean popular music since 2007 through the 'The Sound of Seoul' series, a project that re-discovers Korean punk, soul, and rock from the 1970s.

'Hyundai Card plays not only an economic role, but also a cultural and social role, through which art creates a reflection of the present.'

Walter Mariotti, Domus Editorial Director

Tell us about your personal experience with the special edition of Domus on Gapado Island. What was your first impression of the project? What made you publish an edition that focuses solely on this remote island?

My first impression confirmed the role of Hyundai Card, which plays not only an economic role, but also a cultural and social role, through which art creates a reflection of the present.

Gapado Island is a project in which a commercial brand and the public good come together. Are there any similar projects in Europe?

In Europe, the idea of public and private collaboration was a frontier for a long time, only broached by enlightened entrepreneurs. In Italy, we recall Enrico Mattei, a CEO at the public company ENI, who created new alliances with the private sector. There was also Adriano Olivetti, the first private entrepreneur to create public solutions, based on the idea of a "spiritual community," to empower individual talent and promote human development, despite economic and social differences. But I'm also thinking about recent examples, such as François Pinault at Museo Punta della Dogana in Venice, or Brunello Cucinelli, the "King of Cashmere," in Solomeo, Umbria. Both were able to realize new alliances between private interests and the public good, which fueled the creation of new meaning and new opportunities for individuals and society.

Have you ever visited or heard of Hyundai Card's branded spaces and offices, other than Gapado Island? If so, what was your impression?

Not only have I heard about them, but I had the opportunity and great privilege to visit the Hyundai Card headquarters and meet with Mr. Ted Chung, who invited me to his office and, over a rare and exquisite meal with him, explained to me his vision, philosophy, and interests. He was an incredible source of knowledge for me. We had planned a meeting in Milan this year, during Design Week, but unfortunately, COVID-19 changed our plans. Nevertheless, I want to reschedule again for next month, and I'm sure we will be able to start a collaboration.

'A space surrounded by LPs is a playground as well as a space of inspiration. Hyundai Card always creates a space based on a deep understanding of music.'

he Quiett, Musician

f the numerous spaces created by Hyunai Card, I'm curious to know; which one ave you visited the most? Please also escribe your experience in the space.

The places are of course the usic Library and Vinyl & Plastic. I have personal collection of about 1,000 LPs, nd the Music Library perfectly embods the collector's dream to a level that is comparable and unprecedented. In the st, whenever I visited Japan, I would first op by Shibuya and listen to LPs for hours large LP stores such as Manhattan Rerds and Tower Records. When you go to space filled with records, you encounter e music you already know, the music you ed to know but have forgotten over time, d the music heard for the first time but in ich you have developed a new interest. the number of good record stores is a decline in New York and Tokyo, I am cited that Vinyl & Plastic exists.

The LP space is potentially more important because you are a hip-hop musician.

Record shops also have the feeling of an exhibition hall. I get a lot of inspiration just by looking at the artwork on album covers. There is a 'vitality' when you see, touch, and take an album out. Old LPs are also the source of my music. For instance, my song 'Very Special' uses a sample from the song 'Good to Go Lover' by Gwen Guthrie, an R&B singer of the 1980s. I like songs released between 1972 and 1976. When working on my last album, I was drawn to Gwen Guthrie's pure and clean sound. A space in which one is surrounded by records always presents stimulation and inspiration.

Is there any specific project you would like to explore at the Music Library or Vinyl & Plastic as a professional musician?

Taking the Understage as an example, Hyundai Card creates spaces based on a profound understanding of music. Once COVID-19 calms down, it will be so good to pick up my favorite vinyl, listen to it with others, or hold an event to sample and compose a song or find a rhythm together with other musicians. I would really like to stage a range of shows and interact with people within this space.

'Whenever I hear that Hyundai Card has worked on something new, I get excited just hearing about it.'

Sukwoo Lee, CEO of SWNA

Of the many spaces created by Hyundai Card, which one have you frequented the most? And what unique impressions have you discovered while there?

The highlight of the Design Library is the courtyard. The courtyard, which has preserved the original structure of the hanok, creates another cozy world inside Bukchon. As it is situated right in front of our studio, I use the space more than anyone else. The courtyard has become my own space of inspiration and refuge. The welcoming feeling of being disconnected from the outside world, in harmony with the spirit of numerous design books, gives me the feeling of being on a trip, albeit for a short while. At the Design Library, the space and books form an inspiring union.

As a designer, what impact has the Design Library had on the industry?

I think that the space has reminded the public of how design is a part of our daily life, creating a culture in which the public can approach design in a more organic, integrated way. The Design Library is often home to those who have not majored in design. As a designer, I welcome this phenomenon. As the space is often given as an environment for experts as well as for dates, we must recognize the space's contribution to an expansion in our understanding of design from the professional realm to popular culture.

Please share your initial impressions of Hyundai Card.

Whenever I hear that Hyundai Card has worked on something new, I get excited just hearing about it. I think it's extremely rare for a company or brand to create such an expectation when working on something new. After all, it is a credit card company! When SWNA studio relocated to Bukchon in 2018, the first thing everyone said was 'It's right in front of the Design Library'. It was also very exciting for me. The Design Library is now a cultural landmark in Bukchon. In 2019, I had the opportunity to hold a small SWNA exhibition there, which led us to appreciate how much the Design Library has contributed to the culture and community of the Bukchon.

'Hyundai Card has enhanced the diversity of the Korean architectural landscape.'

ihoon Lee, Principal of SoA

the many spaces created by Hyundai rd, which one have you frequented the st? And what unique impressions have u discovered while there?

I usually visit the Music Library, rage, Vinyl & Plastic, and the Cooking rary. Of these, I am particularly fond the pavilion-like exterior design of the sic Library. The intelligence of creating unfamiliar exterior through conventional hitectural elements is made strikingly parent. The simple and clear concept hind this space using the longitudinal ography of Itaewon street is a particular ue of the Music Library. It is an archi-ture that fully embraces urbanity.

As an architect, what impact have Hyundai Card's architecture and space projects had on the city, the public realm, and the Korean architectural scene?

In terms of spatial planning, Hyundai Card's projects have no shortcomings in the following three areas: architecture, location, and operation. It has set the precedents that good planning and efficient operations are as important as great architecture to create valuable spaces. The spaces created and operated by Hyundai Card have ensured that our society privileges individual architecture of the finest level. Moreover, activities like the Young Architects Program (YAP) have enriched the diversity of the Korean architectural landscape. I acknowledge the considerable contributions Hyundai Card has made in expanding the points at which good architecture meets its public.

Please share your initial impressions of Hyundai Card.

I thought its planning capability was amazing, as exemplified by the Super Concerts inviting world-renowned artists to perform in Korea. It is also closer to an IT company than a credit card company. Last but not least, it is a company that knows how to draw on pertinent design principles.

'It has successfully fulfilled its role in raising the bar for Korean design.'

Yeongkyu Yoo, Founder of Cloudandco

Of the many spaces created by Hyundai Card, which one have you visited the most? And what unique impressions have you discovered while there?

Whenever I come to Korea, I make sure to visit the Design Library. It is also one of the places I tend to recommend to designers overseas. It is a modern yet reserved space with the peacefulness of Korean traditional architecture. The atmosphere naturally you take out your notebook and to become immersed in your own ideas.

As a designer, what impact have Hyundai Card's architecture and space projects had on the design industry, the city, and even on the public realm?

I had a chance to participate in a global project that sought new business opportunities based on collaborations between Japanese artisans and overseas designers. It was a project led by the private sector and aimed to help Japanese designers enter the overseas market through high-quality works and exhibitions. It was a very successful project, after which I received scores of calls expressing interest in my work. I believe Hyundai Card is the only outfit in Korea that is presiding over such a project. As much as 15 years ago, I helped Korean designers enter the overseas market in collaboration with MoMA, New York. Considering that this was an initiative directed by a private company, it was groundbreaking when you think about it. In particular, the Gapado project, conducted by Hyundai Card based on its commitment to the desires and needs of local residents and governments, presented a whole new vision for regional regeneration. Various global programs have been planned, following the leads of public institutions or governments. Yet, there is still room for improvement. Hyundai Card continues to invest unwavering effort in such projects. As a designer, I am truly thankful for their efforts. I expect Hyundai Card will play a significant role in raising the overall standard of design in Korea.

Please share your initial impressions of Hyundai Card.

Not many companies act with such commitment to social engagement projects that present a new vision for regional regeneration, as with the Gapado project. I consider Hyundai Card to be a mentor, who raises the bar for the Korean design scene.

'Hyundai Card has proved how architecture, as a corporate strategy, can mature an architectural culture in the public realm.'

yungmin Pai, Professor at the University
Seoul, Architecture Curator

f the many spaces created by Hyundai ard, which one have you visited the most? nd what unique impressions have you scovered while there?

Of the library series of Hyundai ard, I consider the Music Library to be the ost significant. The space shows, simply d powerfully, how the architecture of private company can contribute to an ban context and culture. Although it is mmon in all Hyundai Card libraries, the proach of making the user, rather than e architecture, the main character is rticularly exceptional in the realization the Music Library.

As an architectural historian, a critic, and a curator, what impact have Hyundai Card's architecture and space projects had on the city and the public realm?

Hyundai Card, together with Amorepacific, has shown how a Korean conglomerate can contribute to the contemporary architectural scene. It has proved how architecture as a corporate strategy can mature an architectural culture and influence the public realm. In that sense, if less widely known compared to Hyundai Card's Library series, the Gapado project is even more important. Notwithstanding the typical roadblocks to project implementation and operation, the Gapado project exemplifies the role architecture should aspire to in society, going beyond a simple measuring of the 'success and failure' of the project

Please share your initial impressions of Hyundai Card.

Hyundai Card deserves your attention because it supports the cultural realm in diverse ways while also pursuing 'aesthetic' marketing, albeit not from within fashionable industries such as fashion, cosmetics, or pop culture-related companies. To be honest, I was at first appalled that all libraries were to be operated based on closed membership except for the external space of the Music Library. Yet, the Gapado project has transformed my perception of Hyundai Card. As it plays a key role in transforming the public's perception of architecture, design, and art, I have been and will continue to be interested in Hyundai Card ventures and activities.

'As a professional athlete, I take pride in training in the finest facility.'

Sungmin Moon, Volleyball Player, Cheonan Hyundai Capital Skywalkers

What distinguishes Castle of Skywalkers from other sports facilities?

Simply put, the space provides an all-in-one integrated system. The first thing that comes to mind is the convenience of the facilities for the players, as everything, except for the hospital, is in one place. Another great advantage is that the training facility is right in front of the dormitory, so we can pop out and engage in individual training at any time. I heard that other teams now hope to create similar spaces, but they've not yet managed to – our space is the envy of other players in other team sports.

Are there any pros and cons only understood by the actual users?

The superb convenience of the restrooms. It is the most frequented place! Toilets, shower booths, washbasins – every element of the space has been tailored to suit the height of the players. The sauna is well equipped, and the spa is used daily. The only downside is that you can hear people training even when you're inside your room. Of course, you can hardly hear them if all of the doors are closed. The last exercise session ends at 9 pm and the lights are turned off afterward, so noise is no longer an issue late at night. The team also constantly checks itself and improves, so there is no real inconvenience in the space. Players have even said that they feel more at home here than in their actual home!

One side of the wall can be opened to connect the inside space with the outside. Does the team often use this function?

We open them once in the morning for ventilation, but never during training. Otherwise, if you look outside while training, you feel peaceful. The outside is well landscaped and there is a reservoir nearby, so we often go for a walk.

What are the direct or indirect influences of the building on the team and the players?

Without a doubt, it is the pride we feel in training in a facility of the highest caliber.

'I was deeply impressed with Hyundai Card's unique approach to the physical experience of a space, instead of going for the cliché services or benefits, such as valet parking services or discounts.'

egyu Bong, Actor

ease share your first impressions of the yundai Card spaces you've visited.

When I first visited the Itaewon usic Library, I was so moved by the space d its potential that I found it difficult to ntain myself. Itaewon and Hannam-dong e densely populated with buildings so at you can see nothing but the builds themselves. Moreover, the stores are ually concealed within specific buildings. e very idea of leaving half of the exterior ace empty in the most crowded and freented place in Seoul was truly striking. sonally, my favorite place is the Cooking rary. I am often amazed that not only the ace and its content but also food and nk, achieve such a high level of quality.

If you could make one further addition to the activities pursued by the Cooking Library, what would it be?

I hope there will be more opportunities to handle the theme of F&B with a little more variety. Using plates that would make the food pop or expanding the tools used for a more advanced experience—these would be interesting additions. For example, even in terms of cooking rice, various details can be explored depending on the type of rice, water, the tools such as pots, pressure cookers, and cast iron, and the heat source. Also of interest would be frequent collaborations with brands, chefs, or restaurants with a long and established history.

If Hyundai Card was to create an entirely new space, what type of a space would you like to see?

Even if the size of the library shrinks a little, it would be great if it could be built in other areas. The era of locality has been flung wide open since the COVID-19 pandemic. As such, it would be great if the library could be expanded beyond the realm of a trendy neighborhood.

'Every time I stop by the Cooking Library, I discover new aspects in the space as if peeling an onion.'

Teo Yang, CEO of Teo Yang Studio

What do you think of Hyundai Card's various spatial projects?

I've visited four in the Library series and Vinyl & Plastic, and I was truly impressed with the selection of location as well as their tailored spatial design. What is even more admirable is their meticulous coordination of details such as the positioning of individual staff members.

I imagine that you visit the Design Library most often considering your profession.

You are right. I consider Gyeongbokgung Palace, Art Sonje Center, Bukchon, MMCA, and the Design Library as spaces of the same flavor and texture. The Design Library is an open yet closed place, in which you simultaneously experience both immersion and openness while reading books in the Design Library. Surrounded by various design elements and genres, it is also a space where one can feel and obtain an idea of design. The Design Library is also relaxing, with the peace achieved by the courtyard. That one can read while sitting on a Jean Prouvé chair also gives me so much joy.

Which book in the Design Library do you enjoy reading the most?

Notwithstanding the vast archive of information known as the internet, the only, ultimate way of obtaining high-quality and credible information is still through books. Whenever I am inside the Design Library, my hands are first drawn to the Domus collection. The collection is bound to be the best reference book for those who deal with space. Particularly, you can observe how architecture and design are interlinked, and how design is integrated with society. I believe it is a valuable archive for architects and designers who possess a strong design philosophy without being swayed by the rise and flow in fashion and beauty trends.

If Hyundai Card was to create another space, what kind of space would you like to see?

Enduring the Japanese colonial period as well as other hardships, Korea has experienced something of a loss of historical identity. Thereafter, during the period of rapid modernization, Korea's unique regionality, crafts, and modern design have not been accurately defined. For me, Hyundai Card libraries are spaces that contain archives and documentation. As such, it would be great to develop another space in which Hyundai Card's unparalleled curatorial approach can record and archive valuable data and materials.

'Among all of Hyundai Card's library series, I think it is the best place to realize the potential of space.'

Wookjeong Lee, Program Director (PD)

What do you think of Hyundai Card's various spatial projects?

I have been to all of the libraries in the Library series. Each library looks different, even from the point of the façade, and is located within symbolic areas. Most particularly, I had high expectations for the Cooking Library, as it is related to my profession, even before its opening. In the digital age, paradoxically, the power of the medium — the printed page — is enormous. Before the Library series, Korea only possessed run-of-the-mill public libraries and bookstores. That is why Hyundai Card's libraries are even more impressive because various experiments take place through the prism of books.

Can you elaborate on your experience in the Cooking Library?

Unlike design and travel, cooking can be performed 'live' inside the library. Of all the premises in Hyundai Card's Library series, it is a space that best realizes the possibilities as yet only dreamed of. The high ceiling and bookshelves on the second floor seize your attention upon entering the space. At first, I wondered how the smell of food could be harmonized with the books. Now I have witnessed how their architectural solution offered an excellent solution without conflict, such as building a separate kitchen space in the basement.

I know that you have traveled the world and collected cookbooks. How would you judge the Cooking Library's collection?

It may be small, but the library is unprecedented in terms of collecting and assembling world-class cookbooks. I especially like the fact that there were a lot of cookbooks, such as art books. I think that cookbooks with good illustrations are of greater value than books with recipes summarised simply. Furthermore, I enjoy cookbooks that partly reveal the chef's personal story, life, and influences – the Cooking Library has a great selection of books in this category. Unfortunately, there are too many spaces in Seoul that use books merely as ornaments. Hyundai Card's Libraries are wonderful spaces in which you can actually access books and take them out to read at any time.

Do you have any suggestions for the Cooking Library?

The Cooking Library, in particular, has pursued a range of ambitious aims, such as offering events that far exceeded the scope and skill level of cooking classes in other similar spaces. One wish I have is that more effort is put into forming a community. I believe that in order for any great content to be sustainable, a community must be formed, a community of fans who always frequent the space with fondness and loyalty.

'The Design Library has a square frame, including a courtyard. When you are in the space, you feel like you are inside a frame. That is why I think the space goes even better with photographs.'

Hasisi Park, Photographer

What do you think of Hyundai Card's Library series?

It is a truly smart way of giving the impression that you are not simply spending money to consume but proposing comprehensive spending on the experience itself. I remember that in my 20s I committed to not owning a credit card before I turned 30. Yet just so I could visit the Design Library, I broke that promise. I am so grateful for these Libraries, which stand as spaces that inspire their users.

Could you tell us more about your experience in the Design Library?

There used to be a café on the ground floor, which now houses the rare book collection. I still have vivid memories of drinking coffee and skimming various design books. I also remember the first time I read rare books on the second floor with gloves on, and they were not simply books but each page was more like a special print or artwork. I especially enjoyed reading magazines including Domus. The Design Library features square-like architectural elements including a courtyard. When you are in the space, you feel like you are inside a frame. That is why I think the space looks even better with photographs.

Do you have any suggestions for the Hyundai Card Libraries?

In light of the COVID-19 crisis, it seems that now is the time to be more alert to climate change. Perhaps, in this wonderful space, we might think of holding activities that address the climate and environmental crises.

‘Hyundai Card Libraries represent the combination of capital, an excellent concept, and innovative design.’

Jaewon Kim, CEO of OrEr

What do you think of Hyundai Card's Library series so far?

I truly sense the power of capital in a good way. Of course, capital alone is not enough to create great content. I believe the Hyundai Card Libraries represent a combination of capital, an excellent concept, and innovative design. As a person who operates and studies spaces, the series provides a superb reference point. In particular, the Cooking Library is the best example of the use of metal in structures in Seoul. The finishing and details are superbly executed.

As a person who runs a space, do you have any suggestions for the operation of the Library series?

The customer services and space operations are truly excellent, with features such as providing menus on iPads in Cooking Library. However, the Libraries fall short on retail functions, as selling items is not a core function of the spaces. For example, the Travel Library would be a much more attractive space if it presented items that are both international and from the local market. In that line, the Saturday Market was recently hosted by the Cooking Library in collaboration with TWL, which was superb. Centering on food, it drew a great deal of attention by showing and selling dishes and tools in the broadest possible sense.

If Hyundai Card was going to create a new space, what do you think it should be?

COEX is the only space in Seoul that can host conferences or fairs. However, it is too wide and divided like a checkerboard, limiting events to dull compositions. In recent years, the number of creative fairs has been on the rise, but there are no spaces in which to fully embrace them. It would be great to introduce a creative space for events, the opposite of COEX, in areas like Seongsu-dong.

‘Hyundai Card was like jewelry for its users.’

Joonwoo Park, Chef and Food Columnist

As a chef, how would you describe your experiences using the Cooking Library?

Since it has a food and beverage store on the first floor, a library on the second floor, a cooking studio, and even a private space, I think that it has a suitable range of content for a ‘cooking’ library. Furthermore, instead of placing the focus on architecture or design, the specialist books and magazines that fill an entire floor of the building further enhance its brilliance.

What impact has the Cooking Library and other related architectural projects had on the industry, on the city, and even on the public realm?

The Cooking Library offers not only a wide assortment of books but also a complex space in which you can engage directly with chefs from different fields. It also provides a range of hands-on experiences to a general public interested in cooking. While it is often the case that the state and local communities devise spaces that fulfill only the most basic of needs, Hyundai Card, as a private company, has built a space that satisfies more nuanced and individualized needs.

What was your first impression of Hyundai Card?

Hyundai Card is not just a credit card but often behaves like jewelry. Its diverse card series including the Red, the Purple, and the Black, as well as the Green, have served as a means of expression for its valued customers.

'We like when a commercial brand creates a cultural and institutional open space for art installations and exhibitions.'

Osgemeos, Street Artist

What is your impression of the Hyundai Card spaces?

We were very impressed to see the space at Storage by Hyundai Card, especially the venue and the Vinyl & Plastic record store. This was the most impressive for us, especially when we saw the JR installation that was so great in that space. The area where the museum is located is incredible. We think this part of the space is the most impressive and amazing. We like when a commercial brand creates a cultural and institutional open space for art installations and exhibitions.

Among all the galleries and exhibition spaces at which you have held a show, which one impressed you the most?

It is difficult to say which one we liked the most because we are lucky to have seen some amazing places. For example, we are back at the Pinacoteca museum in Sao Paulo for our retrospective show. For us, it is a beautiful experience to return there and do this exhibition at a great institution. The Pinacoteca is one of the most special places in the world. There are hundreds of other special and beautiful institutions, but it is difficult to say which one has impressed us the most. Each one has its own special aspects, especially when the museum is doing things the right way and respects the artist and the art.

How was Hyundai Card as a partner for your recent exhibition?

Working with Hyundai Card as a partner for the exhibition was a great experience. Unfortunately, we could not be there for the process of putting together the show or the opening and educational workshops due to COVID-19. However, Hyundai Card did a really nice job because we did everything remotely by internet and e-mail, and the team worked well together. We were impressed by the city, the town, and the neighborhood where Storage is located. I wish we could go back there when everything is better and do more projects in the city. We would love to create a big wall mural at some point in the future.

Designs of Hyundai Card

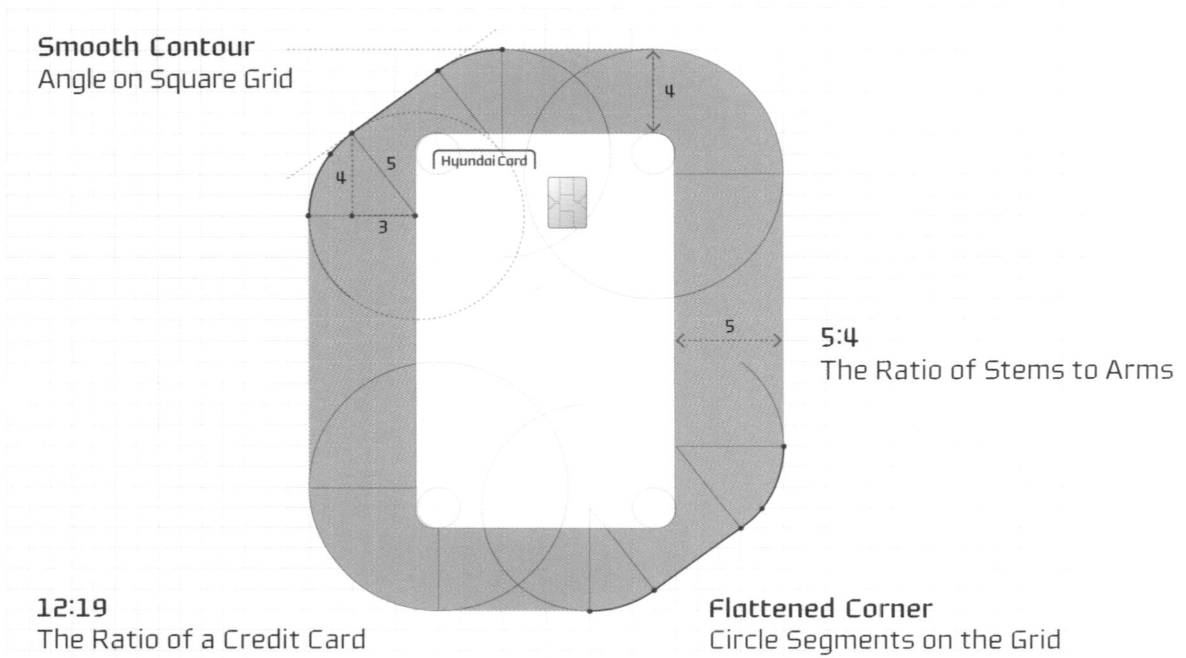
Smooth Contour
Angle on Square Grid
Hyundai Card
4
5
4
3
5
5:4
The Ratio of Stems to Arms
12:19
The Ratio of a Credit Card
Flattened Corner
Circle Segments on the Grid

Youandi New Title

abcdefghij
Short ascenders and descenders

Youandi New Text

Larger radius
Double bars
Larger details
abcdefghij
Longer ascenders and descenders
Double storey a + tail to aboid confusion with an o
Larger overall spacing
Optically corrected weights
Larger counters
Less sharp angles → Bolder strokes

Hyundai Card Youandi Modern Typeface inspired by the shape of a credit card. The Typeface has been redesigned as variable fonts in 2020, which could be varied in an infinite number of families. The new design solved multiple problems: the imbalance between symbols/shapes/heights/sizes, the limited number of families, reduced readability in apps and on the web.

Youandi New Title Condensed Thin	ABCDEFGHIJKLMNOPQRSTUVWXYZ abcdefghijklmnopqrstuvwxyz
Youandi New Title Condensed Regular	ABCDEFGHIJKLMNOPQRSTUVWXYZ abcdefghijklmnopqrstuvwxyz
Youandi New Title Condensed Bold	**ABCDEFGHIJKLMNOPQRSTUVWXYZ abcdefghijklmnopqrstuvwxyz**
Youandi New Title Thin	ABCDEFGHIJKLMNOPQRSTUVWXYZ abcdefghijklmnopqrstuvwxyz
Youandi New Title Thin Italic	*ABCDEFGHIJKLMNOPQRSTUVWXYZ abcdefghijklmnopqrstuvwxyz*
Youandi New Title Light	ABCDEFGHIJKLMNOPQRSTUVWXYZ abcdefghijklmnopqrstuvwxyz
Youandi New Title Regular	ABCDEFGHIJKLMNOPQRSTUVWXYZ abcdefghijklmnopqrstuvwxyz
Youandi New Title Regular Italic	*ABCDEFGHIJKLMNOPQRSTUVWXYZ abcdefghijklmnopqrstuvwxyz*
Youandi New Title Medium	ABCDEFGHIJKLMNOPQRSTUVWXYZ abcdefghijklmnopqrstuvwxyz
Youandi New Title Bold	**ABCDEFGHIJKLMNOPQRSTUVWXYZ abcdefghijklmnopqrstuvwxyz**
Youandi New Title Bold Italic	***ABCDEFGHIJKLMNOPQRSTUVWXYZ abcdefghijklmnopqrstuvwxyz***
Youandi New Title Extra Bold	**ABCDEFGHIJKLMNOPQRSTUVWXYZ abcdefghijklmnopqrstuvwxyz**
Youandi New Title Black	**ABCDEFGHIJKLMNOPQRSTUVWXYZ abcdefghijklmnopqrstuvwxyz**
Youandi New Title Expanded Thin	ABCDEFGHIJKLMNOPQRSTUVWXYZ abcdefghijklmnopqrstuvwxyz
Youandi New Title Expanded Regular	ABCDEFGHIJKLMNOPQRSTUVWXYZ abcdefghijklmnopqrstuvwxyz
Youandi New Title Expanded Bold	**ABCDEFGHIJKLMNOPQRSTUVWXYZ abcdefghijklmnopqrstuvwxyz**

General Purpose Credit Card (GPCC). GPCC represents Hyundai Card's exclusive and proprietary credit card products. From the top left, listed are the premium card lines: the Black, the Purple osée (Jade, Versailles, Seine), the Red, the Green, the Pink (Glossy, Stranger, Lollipop, Little Black Dress, Satin), MX BOOST (Bubble Wrap, Gummy Bear, Canvas, M Charger, X Charger, M Fluffy, X Fluffy, M Steel, X Steel, the Gear, the Coil, the Can), M, M2, M3, X, X2, X3, T3, Z (Family Z, Work Z, Ontact Z, Clap Z, Rolling Z, Sweet Z, Block Z, Slate Z, Laser Z), DIGITALLOVER (Jean Crush, Rusty Robot, Star Bomb, Foggy Planet), ZERO Limited Edition (Black, Gold), and Chameleon

PLCC (Private Label Credit Card). PLCC means 'business operator marked credit card'. PLCC provides optimized benefits to clients who use certain brands or corporations. From the top right, the listed credit cards are Hyundai Mobility Card (Origin, N Line, Lounge, My PONY, UFO, Blue Night), Hyundai Motors EV Card (Pixel Racer, Pixel Light, I'm on it, Now Loading), MUSINSA Hyundai Card (MUSINSA Magazine, Sold Out, Orange Tag, Care Label, Shoebox-Blue, Shoebox-Red, Green Button, Denim), SOCAR Card (MAP, STRAIGHT, RIDER, 88, PIN), Baemin Hyundai Card (On My Way, Sunglasses, The More The Better, Delivery Bag, Sunny Side Up, Seaweed, Tteokbokki, Omega 3), Starbucks Hyundai Card (Mystical, Caution!, Midnight, Sparkle, Starry), Korean Air Card (the Craft, the Wing, the Pass, the Tag, the Wing Metal, the Pass Metal, Sunrise, Aurora, Sunset), Energy Plus Card, SSG.COM Card, Costco Rewards Hyundai Card, Smile Card (MATRIX, CHAPLIN, SKY, SMILE CLOCK, EINSTEIN, HERITAGE), Kia Members Credit Card, and E-Mart e-card.

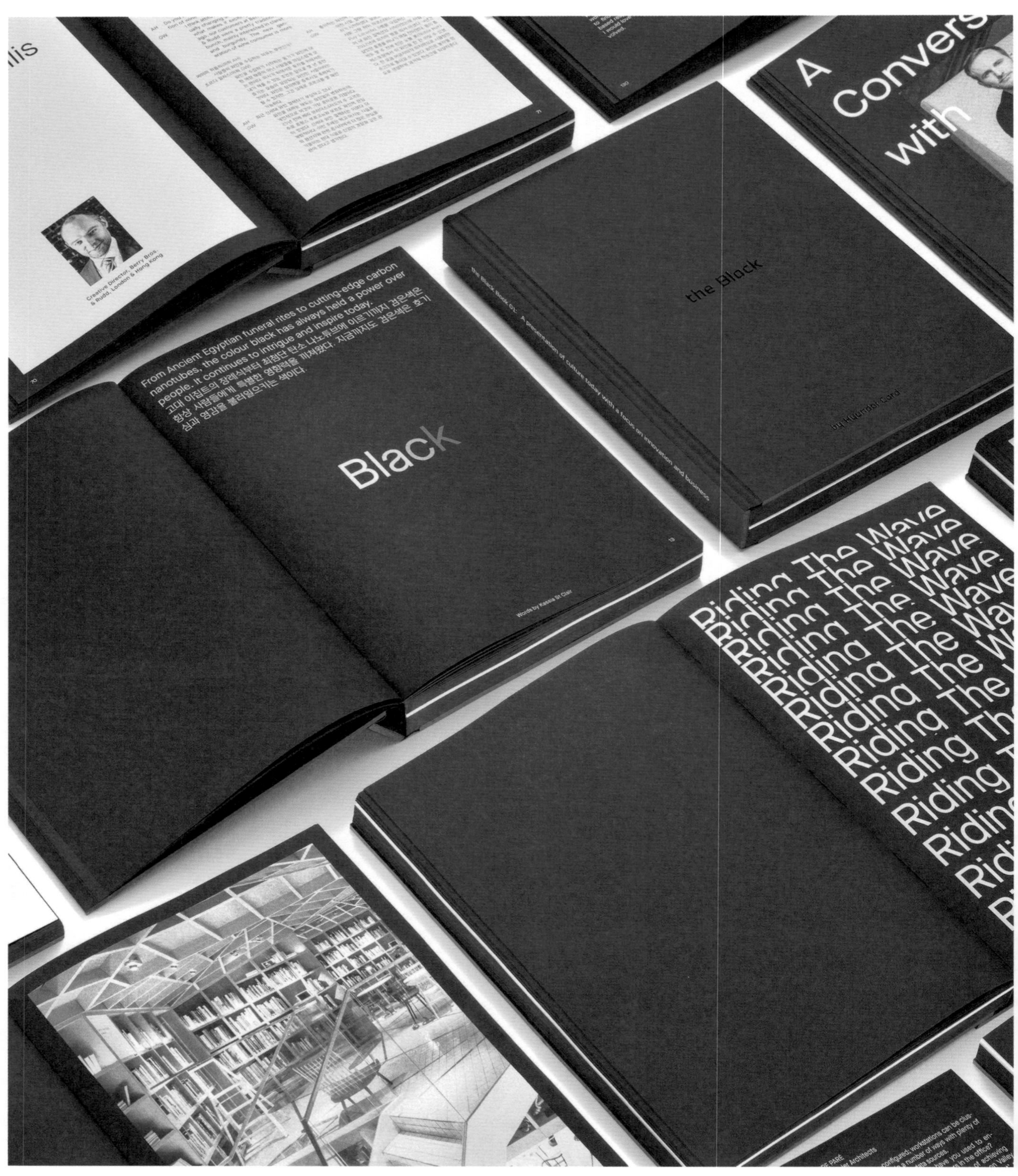
From Ancient Egyptian funeral rites to cutting-edge carbon nanotubes, the colour black has always held a power over people. It continues to intrigue and inspire today.
고대 이집트의 장례식부터 최첨단 탄소 나노튜브에 이르기까지 검은색은 항상 사람들에게 특별한 영향력을 끼쳐왔다. 지금까지도 검은색은 호기심과 영감을 불러일으키는 색이다.
Black
Words by Kassia St Clair
the Black
the Black Book 01. A presentation of culture today with a focus on innovation and business
Creative Director, Berry Bros. & Rudd, London & Hong Kong
A Conversation with
Riding The Wave

yundai Card the Book. A new release of e premium credit card line package—e Black, the Purple, and the Red—is presented under the theme, 'the Book'. This a lifestyle book series with the persona Hyundai Card premium cards: Business nd innovation for the Black, design and avel for the Purple, art and fashion for e Red. With the motto of 'See Color Differently', what color connotes has been defined and expressed via cultural contents and images. All designs and layouts so focused on conveying the identity of olor. The way of explaining the benefits information of the card is replaced by e indirect yet robust method of filling e pages with contents tailored to each olor and the lifestyle pertaining to it. Of ourse, it is unforgotten that this book is 'package for credit cards'. Before the ader meets the card, a few hint pages ppear. After the pages, there is rather thick page containing the actual card. his is to let the readers naturally find e card as the thick page holding the ard stops the hand flipping through the ook. It was created in collaboration with itish design studio Made Thought and ondon-based writer and consultant Jonhan Openshaw.

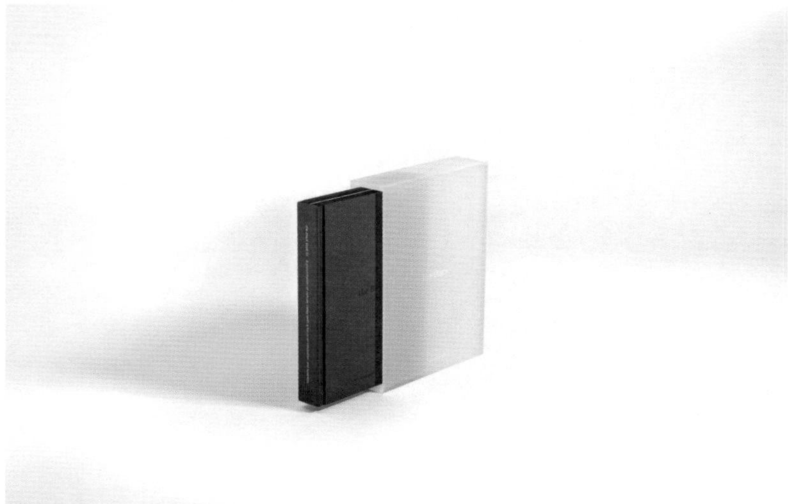

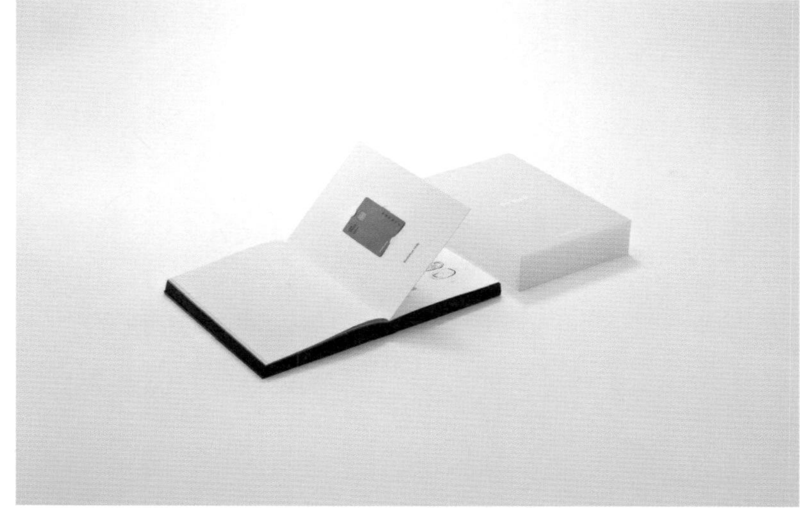

the
Purple
the Purple

the Red
My Hot
Luxury

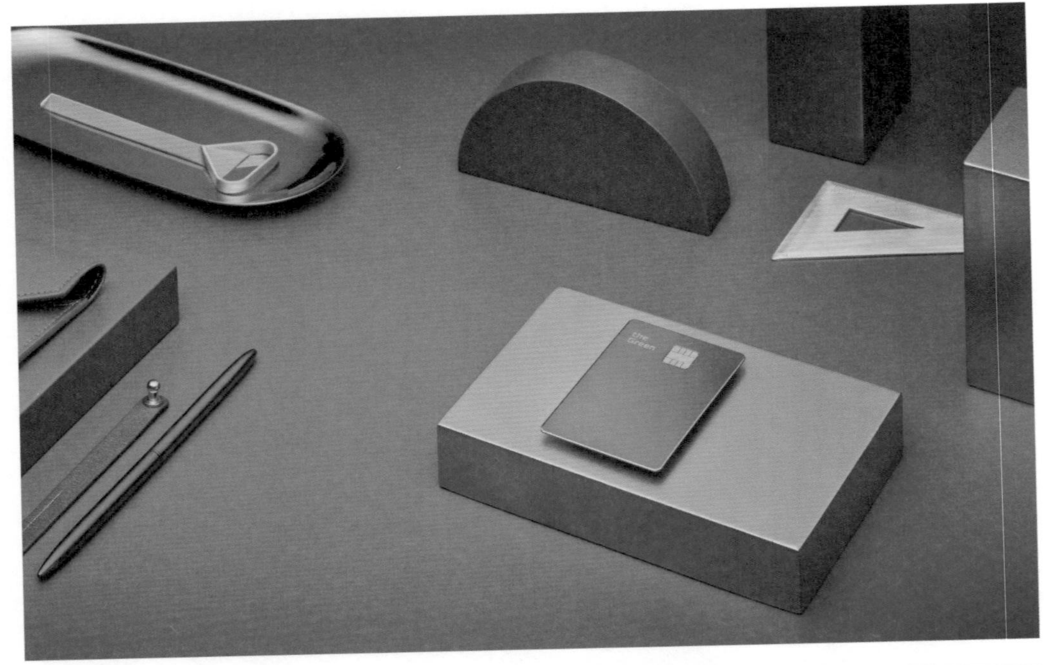
the
Green

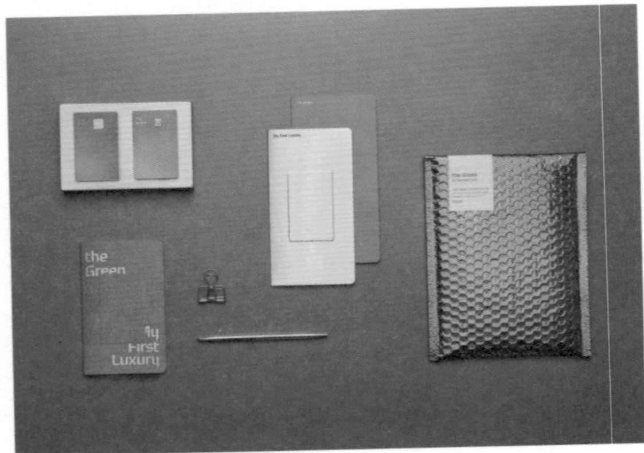
the
Green
1y
First
Luxury

Hyundai Card's new premium card launched four years after the Green's release in 2017. The Pink credit card, created with the slogan of 'My First Seduction', reflects today's trend, breaking away from the calm ambiance of traditional premium cards. Breaking away from the standard shade of 'pink', the Pink distinguishes itself. Reflecting on the shopping-centered benefits of the card, a recyclable bag-shaped package has been developed.

Hyundai Card MX BOOST card design. Customers can choose from a total of 12 unique card design options. By boldly breaking away from the minimalism that Hyundai Card used to adhere to, the design uniquely employs a variety of styles and senses. Along with the launch of the credit card, Hyundai Card also designed BOOSTTEM, items to 'boost' daily lives. BOOSTTEM consists of 21 types, 50 items in total.

MX BOOST
MX BOOST
MX BOOST
MX BOOST

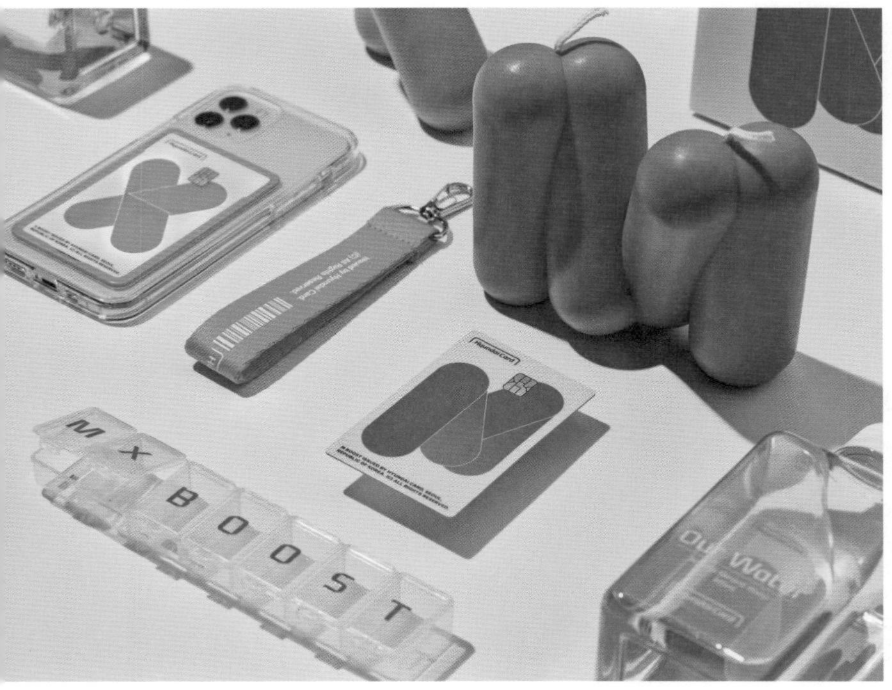
M X B O O S T

BENEFITS
STICKERS
CARD
MX BOOST MX
HYUNDAI CARD
MORE OR LESS
MX
X
MX BOOST

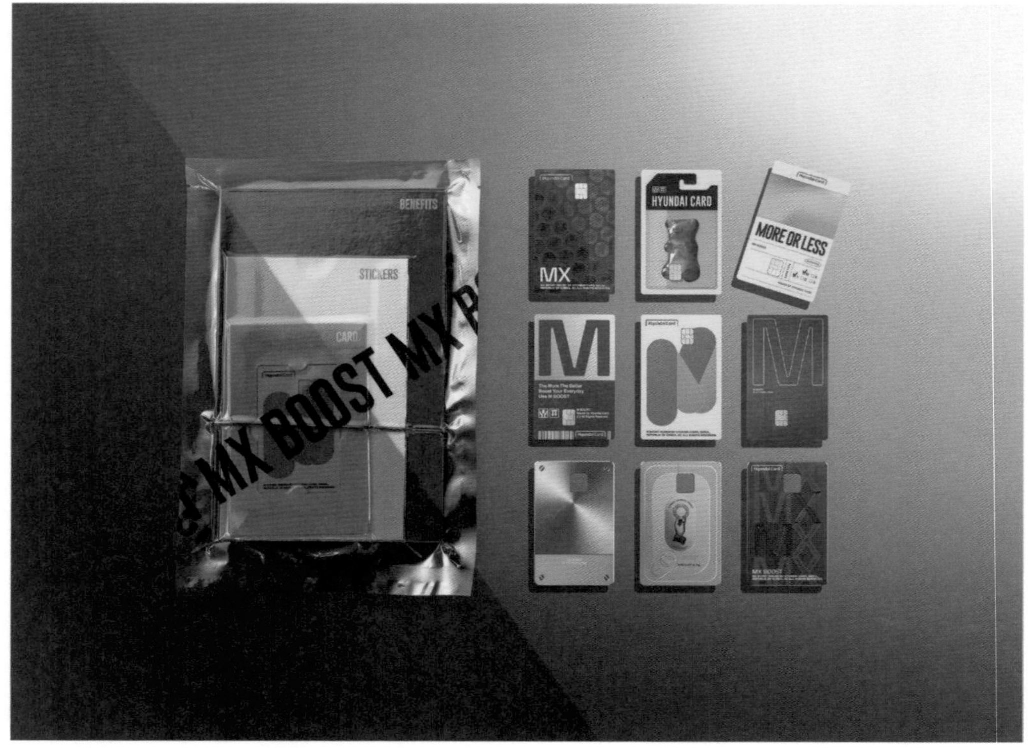
BENEFITS
STICKERS
CARD
MX BOOST MX
HYUNDAI CARD
MORE OR LESS
MX
M
MX BOOST

yundai Card Z. As 'the ultimate discount ard', the keywords are Ultimate and Game hanger. The series features card designs at embody extreme and provocative artorks symbolizing 'the key player with ultiate discount benefits.'

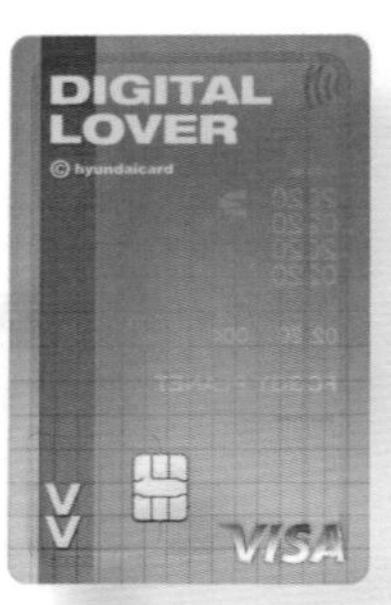

The credit card design of Hyundai Card DIGITAL LOVER. Characterized by the thorough appeal of the brand's direction via images, this card is tailored for digital natives who create their own universe in digital realm. Four versions were released: Jean Crush, Rusty Robot, Star Bomb, and Foggy Planet. The designs are derived from space or spaceships, such as Jean Crush expressing the fragments of a corroded spaceship with a metal sheet material, and the translucent plate Foggy Planet being reminiscent of a card key used in a spaceship. This design has won the iF Design Award and IDEA award. For the DIGITAL LOVER and ZERO Limited Edition, each plate design motif was also reflected in the package. For DIGITAL LOVER, the envelope shows a photo of the plate. ZERO Limited Edition designed gold-looking stickers and other printed materials under the theme of 'Fake Luxury.'

DIGITAL
LOVER
DIGITAL
LOVER

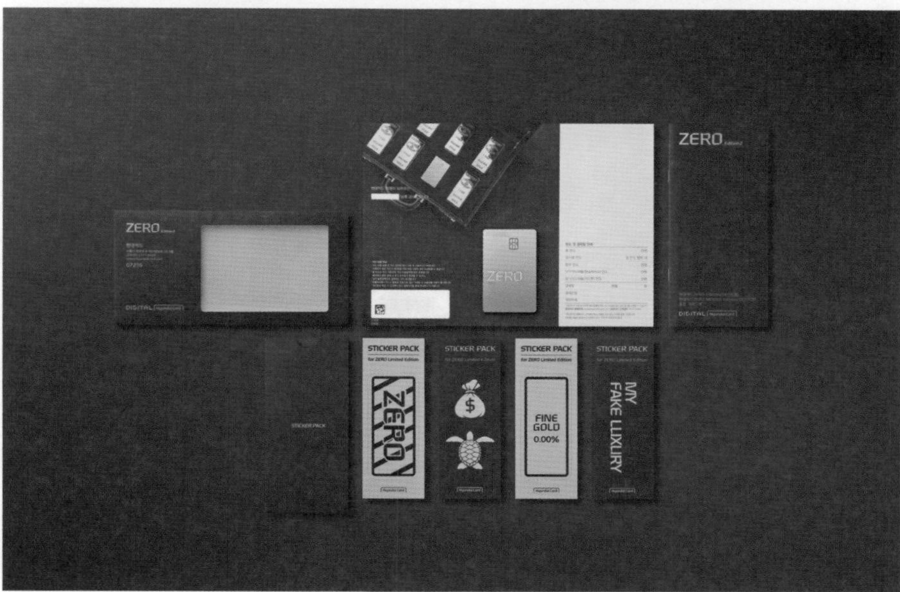
ZERO
ZERO
ZERO
STICKER PACK
STICKER PACK
STICKER PACK
STICKER PACK
ZERO
$
FINE
GOLD
0.00%
MY
FAKE LUXURY

The card was introduced by the PLCC partnership between Starbucks and Hyundai Card. The core elements of the Starbucks brand are reinterpreted from the perspective of Hyundai Card. The design process started with utilizing the Siren logo, the most loved by loyal customers of Starbucks. From there, Hyundai Card reinterpreted the logo into various forms and added more designs with stars, the symbol of the Starbucks reward system.

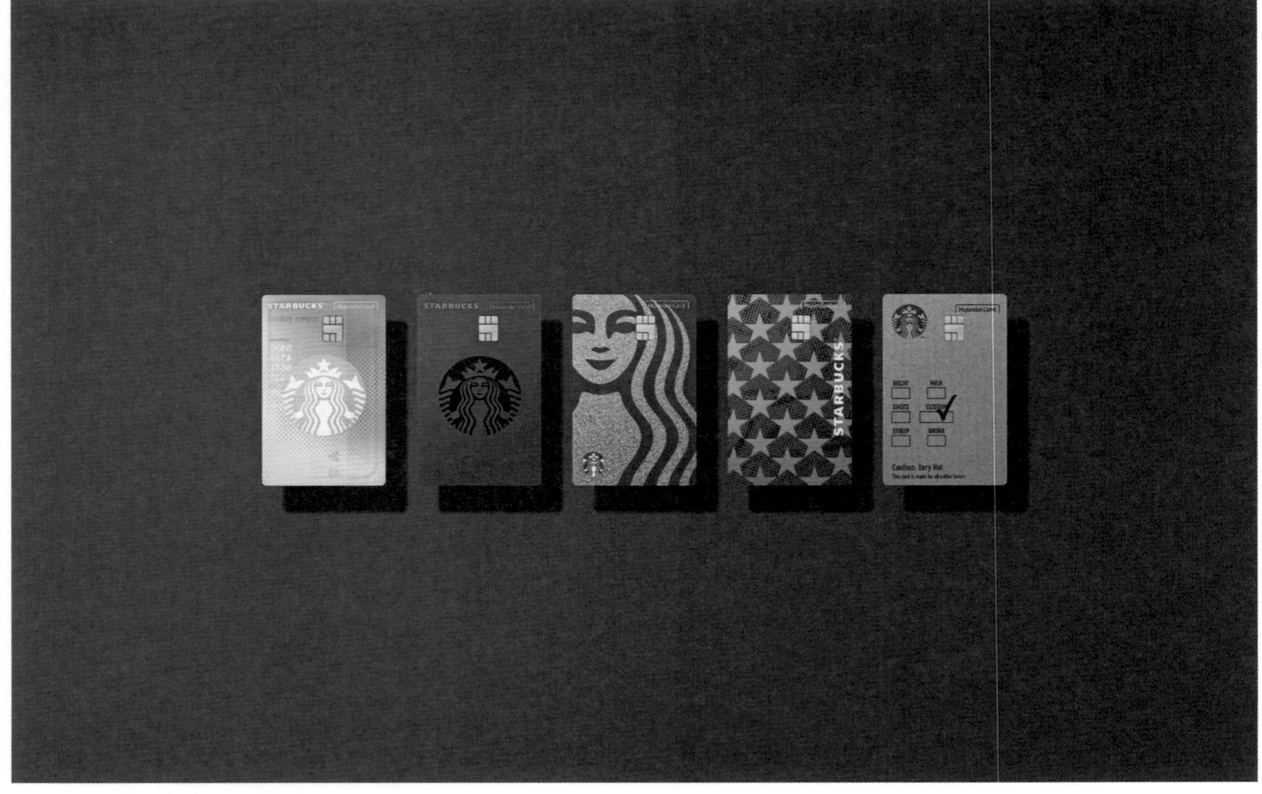

The card was launched by the PLCC partnership between Korean Air and Hyundai Card. The card offers uniquely optimized benefits for travel. Of four types in total, the series consists of 030, 070, and 150 cards reminiscent of flight numbers, and 'the First', available only to Korean Air premium members with Morning Calm or higher. To reflect the excitement of travel on the card plate, the design shows airplane tickets, baggage tags, airplane fuselage and design patterns on airplane wings. Also, for the premium card 'the First', images of scenery viewed from the plane, sunrise, sunset, and aurora were utilized. Customers can choose from six plastic and three metal card plates offered for each card. Further, the package includes passport cases, passport-type notes, and a book on various destinations served by Korean Air, allowing the customers to feel as if they are preparing for travel.

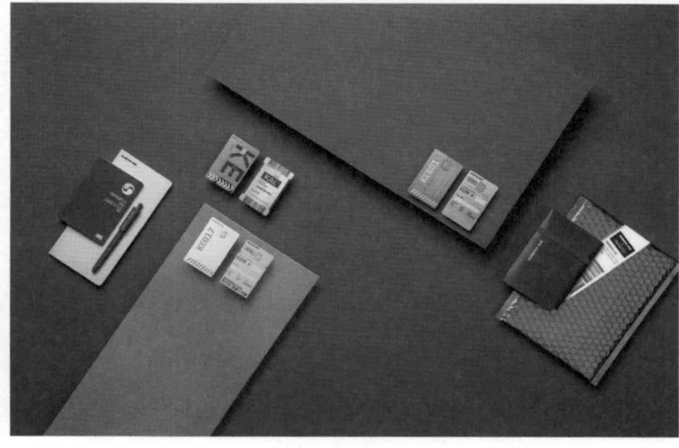

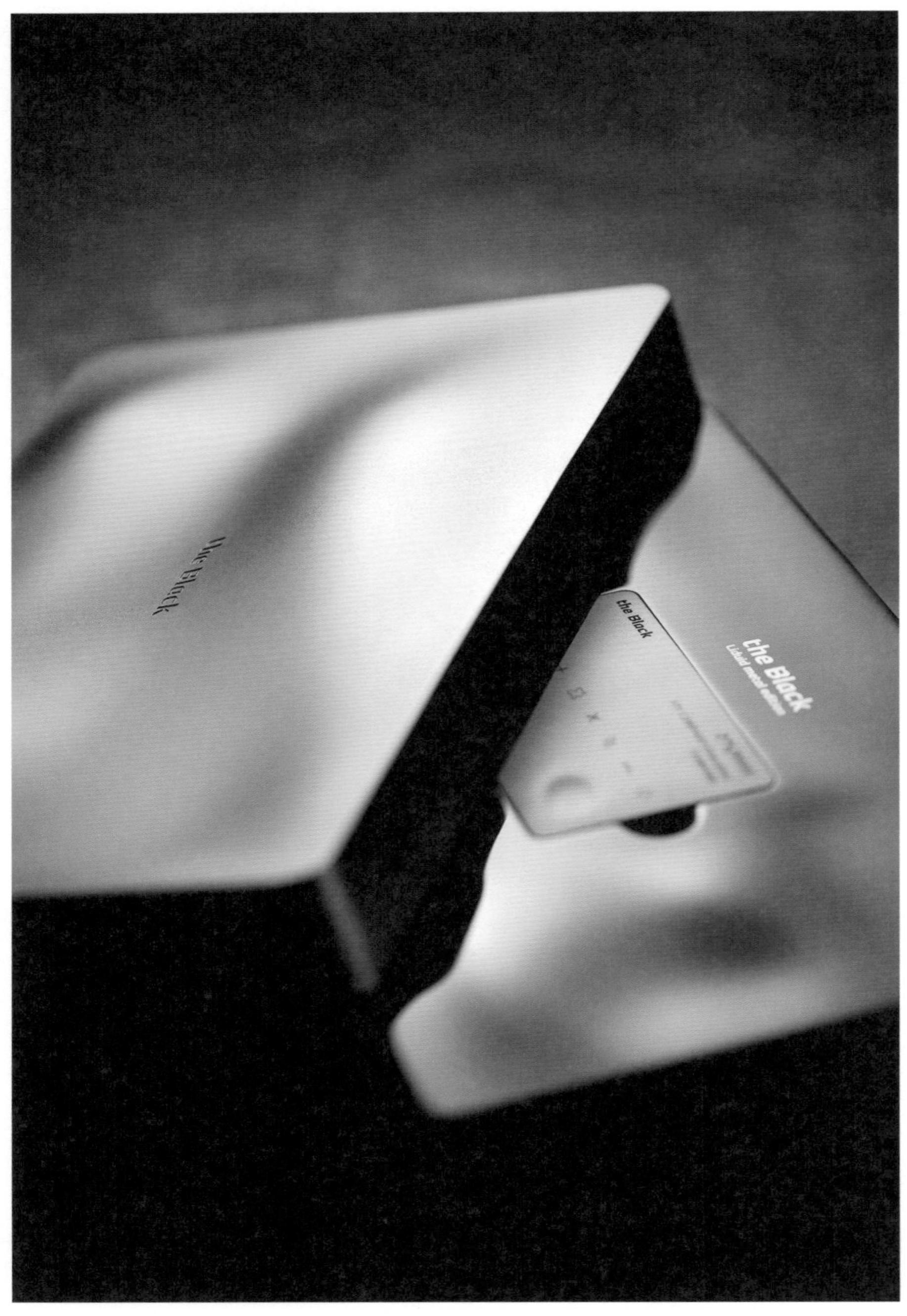

The design of old the Black, the Purple, the Red package. They are no longer in use. Yet, the design provided a luxurious package with a mixture of new techniques and materials. The Black package presented atypical and perfect details suitable for its liquid metal material. The Red and the Purple package stemmed out of balance between luxurious sensibility and eco-friendliness. The Red package employed an easily disposable pulp material while the Purple utilized a can-shaped design, reusable for other purposes than holding credit cards.

The card-issuance envelope and package design for M Card. Instead of vinyl envelopes, which were often criticized for potential security issues, paper envelopes sewn with threads and an authentication sticker were utilized. Added here was the unique and brand-new scent of Hyundai Card, providing a new experience upon opening the envelope. The Platinum Card 3 Series was the first to use vacuum packaging. The design stimulates curiosity and prevents any damages or security loopholes.

M3
M3
M3
M3
DIGITAL

M·M2·M3·T3
M·M2·M3·T3

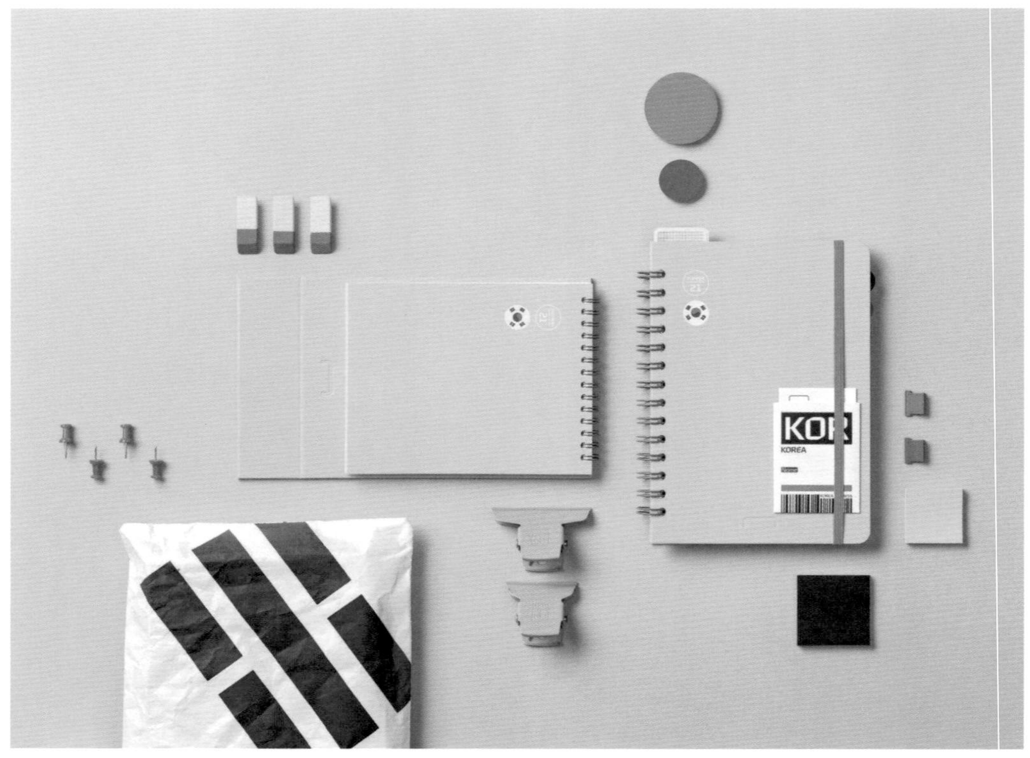

Hyundai Card, Hyundai Capital, and Hyundai Commercial established themselves as global companies with 13 global offices worldwide. Considering the expansion and employees worldwide, the Global Edition Diary and Calendar have been issued annually to share the corporate culture and philosophy. The design of the Diary and Calendar contains Hyundai Card's unique design interpretation under the theme of one of the countries where the company operates. Starting with Germany in 2016, the cover of the Diary and Calendar have reflected the colors and icons inspired by the national flags of each country. The page dividers and the sticker on the back page also facilitate a secondhand experience of the themed country. On the front pages, a brief introduction and photos of each office are included

MyD is an employee ID designed by Hyundai Card Design Lab, breaking away from the old employee ID format. Removing the built-in ID card, the outer design is a simple white piece. The design seems more systemized by secluding one information, name of the employee, form another, a portrait. The design won the iF Design Award in 2021.

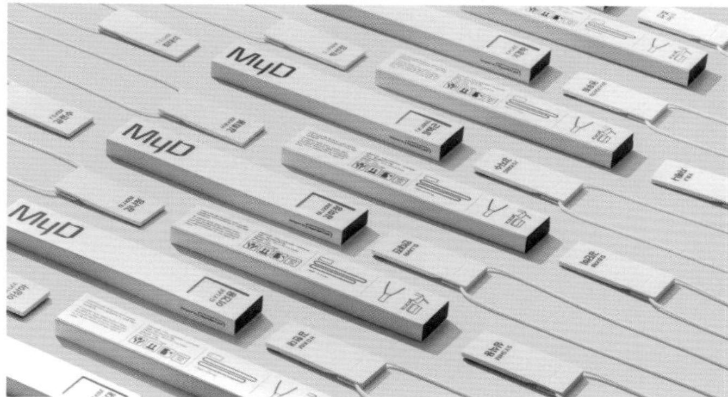

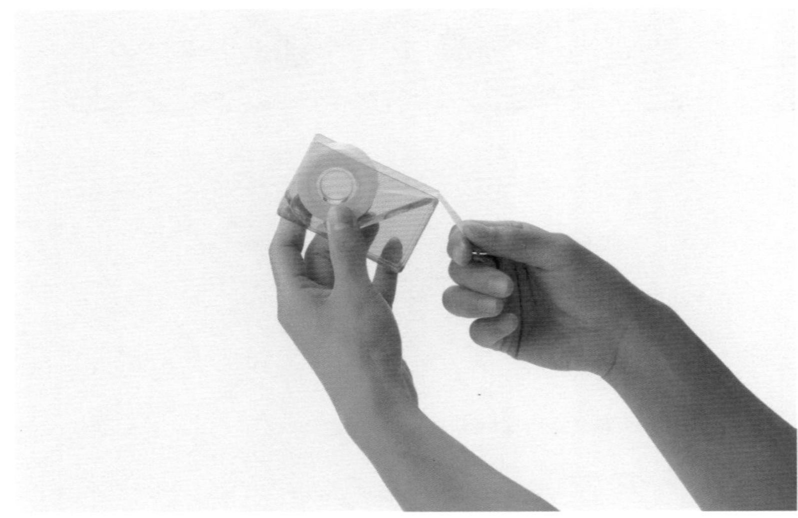

Our Tools were designed after the card plate, the core of Hyundai Card's design, and utilized a card ratio of 1:1.58. Included here are a total of 15 products including office supplies and daily necessities such as an alarm clock, a battery charger, and a money clip. It also features a modular design, allowing employees to make their own sets according to the usage.

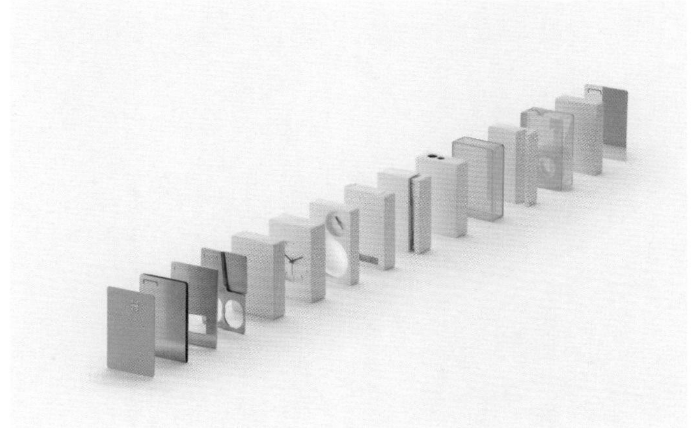

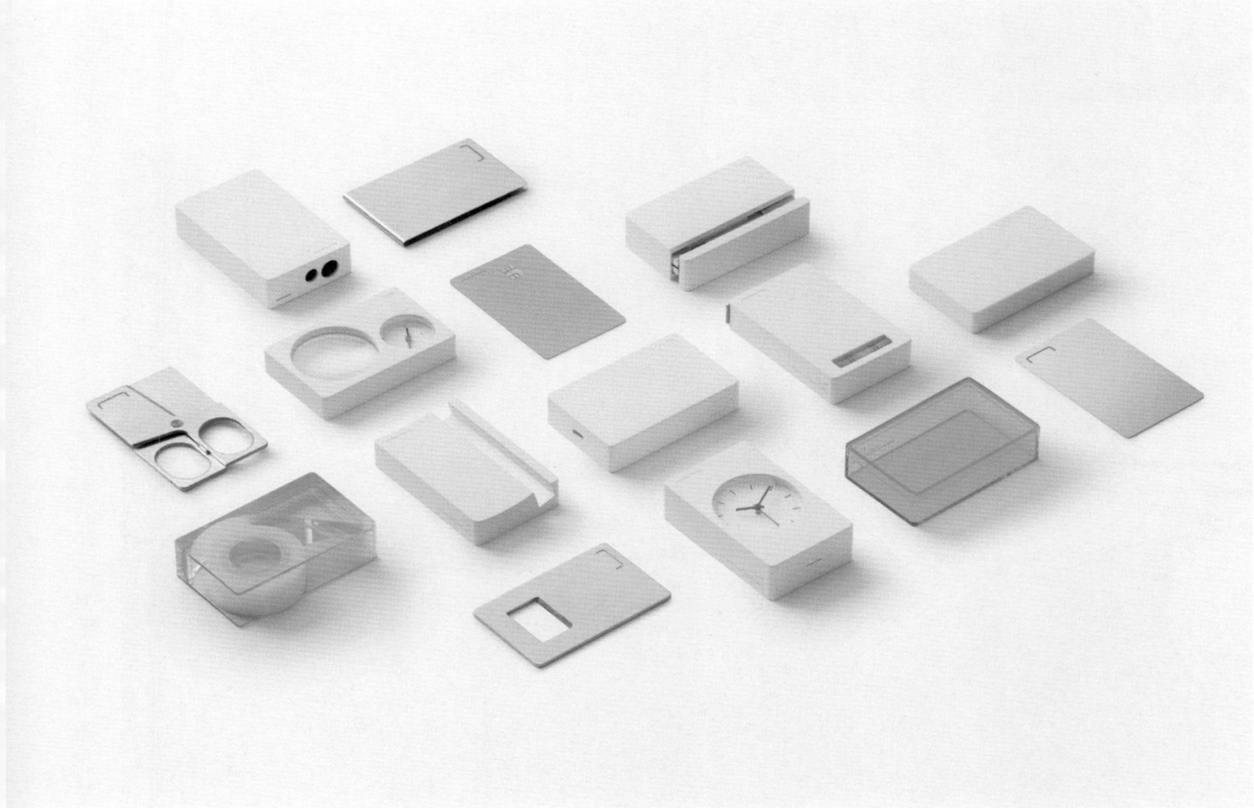

The Table and Ping-Pong Table. The Table and Ping-Pong Table installed in the break rooms of Yeouido Headquarters allow employees to have unique experiences while at leisure. The Table, modeled after Yeouido, also has a space for Korean chess.

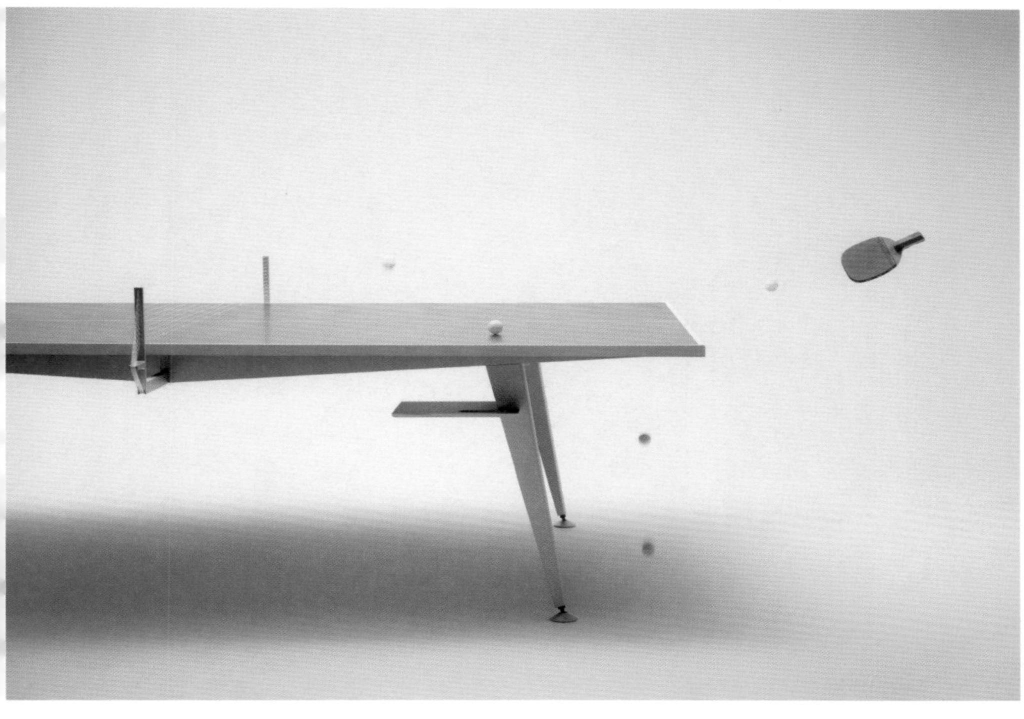

Parking signage. Made by folding a single iron plate, it is easy to transport and store when necessary. Also, it is highly visible with the emphasized icon graphic made of light reflective sheets.

Our Water. The credit card design was a key motif.

The Oyster Project turned everyday kitchen appliances, which used to be negligible, into must-haves. In line with Oyster's unique sensibility of being 'inornate', a neutral and functional design was employed. Oyster suits people who pursue both rational consumption and style. For instance, the rubber gloves are offered in orange, navy, and beige, not in typical red or pink. With an apron, dishcloth, an oven glove, and a cooking toolset that can be stored efficiently even in a narrow space, Oyster enhances the visual perfection of the kitchen space.

The Emergency Card, a card that stores emergency funds for unexpected circumstances in daily life. As a prepaid card that can be immediately used as necessary, the card is delivered sealed in a can package reminiscent of disaster relief supplies. The design consists of fluorescent green colors and diagonal graphics used in disaster relief goods. Added on the front of the card is the phrase 'Emergency Use Only.' At the top of the can package, instructions for usage and panic prevention are also included.

My Taxi was a concept car project partnered with Kia Motors. It was designed based on the Ray, a compact electric vehicle without CO^2 emission. Inside, the passenger seat can be folded to expand the space for baby strollers or large luggage. Additionally, the My Taxi app supported reservations, location tracking, and payment in real-time. The design won the gold prize in the communication category at the iF Design Award 2014.

462

Based on the traditional gate design of Jeju island, 'Jeong Ju-seok' and 'Jeong-nang', Hyundai Card redesigned bus stops of the island. Evidently, the new bus stop design was able to portray the charm of Jeju. The design elements were minimized to harmonize with the beautiful landscape of Jeju. The design was offered in three types—a sign type, a bench type, and a shelter type—to be applied according to the diverse environment of Jeju. Further, its modular design reduced manufacturing and maintenance costs as different types can use the same parts.

Hyundai Capital Skywalkers uniform and court design renewal. The design aimed to express a stronger team by changing the main design to solid colors instead of stripe patterns. In particular, the neckline design and the detail of the sewing line were upgraded. Also employed is a functional material that quickly absorbs and dries sweat during games. The team's home stadium court uses navy and orange, instead of light green and orange, for the first time in professional volleyball history. This court design renewal considered harmony with the uniform design. The dramatic contrast between the court and the uniform completes an extraordinary design that makes players shine the most on the court.

2013 London Design Festival Exhibition. Hyundai Card participated in the London Design Festival, presenting an exhibition titled 'Money' examining the evolution of credit cards as a currency through different materials and design aspects. The Exhibition introduced the changes and development of credit cards that Hyundai Card has achieved along with a variety of projects.

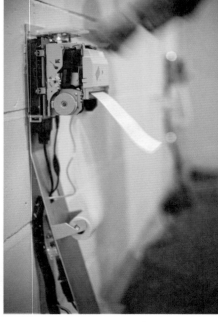

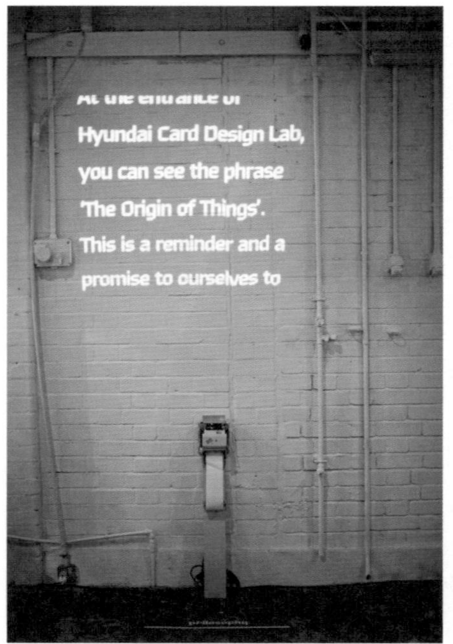

Hyundai Card Design Lab,
you can see the phrase
'The Origin of Things'.
This is a reminder and a
promise to ourselves to

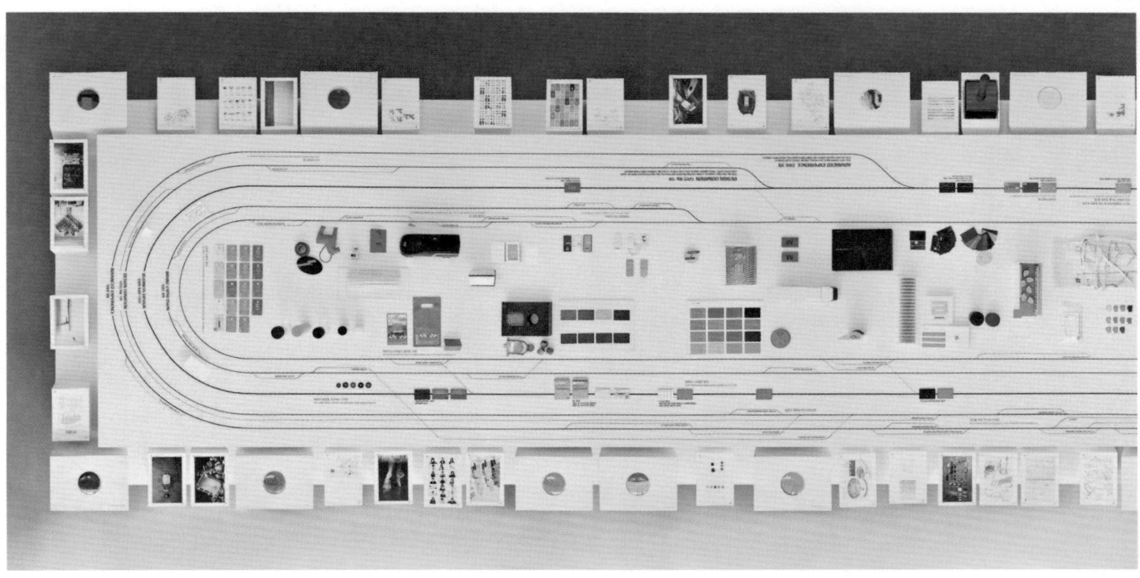

Design Library Exhibition. Based on Hyundai Card's brand and corporate philosophy, the exhibition presented how the design of Hyundai Card has been developed not only in terms of a credit card but also in other projects.

<Traces> Exhibition. As the opening exhibition of Hyundai Card Storage, introduced were a wide variety of design projects beyond the card design stirved by Hyundai Card Design Lab.

The book <Design without Words> was published by Hyundai Card. As the title suggests, it covers the processes, sketches, and outputs of various Hyundai Card-led projects, under the proposition that 'sometimes a single photo speaks louder than detailed descriptions.'

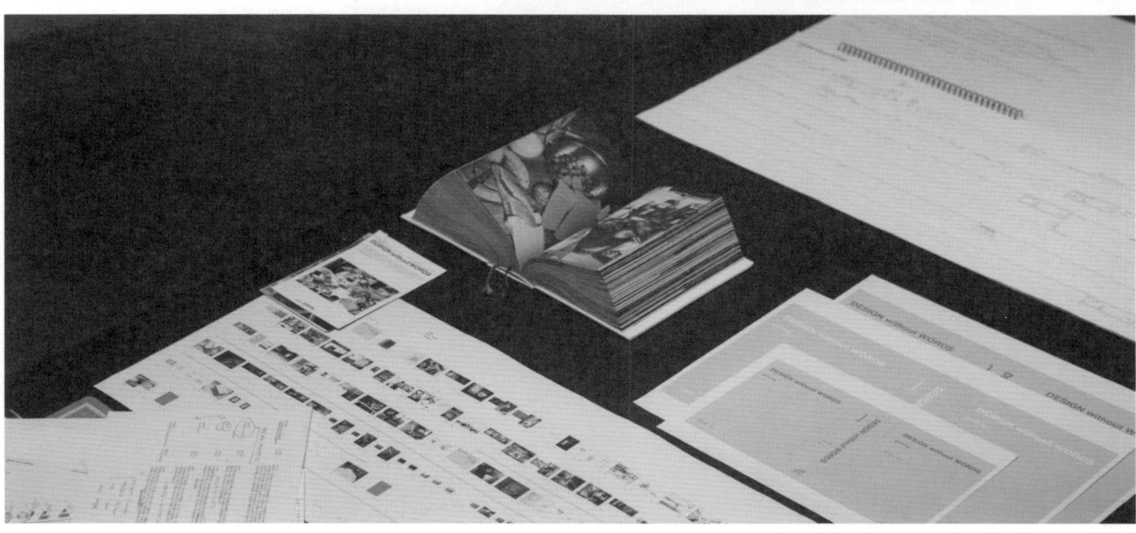

The Understage design identity took its motif from the huge open steel façade of the space. The brochures and design materials of the Music Library tried to visualize the depth of the musical experience through the form of albums stacked on top of each other for many years. The color variations express diverse genres.

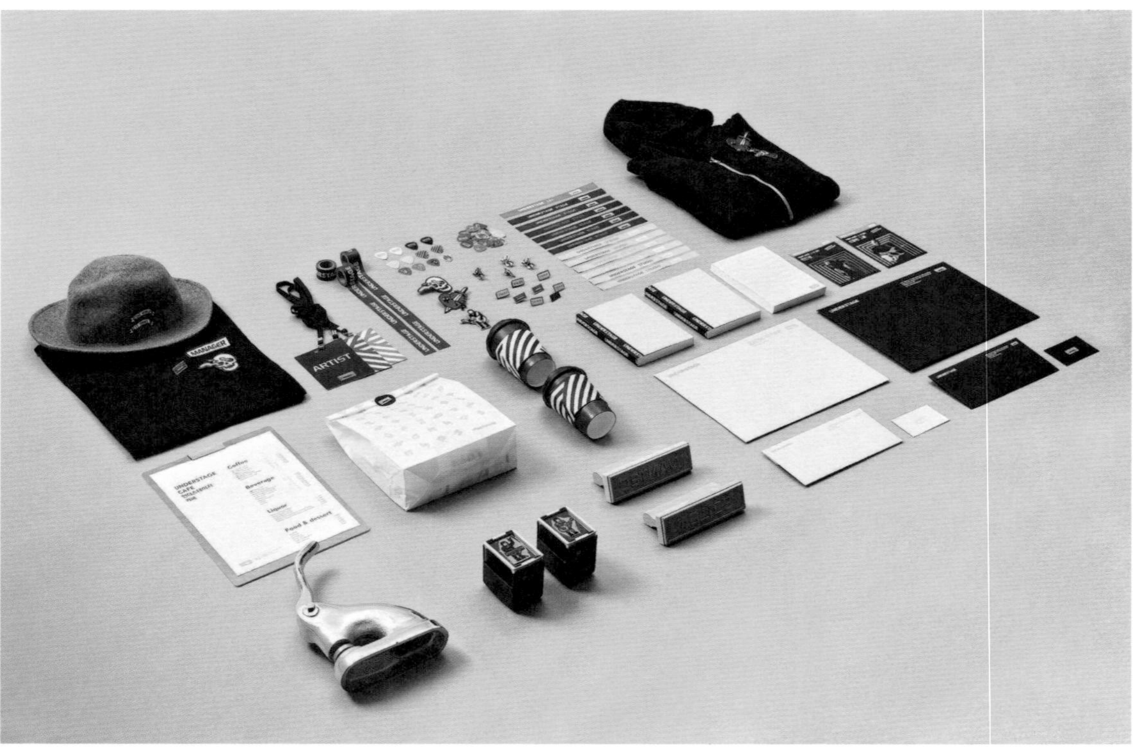

Various items are used and sold in Gapado. Instead of pursuing perfection of the design itself, Hyundai Card Design Lab came up with designs that enable Gapado Island residents to expand and utilize thereafter.

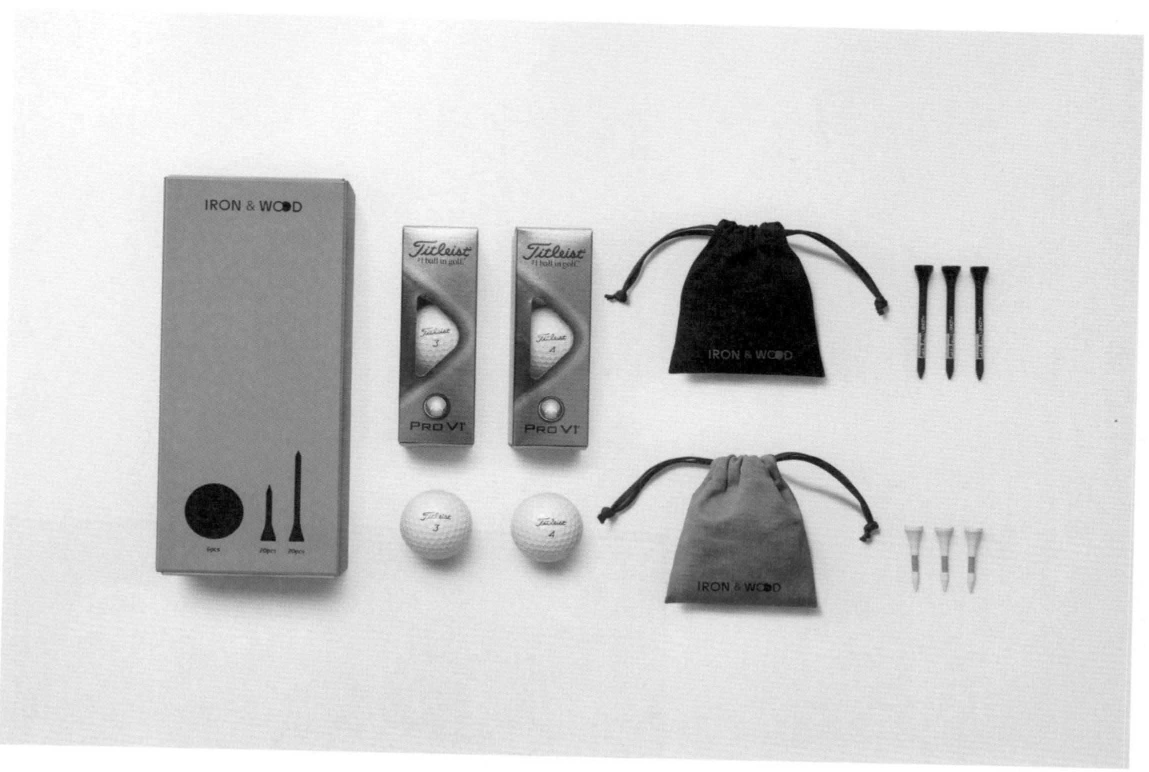

IRON & WOOD is an exclusive space for Hyundai Card Premium members. Here, the members could improve their golf techniques through data-based systematic training services. The nation's top professionals offer coaching services. Also delivered are club customization services based on professional fitters' analysis, as well as effective putting practice opportunities using advanced putting simulators. The BI design has a green reminiscent of fairways and the word mark design is based on the movement of a golf ball.

Hyundai Card App 3.0, which won the top prize in the Communication/Apps category at the iF Design Award 2021, was highly praised for its innovative structure and user convenience. Hyundai Card App 3.0, proprietarily developed by Hyundai Card in August 2020, has moved away from the common method of financial apps that lists all menus horizontally. Instead, the main menu, most frequently used by users, was reconstructed as 'Dual Home'. Also, artificial intelligence (AI) technology was utilized, providing tailored content for each user.

Hyundai Card DIVE is an app that focuses on culture and lifestyle. Its main feed presents major trends, news, and recommended articles on domestic and foreign lifestyles. The proprietary algorithm recommends articles that fit the user's taste. In addition, it serves as a multi-platform for space reservations and participation in major events hosted by Hyundai Card.

Hyundai Card's International Design Awards 2010 - 2021

Since the formation of the in-house design team in 2009, Hyundai Card has submitted various types of designs to the established international design awards. In that respect, Hyundai Card prides itself to be much closer to a cultural brand than to a financial company. The designs submitted to the global awards are not limited in scope, crossing boundaries from credit card design as Hyundai Card's main business, designs that predict future taxi services, and to designs emphasizing regional characteristics.

IDEA (International Design Excellence Awards) is a design award of global authority. Winners are selected based on five criteria: design innovation, user experience, user benefits, social responsibility, and aesthetics.

The iF Design Award (International Forum Award) is a design award with top-notch recognition and reliability. The evaluation comprehensively examines design materials and innovation of five categories: product, communication, package, interior, architecture, and concept.

Along with the iF Design Award and IDEA, the Red Dot Design Award is one of the world's top three design awards. Winners are selected based on nine rigorous judging criteria including design innovation, possibility, quality, human, engineering considerations, durability, etc.

Year	Project Title	Award	Level
2010	Seoul Station Media Art Shelter	IDEA	Gold
		iF Design Award	winner
		Red dot Design Award	winner
2011	The Dream Project	IDEA	Gold
2013	Design Library	DFAA	Grand
2014	My Taxi	iF Design Award	Gold
2015	Hyundai Card Weather	Red dot Design Award	winner
2016	Hyundai Card Weather	iF Design Award	winner
2017	Vertical Type Credit Card	iF Design Award	winner
2018	The Gapado project	Korea Design Award	Grand
2019	the Green	IDEA	winner
	The Gapado project	iF Design Award	winner
	Hyundai Card Culture (DIVE App)	iF Design Award	winner
2020	DIGITAL LOVER	IDEA	winner
2021	DIGITAL LOVER	iF Design Award	winner
	MyD	iF Design Award	winner
	Hyundai Card App 3.0	iF Design Award	winner

Seoul Station Media Art Shelter(2010)

Vertical Type Credit Card(2017)

Index

Profile

Architecture Studios

101 Architects

CEO Wook Choi

Founded in 2002 by Wook Choi, who graduated from the architecture department of Hongik University and Università Iuav di Venezia (dottore in arch.), 101 Architects pursue spaces made up of scenery and heartfelt emotions instead of the form or concept of objects. It is currently working on projects that seek to discover the value of time within a location and cultural scenery, and it bases its value on humanistic records. Wook Choi's main works include the Hakgojae Gallery, Dugahun, Mapo Bridge Plaza, and the Hanyangdoseong Visitor Center. He also publishes Domus Korea.

Hyundai Card projects

Yeouido headquarters building 3, Yeongdeungpo office, Busan office, Design Library, Cooking Library, Gapado project He won the 2013 DFAA (Design for Asia Awards) Grand Prize with the Design Library and the 2014 Kimm Jong Soung Award with the Yeongdeungpo office.

Gensler

CEO Diane Hoskins, Andy Cohen

It was founded in San Francisco in 1965 by M. Arthur Gensler Jr., his wife Rue Gensler, and James Follett. Currently, it has grown into a global design company with 6,000 employees in 48 cities in 16 countries. Philippe Paré, who was in charge of the Hyundai Card project, oversees its Paris office. Sabu Song participated in the interior design of the Music Library. Major operations include the Nvidia headquarters, Hyatt Foundation headquarters, Bank of California Stadium, Shanghai Tower, Shanghai Johnson Controls headquarters, Philippines Finance Center Tower, and Incheon International Airport Terminal 2, among others.

Hyundai Card projects

Yeouido Headquarters' Convention Hall, Workspace, UX Lab, HCA, HCUK, HCBE, BHAF, Music Library, Studio Black

Spackman Associates

CEO Mary Spackman

It was founded in 2002 by Mary Spackman, who graduated from Smith College's Department of Political Science and the New York School of Interior Design. CEO Mary Spackman worked at Rockwell Group and Space, carrying out various projects including the first W Hotel in New York, office building, and restaurant. Spackman Associates has a reputation for creating unique corporate identities through in-depth communication. The CEO's main works include the offices of Amorepacific, Samyang, Shinyoung, Dongwha, SK, Kim & Chang, Yulchon, as well as the space design for Naver and Somerset Palace Seoul.

Hyundai Card projects

Yeouido Headquarters' Convention Hall, Auditorium, Lecture Room, the Box, Café & Pub, Vinyl & Plastic, Storage, Air Lounge

Ateliers Jean Nouvel

CEO Jean Nouvel

It was founded in 1994 by Jean Nouvel, a graduate of the École de Beaux-Arts in Paris. All projects are original enough to be masterpieces, and they always employ cutting-edge technology and materials. It stems from a unique concept for unifying people, place, and time, rather than a consideration of style or ideology. The basis of the urban humanistic context and the provocative nature of contemporary architecture infuses a real uniqueness into the projects that it undertakes. Various awards prove its worldwide reputation. It received the Aga-Khan Prize from the Arab World Institute in Paris in 1989, Golden Lion Award at the Venice Architecture Biennale in 2000, Royal Institute of British Architects (RIBA) Gold Medal in 2001, Japan Art Association Premium Imperial, Lucerne Cultural Center Borromini Award, and Pritzker Prize in 2008. Jean Nouvel's major works include the Arab World Institute, Cartier Foundation, Qué Branly Museum, Philharmonic Concert Hall, Lyon Opera House, Barcelona Glory Tower, London One New Challenge, Doha Tower, National Museum of Qatar, Louvre Abu Dhabi, and Leeum Museum

Hyundai Card project

Yeouido Headquarters' Design Lab

Wonderwall

CEO Masamichi Katayama

Wonderwall, a global interior design company, was founded in 2000 by Masamichi Katayama, a graduate of Osaka Designer College. While respecting Japanese traditions and design philosophies, he is attracting attention for his unconstrained approach to turning concepts into reality. "We have created an environment that appeals to the subconscious of our visitors by conceptualizing a unique space that enhances the branding and vision the customers aspire to." (Masamichi Katayama) With a focus on interiors, Wonderwall's work spans creative and architectural disciplines, therefore, its portfolio includes projects from around the world, including fashion boutiques, branding spaces, and large commercial establishments. Katayama has designed spaces for various global brands, such as Nike, Thom Browne, Uniqlo, Diesel, Dean & Deluca, Pierre Herme, and Samsung.

Hyundai Card project

Travel Library

Doojin Hwang Architects

CEO Doojin Hwang

It was founded in 2000 by Doojin Hwang, who graduated from Seoul National University and Yale University. He gained attention for a series of works that reinterpreted the hanok from the perspective of modern architecture based on his understanding of history and culture. As seen in the case of the Rainbow Cake Architecture (middle-rise, high-density residential complex) as a solution to the Korean urban problem, he aims for architecture that encompasses various issues related to structure and form, function and efficiency, as well as science and humanities. His major works include the Korean Gallery, Museum of Far Eastern Antiquities, Stockholm, Sweden, Choonwongdang Hospital of Korean Medicine & Museum, Gahoeheon, Han River Bridge Pedestrian Facilities(Jamshil, Hannam), Gallery Artside, Open Books, Won & Won 63.5, Mookas HQ, and North Terrace.

Hyundai Card project

Castle of Skywalkers.

Ga.A Architects

CEO Koh Daekon

It was established by Moongyu Choi after graduating from Yonsei University and Columbia Graduate School of Architecture in 1999 and gaining experience at Toyo Ito Architects, Hanul Architecture, and Group See. Through systematic and consistent design, Moongyu Choi pursues rational and future-oriented creation that meets the needs of modern society, ultimately making life more valuable and beautiful. His major works include Maison de La Corée in Paris, Veritas Hall Yonsei Campus, Y Study House, UOS Centennial Memorial Hall, SSU Student Union, Heyri K Gallery, Arumdri Media, Taehaksa Publishing, Booksea Publishing, Ssamziegil, I like Dalki, and Hansook Cheong Memorial.

Hyundai Card projects

Music Library, Understage

Samuso Hyojadong

CEO Seungmo Seo

Samuso Hyojadong, an architectural firm, was founded in 2004 by Seungmo Seo, who studied at Kyungwon University and Tokyo University of the Arts. The firm has carried out a number of hanok remodeling projects centered on Seochon, and its work is characterized by calm and serene spaces and delicate details, consistently assuming the form of caring for people. Major works include renovation of the façade of the Theory flagship store, DDP Design Library, the exhibition design of Ryuichi Sakamoto at Piknic, and Namhae 613 Inn.

Hyundai Card project

Vinyl & Plastic façade renovation

Writers

Jiho Park

He worked for 17 years for fashion and lifestyle magazines, such as Esquire and Arena. For seven years, he served as an editor for Arena. Since 2016, he has been working on a number of projects that unfold content based on sensitive spaces, such as Jiho Park's late-night bookstore and late-night salon. In 2019, he collaborated with the Seoul Metropolitan Government to carry out the Urban Space Odyssey(USO) project, which planned to renovate the 50-year-old Jungnim Warehouse as a cultural space. In 2021, in the 90-year-old Shin-A Memorial Hall located on Jeongdong-gil, he opened the library of inspiration 102, a space for curating experiential content every month. Currently, he is working as a special editor for Hyundai Card DIVE, a director of the Daelim Cultural Foundation, and an ambassador for Seoul. His books include Inside Hyundai Card and Leica, Tools of Inspiration.

Eunkyung Jeon

She is the director of the Monthly DESIGN magazine. Working as a design journalist, and editor in chief for 17 years, she interviewed domestic and foreign designers, managers, and marketers, writing and planning articles on various architectural and design projects as well as design and lifestyle trends. She served as the director of the Seoul Design Festival and the Audi Design Challenge, as well as a jury member for the iF Design Award and Korea+Sweden Young Design Award. She launched Work Design, a content on work with Seoul Work Design Week. Her books include Design Power(author) and Consumer Evolves(co-author), and she has participated in design consulting for various companies.

Yoonkyung Bae

He lectures on architectural design and theory at Dankook University and Yonsei University, and writes architecture-related columns for various media outlets. He is the author of Amsterdam Architecture Journey, A Delightful World Architecture Tour for Children, DDP Metonymic Landscape(co-author), Urban Issue: Proposals for a City to Live Together(co-author), Small Shining, and Amorepacific New Beauty Space(co-author).

Ted Chung

The Way We Build :
A Journey Through the Spaces of Hyundai Card

Publication of 1st Edition (English Edition)
March 17, 2022

Chief director Ted Chung
Writer Jiho Park, Eunkyung Jeon, Yoonkyung Bae
Translator Jane Hong
Design by Made Thought
Photo by Textureontexture, Kyongseop Shin,
Hojeong Chun, Yongjoon Choi,
Myeongsik Kim, Sooyeon Yoon,
Dongwook Chung, Chanwoo Park (Studio Zip),
courtesy of Gensler
Planning Inspiration 102

Published by JoongangilboS
100 Seosomun-ro, Jung-gu, Seoul, Korea
Jbooks@joongang.co.kr

ISBN 978-89-278-1283-8 03540